Frontiers in Mathematics

Advisory Editors

William Y. C. Chen, Nankai University, Tianjin, China
Laurent Saloff-Coste, Cornell University, Ithaca, NY, USA
Igor Shparlinski, The University of New South Wales, Sydney, NSW, Australia
Wolfgang Sprößig, TU Bergakademie Freiberg, Freiberg, Germany

This series is designed to be a repository for up-to-date research results which have been prepared for a wider audience. Graduates and postgraduates as well as scientists will benefit from the latest developments at the research frontiers in mathematics and at the "frontiers" between mathematics and other fields like computer science, physics, biology, economics, finance, etc. All volumes are online available at SpringerLink.

Alexander Iksanov • Alexander Marynych •
Andrey Pilipenko • Ihor Samoilenko

Locally Perturbed Random Walks

# Birkhäuser

Alexander Iksanov
Faculty of Computer Science & Cybernetics
Taras Shevchenko National University of Kyiv
Kyiv, Ukraine

Alexander Marynych
Faculty of Computer Science & Cybernetics
Taras Shevchenko National University of Kyiv
Kyiv, Ukraine

Andrey Pilipenko
Institute of Mathematics
Ukrainian National Academy of Sciences
Kyiv, Ukraine

Ihor Samoilenko
Faculty of Computer Science & Cybernetics
Taras Shevchenko National University of Kyiv
Kyiv, Ukraine

ISSN 1660-8046   ISSN 1660-8054   (electronic)
Frontiers in Mathematics
ISBN 978-3-031-83918-4   ISBN 978-3-031-83919-1   (eBook)
https://doi.org/10.1007/978-3-031-83919-1

© The Editor(s) (if applicable) and The Author(s), under exclusive license to Springer Nature Switzerland AG 2025

This work is subject to copyright. All rights are solely and exclusively licensed by the Publisher, whether the whole or part of the material is concerned, specifically the rights of translation, reprinting, reuse of illustrations, recitation, broadcasting, reproduction on microfilms or in any other physical way, and transmission or information storage and retrieval, electronic adaptation, computer software, or by similar or dissimilar methodology now known or hereafter developed.
The use of general descriptive names, registered names, trademarks, service marks, etc. in this publication does not imply, even in the absence of a specific statement, that such names are exempt from the relevant protective laws and regulations and therefore free for general use.
The publisher, the authors and the editors are safe to assume that the advice and information in this book are believed to be true and accurate at the date of publication. Neither the publisher nor the authors or the editors give a warranty, expressed or implied, with respect to the material contained herein or for any errors or omissions that may have been made. The publisher remains neutral with regard to jurisdictional claims in published maps and institutional affiliations.

This book is published under the imprint Birkhäuser, www.birkhauser-science.com by the registered company Springer Nature Switzerland AG
The registered company address is: Gewerbestrasse 11, 6330 Cham, Switzerland

If disposing of this product, please recycle the paper.

To our wives

Natalka, Olena, Nataliya and Tetiana

Preface

A standard random walk on the line is the sequence of successive sums of independent identically distributed real-valued random variables. Standard random walks are among the simplest probabilistic objects in discrete time, and by now the theory of standard random walks has achieved maturity which particularly means that most of their properties are well-understood. Modern Probability Theory pays considerable attention to perturbed random walks attempting to find out to what extent a slight modification of a standard random walk influences properties of the modified object. We note that "perturbed" is used as a synonym for "spoiled."

Like in the proverb saying "it's easier to break than to make" there are a number of ways to produce a perturbed random walk. The book [68] is concerned with a particular instance of "globally perturbed random walks" for which the perturbations have a global impact and change transition rates of a standard random walk in all states, making, in most cases, the walk non-Markovian. Another class of globally perturbed random walks is discussed in Chapter 6 of [60].

In this book we shall focus on a completely different class of perturbed random walks which we call "locally perturbed random walks" and whose source of perturbation is a local impurity of the underlying state space, that only changes the behavior of a random walk upon visiting those "spoiled" states. The book offers some elements of an asymptotic theory of such random walks. The introductory Chap. 1 starts with a motivation, providing examples of locally perturbed random walks. It proceeds with a brief overview of the mathematical toolbox and approaches that will be utilized throughout the book. Chapter 2 is concerned with constructing various stochastic processes that serve as scaling limits for locally perturbed random walks. We call these processes "Lévy-type processes with singularities." In particular, we discuss in considerable detail reflected and skewed processes. We review both old and new facts about a skew Brownian motion and present the first rigorous definition of a skew stable Lévy process. These processes serve as scaling limits for a wide class of locally perturbed random walks known as random walks with membranes. There are different approaches to the analysis of locally perturbed random walks available in the literature. However, no general theory exists for deriving singular diffusions as scaling limits of locally perturbed random walks. In Chap. 3 we undertake the first attempt toward this goal by proving various limit theorems for locally perturbed random walks.

The book is largely self-contained, with many parts accessible to students familiar with probability and stochastic processes on the level of the "second course in probability." The Appendix collects more advanced background material, which is usually not included in textbooks.

The book presents, for the most part, an outcome of research done in 2020–2023 within the project 2020.02/0014 "Asymptotic regimes of perturbed random walks: on the edge of modern and classical probability" funded by the National Research Foundation of Ukraine. The authors gratefully acknowledge generous financial support from the National Research Foundation of Ukraine.

Kyiv, Ukraine
Alexander Iksanov
Alexander Marynych
Andrey Pilipenko
Ihor Samoilenko

List of Notation

General Notation

$\mathbb{N} := \{1, 2, \ldots\}$
$\mathbb{N}_0 := \{0, 1, 2, \ldots\}$
$\mathbb{Z} := \{0, \pm 1, \pm 2, \ldots\}$
$x \wedge y := \min(x, y)$, $x \vee y := \max(x, y)$
$x_+ := \max(x, 0)$, $x_- := -\min(x, 0)$
$\lfloor x \rfloor$ – the floor function defined by $\lfloor x \rfloor = \sup\{n \in \mathbb{Z} : n \leq x\}$
$\Longrightarrow$ – weak convergence of probability measures, convergence in distribution of random processes
$\xrightarrow{d}$ – convergence in distribution of real-valued random variables
$\stackrel{d}{=}$ – equality of distributions
$\mathbb{1}_A$ – the indicator of an event A, which is equal to 1 if A occurs, and is equal to 0 otherwise
Càdlàg function – right continuous function (defined on a subset of $\mathbb{R}$) with finite limits from the left
$F^{\leftarrow}$ – the right-continuous generalized inverse of a nondecreasing càdlàg unbounded function F, see p. 217
F^{-} – the left-continuous generalized inverse of a nondecreasing càdlàg unbounded function F, see p. 217
Γ – the Euler gamma function defined by $\Gamma(x) := \int_0^\infty y^{x-1} e^{-y} dy$ for $x > 0$
LEB – Lebesgue measure
$\|f\|_\infty = \sup_{x \in E} |f(x)|$ for a function f bounded on a set E
a.s. – almost sure or almost surely
$\langle x, y \rangle$ – the standard inner product of vectors x and y from a Euclidean space
$\langle \nu, f \rangle = \int f(x) \nu(dx)$, where ν is a measure and f is a measurable function
$\mathcal{B}(E)$ – the Borel sigma-algebra of a metric space E

Functional Spaces

$C_b(E)$ – the space of bounded continuous functions on a metric space E

$C_0(E)$ – the space of continuous functions on a locally compact metric space E which vanish at infinity

$C_+([0, \infty))$ – the space of continuous functions f defined on $[0, \infty)$ with $f(0) \geq 0$

$D([0, \infty), E)$ – the Skorokhod space of càdlàg functions defined on $[0, \infty)$ with values in a metric space E

$D([0, \infty)) := D([0, \infty), \mathbb{R})$; throughout the book we assume that $D([0, \infty))$ is endowed with the J_1-topology

$(D([0, \infty)))^k$ – the Cartesian product of k copies of $D([0, \infty))$ endowed with the product J_1-topology

$D_+([0, \infty))$ – the space of càdlàg functions f defined on $[0, \infty)$ with $f(0) \geq 0$

$D_{0,\uparrow}^u([0, \infty))$ – the space of nondecreasing unbounded càdlàg functions f defined on $[0, \infty)$ with $f(0) = 0$

$D_{0,\uparrow\uparrow}^u([0, \infty))$ – the space of strictly increasing unbounded càdlàg functions f defined on $[0, \infty)$ with $f(0) = 0$

Stochastic Processes

S_ξ – a standard random walk with jumps distributed like ξ, that is, $S_\xi(n) = S_\xi(0) + \xi_1 + \cdots + \xi_n$ for $n \in \mathbb{N}$

$W := (W(t))_{t \geq 0}$ – a standard Brownian motion on $\mathbb{R}$

$U_\gamma^{\mathfrak{s},\beta,\mu}$ – a stable distribution with index of stability $\gamma \in (0, 2]$, scale parameter $\mathfrak{s} > 0$, skewness parameter $\beta \in [-1, 1]$, shift parameter $\mu \in \mathbb{R}$, and the characteristic function given in (4.11)

$U_\alpha := (U_\alpha(t))_{t \geq 0}$ – a symmetric α-stable Lévy process with $\alpha \in (1, 2)$ and the characteristic function $\mathbb{E} \exp(i z U_\alpha(t)) = \exp(-t|z|^\alpha)$ for $z \in \mathbb{R}$ and $t \geq 0$

$\mathcal{S}_\alpha := (\mathcal{S}_\alpha(t))_{t \geq 0}$ – a context-dependent notation for an α-stable Lévy process with the characteristic function of $\mathcal{S}_\alpha(1)$ (or the Lévy-Khintchine triplet) specified on every occurrence

$\mathcal{V}_\beta := (\mathcal{V}_\beta(t))_{t \geq 0}$ – a drift-free β-stable subordinator with $\beta \in (0, 1)$ and the Laplace transform $\mathbb{E} \exp(-s \mathcal{V}_\beta(t)) = \exp(-t s^\beta)$ for $s \geq 0$ and $t \geq 0$

$W_\gamma^{\text{skew}} = (W_\gamma^{\text{skew}}(t))_{t \geq 0}$ – a skew Brownian motion with permeability parameter $\gamma \in [-1, 1]$, see Definition 2.2.1

$W_\mathbf{p} = (W_\mathbf{p}(t))_{t \geq 0}$ – a Walsh Brownian motion with parameter $\mathbf{p} = (p_1, \ldots, p_d)$, see Definition 2.2.2

Miscellaneous

R_λ^Z or R_λ – the resolvent of a Markov process Z, see formula (4.14)

V_λ^Z or V_λ – the resolvent of a killed Markov process Z, see formula (4.17)

List of Notation

$V_\lambda 1$ is understood as $V_\lambda f$ with $f(x) = 1$ for $x \in \mathbb{R}$

$\mathcal{SR}(y)$ – the Skorokhod reflection problem for a function y, see p. 14

$\mathcal{GSR}(y, h)$ – the generalized Skorokhod reflection problem for a noise y and a reflection governed by h, see p. 19

$\mathcal{G}_\delta(y, F)$ – a solution to the switch problem with a noise y, a regulator F and a gap δ, see p. 91

$\langle Z_1, Z_2 \rangle$ – the predictable quadratic covariation of semimartingales Z_1 and Z_2; in particular, $\langle Z, Z \rangle$ is the predictable quadratic variation of Z

Contents

1   **Introduction** 1

2   **Lévy-Type Processes with Singularities** 11
    2.1   Reflected Processes 11
        2.1.1   The Skorokhod Reflection Problem 11
        2.1.2   The Generalized Skorokhod Reflection Problem 19
        2.1.3   Properties of a Reflected Brownian Motion with Jump-Type Exit from 0 25
    2.2   The Skew Brownian Motion and Its Generalizations 34
        2.2.1   The Skew Brownian Motion 34
        2.2.2   The Walsh Brownian Motion 42
    2.3   The Skew Stable Lévy Process 47
        2.3.1   Definition 47
        2.3.2   Proof of Theorem 2.3.1 51
        2.3.3   An Equation for the Skew Stable Lévy Process 59

3   **Functional Limit Theorems for Locally Perturbed Random Walks** 73
    3.1   Preliminaries 73
        3.1.1   Definition of Random Walk with Membrane 73
        3.1.2   Examples of Random Walks with Reflection 79
    3.2   Functional Limit Theorems for Random Walks with Reflection 82
        3.2.1   Perturbations with Finite Mean 82
        3.2.2   On Two-Stage Models and the Generalized Reflection 90
        3.2.3   Perturbations with Regularly Varying Distribution Tails .. 107
        3.2.4   Perturbations with Slowly Varying Distribution Tails 117
    3.3   Random Walks with Membrane and a Skew Brownian Motion 121
        3.3.1   Convergence to a Walsh Brownian Motion 124
        3.3.2   Examples 132
        3.3.3   Proof of Theorem 3.3.2 139
        3.3.4   Proof of Theorem 3.3.3 155
        3.3.5   Proof Corollary 3.3.1 162
    3.4   Limit Theorems for Heavy-Tailed Random Walks with Membrane at 0 164

		3.4.1	Main Result	164
		3.4.2	Proof of Theorem 3.4.1(a)	166
		3.4.3	Proof of Theorem 3.4.1(b)	177
	3.5	Multidimensional Random Walks with Membranes		180
		3.5.1	Two-Dimensional Random Walks with Finite Membranes	183
		3.5.2	Multidimensional Random Walks with Periodic Membranes Concentrated on Hyperplanes	194
4	**Auxiliary Results**			205
	4.1	Probability Measures and Weak Convergence		205
	4.2	The Space of Continuous Functions $C([0, \infty), \mathbb{R}^d)$		209
		4.2.1	Weak Convergence and the Locally Uniform Topology	209
		4.2.2	The Wiener Measure and Donsker's Invariance Principle	210
	4.3	The Skorokhod Space $D([0, \infty), \mathbb{R}^d)$		211
		4.3.1	The J_1-Topology	211
		4.3.2	Convergence of Probability Measures on $D([0, \infty), \mathbb{R}^d)$	213
		4.3.3	Continuity of Mappings on $D([0, \infty), \mathbb{R}^d)$	213
		4.3.4	Generalized Inverse Functions	216
	4.4	Stable Distributions		219
		4.4.1	Domains of Attraction of Stable Distributions	220
	4.5	Convergence of Markov Processes		222
	4.6	Itô's Excursion Theory		225
Bibliography				239
Index				247

Introduction

The scariest moment is always just before you start.

Stephen King, *On Writing: A Memoir of the Craft (2000)*

Abstract In this introductory chapter we discuss various models of pure and applied probability, which lead to locally perturbed random walks. We discuss mathematical tools and approaches to be utilized throughout the book, and provide a detailed overview of the book's structure.

Let $S_\xi := (S_\xi(n))_{n\geq 0}$ be a one-dimensional *standard random walk* defined by

$$S_\xi(0) := x, \quad S_\xi(n) := S_\xi(0) + \xi_1 + \cdots + \xi_n, \quad n \in \mathbb{N},$$

where $x \in \mathbb{R}$, and the increments $\xi_1, \xi_2, \ldots$ are independent copies of a real-valued random variable ξ.

Throughout the book we assume that the Skorokhod space $D([0,\infty))$ is endowed with the J_1-topology. Relevant properties of this space are discussed in Sect. 4.3 of the Appendix. It is known that if the distribution tail of ξ exhibits a "nice" asymptotic, then the sequence of processes $((S_\xi(\lfloor nt \rfloor))_{t\geq 0})_{n\geq 1}$, properly centered and normalized, converges in distribution on $D([0,\infty))$ to a Lévy process. For instance, if $\mathbb{E}\xi = 0$ and $\sigma^2 := \operatorname{Var}\xi \in (0,\infty)$, then Donsker's invariance principle tells us that

$$\left(\frac{S_\xi(\lfloor nt \rfloor)}{\sigma\sqrt{n}}\right)_{t\geq 0} \Longrightarrow (W(t))_{t\geq 0}, \quad n \to \infty, \tag{1.1}$$

where $W = (W(t))_{t \geq 0}$ is a standard Brownian motion.

Consider now a Markov chain $X := (X(n))_{n \geq 0}$, whose transition probabilities coincide with those of S_ξ everywhere except on a given set A. We say that X is a *random walk with membrane A*. Although any Markov chain on $\mathbb{R}$ can be regarded as a random walk with membrane upon setting $A = \mathbb{R}$, such trivial cases will be excluded in what follows. Random walks with membranes form a subclass of *locally perturbed random walks*. Their source of perturbation is a local impurity of the underlying state space, which changes the behavior of a random walk upon visiting those "spoiled" states. In this book we are mainly interested in the distributional limit behavior of locally perturbed random walks under the normalization which appears in the classical distributional limit theorems for S_ξ.

To get a better feeling of the expected results we consider a few warm-up examples. Let $\mathbb{P}\{\varepsilon = \pm 1\} = 1/2$, so that S_ε is a simple symmetric random walk on $\mathbb{Z}$ starting at $x \in \mathbb{Z}$. Assume that the Markov chain X satisfies $X(0) = |x|$ and has the same transition probabilities as S_ε everywhere except on $A = \{0\}$, whereas the transition probability at $\{0\}$ is defined by $\mathbb{P}\{X(1) = 1 | X(0) = 0\} = 1$. Then X is a simple symmetric random walk on $\mathbb{N}_0$ with reflection at 0. According to Donsker's invariance principle (1.1),

$$\left(\frac{S_\varepsilon(\lfloor nt \rfloor)}{\sqrt{n}}\right)_{t \geq 0} \Longrightarrow (W(t))_{t \geq 0}, \quad n \to \infty.$$

Since $(X(n))_{n \geq 0}$ has the same distribution as $(|S_\varepsilon(n)|)_{n \geq 0}$, an application of the continuous mapping theorem yields

$$\left(\frac{X(\lfloor nt \rfloor)}{\sqrt{n}}\right)_{t \geq 0} \Longrightarrow (|W(t)|)_{t \geq 0}, \quad n \to \infty.$$

We continue with a slight modification of the previous example which often pops up in queuing theory. Let $\xi_1, \xi_2, \ldots$ be independent copies of a random variable ξ with zero mean and variance $\sigma^2 \in (0, \infty)$, and $\eta_1, \eta_2, \ldots$ independent copies of a nonnegative random variable, the two sequences being independent. Define now locally perturbed random walks X_1 and X_2 by

$$X_1(n+1) := (X_1(n) + \xi_{n+1})_+ = \begin{cases} X_1(n) + \xi_{n+1}, & \text{if } X_1(n) + \xi_{n+1} > 0, \\ 0, & \text{if } X_1(n) + \xi_{n+1} \leq 0 \end{cases}$$

and

$$X_2(n+1) := \begin{cases} X_2(n) + \xi_{n+1}, & \text{if } X_2(n) + \xi_{n+1} > 0 \text{ and } X_2(n) > 0, \\ 0, & \text{if } X_2(n) + \xi_{n+1} \leq 0 \text{ and } X_2(n) > 0, \\ \eta_n, & \text{if } X_2(n) = 0. \end{cases} \quad (1.2)$$

1 Introduction

Here, as usual, $x_+ = \max(x,0)$ for $x \in \mathbb{R}$. We interpret X_1 and X_2 as "an amount of goods in a warehouse" or "the number of requests in a queue" at time n. The random variable ξ_n is the difference between "arrivals" and "departures" at time n, and the behavior of X_1 and X_2 at zero encodes the rules that are applied when a stock or a queue is empty. Note that X_1 and X_2 can be thought of as random walks with membranes located on the negative halfline.

There is no simple relation between the distributions of $(X_1(n))_{n\geq 0}$ and $(|S_\xi(n)|)_{n\geq 0}$, unless the distribution of ξ is symmetric. Nevertheless, the distributional limit of $(\sigma^{-1} n^{-1/2} X_1(\lfloor nt \rfloor))_{t \geq 0}$ as $n \to \infty$ is again $(|W(t)|)_{t \geq 0}$. As we shall see in Sect. 2.1, the sequence $(X_1(n))_{n \geq 0}$ is the image of $(|S_\xi(n)|)_{n \geq 0}$ under the Skorokhod reflection map and, by continuity of the latter, converges in distribution to the image $W^{\text{refl}} = (W^{\text{refl}}(t))_{t \geq 0}$ of the Skorokhod reflection map applied to W. Adhering to this viewpoint, the process W^{refl} can be defined as a unique strong (pathwise) solution to the equation

$$W^{\text{refl}}(t) = W(t) + L_{W^{\text{refl}}}(t), \quad t \geq 0, \tag{1.3}$$

where $L_{W^{\text{refl}}}$ is a nondecreasing continuous process with $L_{W^{\text{refl}}}(0) = 0$, which only increases at times when W^{refl} hits 0. Furthermore, the sum $W(t) + L_{W^{\text{refl}}}(t)$ is nonnegative for all $t \geq 0$. It is known that

$$(W^{\text{refl}}(t))_{t \geq 0} \stackrel{d}{=} (|W(t)|)_{t \geq 0}, \tag{1.4}$$

and that $L_{W^{\text{refl}}}$ in (1.3) is a symmetric local time of W^{refl} at 0. The process W^{refl}, as well as $|W|$, is called a *reflected Brownian motion*.

Equation (1.3) is called *Skorokhod reflection equation*. Under mild assumptions there exists a unique solution to (1.3) with a process W which is not necessarily a Brownian motion. Furthermore, this solution which we call a *reflected process* is a continuous function of the process W. One may expect that formula (1.3) is an efficient tool for guessing a reflected process which could be a scaling limit of a reflected random walk. We stress that distributional equality (1.4) which holds true whenever W is a Brownian motion is an exception rather than a rule.

The distributional limits of $(\sigma^{-1} n^{-1/2} X_2(\lfloor nt \rfloor))_{t \geq 0}$ as $n \to \infty$ are of a more interesting nature, especially when $\mathbb{E}\eta = \infty$. If $\mathbb{E}\eta \in (0, \infty)$, the distributional limit is again $|W|$. It will be shown in Sect. 3.2.3 that if the distribution of η belongs to the domain of attraction of a β-stable distribution with $\beta \in (0, 1)$, then the distributional limit is rather nonstandard and given by

$$W^{\text{refl},\beta}(t) := W(t) + \mathcal{V}_\beta(\mathcal{V}_\beta^{\leftarrow}(\max_{s \in [0,t]}((W(s))_-))), \quad t \geq 0, \tag{1.5}$$

where $\mathcal{V}_\beta$ is a β-stable subordinator which is independent of a Brownian motion W, $\mathcal{V}_\beta^{\leftarrow}$ is its generalized inverse, and $x_- = \max(-x, 0)$ for $x \in \mathbb{R}$. Of course,

$\max_{s \in [0, t]}((W(s))_-) = -\min_{s \in [0, t]}(W(s))$ provided that $W(0) = 0$. The nondegenerate "reflecting" term $\mathcal{V}_\beta(\mathcal{V}_\beta^\leftarrow (\max_{s \in [0,t]}((W(s))_-)))$ is, roughly speaking, the scaling limit of the composition of S_η and the number of visits of $(X(k))_{0 \leq k \leq n}$ to 0. Even though neither of the composed processes converges in distribution under Donsker's scaling, their composition does indeed exhibit growth at the "magic" square-root rate. We interpret the process $W^{\mathrm{refl},\beta}$ as a Brownian motion with jump-type exit from 0 and reflection governed by the β-stable subordinator $\mathcal{V}_\beta$.

Consider now a non-reflecting perturbation of a simple random walk S_ε at 0 such that the resulting locally perturbed random walk may hit negative integers. Namely, let $X = (X(n))_{n \geq 0}$ be a Markov chain on $\mathbb{Z}$ with the transition probabilities $\widetilde{p}_{i,i\pm 1} = 1/2$ for $i \neq 0$, $\widetilde{p}_{0,1} = p$ and $\widetilde{p}_{0,-1} = 1 - p$, where $p \in (0, 1)$. With the help of an argument similar to Andre's reflection principle, it can be checked that the n-step transition probabilities of X are given by

$$\widetilde{p}_{i,j}^{(n)} = p_{i,j}^{(n)} + \gamma \, \mathrm{sgn}\,(j) p_{|i|,-|j|}^{(n)}, \tag{1.6}$$

where $\gamma = 2p - 1$ and $p_{i,j}^{(n)} := \mathbb{P}\{S_\varepsilon(n) = j \mid S_\varepsilon(0) = i\}$ is the n-step transition probability of a simple random walk. Therefore, it comes as no surprise that Donsker's scaling of X converges in distribution to a Markov process W_γ^{skew} on $\mathbb{R}$ whose transition probability density function is equal to

$$p_t(x, y) = \varphi_t(y - x) + \gamma \, \mathrm{sgn}\,(y)\varphi_t(|x| + |y|), \quad x, y \in \mathbb{R}, \quad t > 0.$$

Here, the function φ_t defined by $\varphi_t(z) := (2\pi t)^{-1/2} e^{-\frac{z^2}{2t}}$, $z \in \mathbb{R}$, $t > 0$, is a density of the normal distribution with mean 0 and variance t. The process W_γ^{skew} is called a *skew Brownian motion* with permeability parameter $\gamma \in [-1, 1]$.

Now, let $X = (X(n))_{n \in \mathbb{N}_0}$ be a perturbation on a finite set A of a general integer-valued standard random walk S_ξ. Under some natural conditions the scaling limit process of X is still a skew Brownian motion (see Sect. 3.3). It is hopeless to obtain a simple formula similar to (1.6) for the transition probabilities of a general locally perturbed random walk and/or to exploit classical methods for investigating scaling limits of X. Furthermore, the skew Brownian motion with $\gamma \notin \{-1, 0, 1\}$ is not a continuous function of the underlying Brownian motion W, in contrast to a reflected Brownian motion. Hence, arguments based on the continuous mapping theorem fail when proving functional limit theorems in this case. On the other hand, it is known that W_γ^{skew} is a unique strong solution to the stochastic differential equation with generalized drift

$$\mathrm{d}X(t) = \mathrm{d}W(t) + \gamma \mathrm{d}L_0^X(t), \quad t \geq 0, \tag{1.7}$$

where L_0^X is a symmetric local time at 0 of the unknown process X.

1 Introduction

If the original standard random walk, properly scaled, centered, and normalized, converges in distribution to a Brownian motion (or a general Lévy process), then the scaling limit of the corresponding locally perturbed random walk with a finite membrane should behave like a Brownian motion (or a Lévy process) outside 0. The main problem is to describe how it exits 0.

We split the exposition into two parts:

- *Lévy-type processes with singularities at* 0, where we describe different types of processes that appear as scaling limits of locally perturbed random walks. For instance, we investigate properties of a reflected Brownian motion, a skew Brownian motion, a Walsh Brownian motion, etc.
- *Functional limit theorems*, where we prove distributional convergence of locally perturbed random walks, properly scaled, centered, and normalized, to processes introduced in the first part.

There are numerous methods to describe a Markov process. One may use transition probabilities, semigroups, resolvents, martingale problems, stochastic differential equations, the Skorokhod problem and its generalizations (for reflected processes), Itô's excursion theory, etc. It is hard to say which approach is better, and each of them has its own merits and drawbacks. Thus, we shall use various methods interchangeably. For instance, the easiest way to define a skew Brownian motion is to give a formula for its transition probabilities. A description of a process from the viewpoint of excursion theory provides us with a transparent probabilistic understanding of its structure. Methods based on a martingale problem or resolvent analysis are most effective for proving functional limit theorems.

We proceed with a brief overview of sections and our main results. We postpone historical notes and comments until the end of each section. Section 2.1 deals with reflected processes. We discuss in Sect. 2.1.1 the classical Skorokhod reflection problem and its applications to a construction of a reflected Brownian motion and reflected diffusions. Also in this section, we identify a reflecting term L of the problem with the local time of the reflected process. A generalization of the Skorokhod problem to the case of jump-type exit from 0 is treated in Sect. 2.1.2. Although this generalization is not as known as the classical Skorokhod reflection problem, its usage is an efficient tool for establishing functional limit theorems, in which the limit processes admit representations similar to (1.5). We investigate in Sect. 2.1.3 some properties of the process $W^{\text{refl},\beta}$ appearing in (1.5). In particular, we justify the Markov property of $W^{\text{refl},\beta}$ and prove that the process $(L_0^{W^{\text{refl},\beta}}(t))_{t\geq 0}$ defined by $L_0^{W^{\text{refl},\beta}}(t) := \mathcal{V}_\beta^\leftarrow(\max_{s\in[0,t]}((W(s))_-)$ for $t \geq 0$ is a local time at 0 of $W^{\text{refl},\beta}$. Representation (1.5) can then be written in an equivalent form

$$W^{\text{refl},\beta}(t) = W(t) + \mathcal{V}_\beta\big(L_0^{W^{\text{refl},\beta}}(t)\big), \quad t \geq 0. \tag{1.8}$$

In Sect. 2.2 we discuss the skew Brownian motion and its natural generalization called the *Walsh Brownian motion*. Several limit theorems are given together with a martingale characterization, to be exploited later in the text. Section 2.3 is concerned with constructing an analogue of the skew Brownian motion in the situation that an underlying Brownian motion is replaced with a symmetric α-stable Lévy process $(U_\alpha(t))_{t\geq 0}$, $\alpha \in (1, 2)$. This is a new object that appeared for the first time in the recent paper [70]. We define the *skew α-stable Lévy process* with parameter $\beta \in (0, \alpha - 1)$ as a distributional scaling limit of small local perturbations of U_α upon visits to 0. It turns out that the skew α-stable Lévy process with parameter β is a *weak* solution, see Definition 2.3.2, to the equation

$$X(t) = X(0) + U_\alpha(t) + \mathcal{S}_\beta(L_0^X(t)), \quad t \geq 0, \tag{1.9}$$

which looks rather similar to (1.5), (1.7), and (1.8). Here, $\mathcal{S}_\beta$ is a β-stable Lévy process of locally finite variation, which is independent of U_α, and L_0^X is a local time of X at 0. It is known that Eq. (1.7) does not have solutions if $|\gamma| > 1$. We prove a similar result for Eq. (1.9). Also, we provide characterizations of the skew α-stable Lévy process with parameter β in terms of resolvents and with the help of Itô's excursion theory.

We state and prove in Chap. 3 functional limit theorems for various locally perturbed random walks. In particular, we show that their scaling limits are the processes discussed in Chap. 2. The most interesting results are obtained for the one-dimensional case, in which we consider two types of local perturbations. The resulting processes are random walks on $\mathbb{N}_0$ reflected to the right upon crossing 0 (see Sect. 3.1.2) and random walks on $\mathbb{Z}$ with a finite membrane, that is, random walks perturbed at a finite number of points (see Sects. 3.3 and 3.4). Our theorems presented in Sects. 3.1.2 and 3.3 depend on the distribution tail behavior of both the increments of the original unperturbed random walk S_ξ and the perturbations. Now we discuss typical results that we obtain for *random walks with reflection*. Assume that $\mathbb{E}\xi = 0$ and $\text{Var}\,\xi \in (0, \infty)$.

(a) If the jumps from 0 have a finite mean, for instance, $\mathbb{E}\eta \in (0, \infty)$ in (1.2), then Donsker's scaling limit of the random walk with reflection is a reflected Brownian motion; see (1.3) or (1.4). This setting is treated in Theorem 3.2.1.
(b) If the distribution of the jumps from 0 belongs to the domain of attraction of a stable distribution concentrated on the nonnegative halfline, then Donsker's scaling limit is a Brownian motion with jump-type exit from 0 and reflection governed by a β-stable subordinator; see (1.5) or (1.8). Details can be found in Theorem 3.2.3.
(c) If the distribution tail of the jumps is slowly varying at infinity, then Donsker's scaling limit does not exist. This is shown in Theorem 3.2.4.

As far as the results from the preceding paragraph are concerned, our main tools are the Skorokhod reflection problem and its generalizations, combined with the continuous mapping theorem. We stress that our technique applies to situations which are more general than the reflection of a Brownian motion.

1 Introduction

Here are our results for integer-valued *random walks with membrane* that consists of a finite number of points.

(i) If $\mathbb{E}\xi = 0$, $\operatorname{Var}\xi \in (0, \infty)$ and the perturbations have a finite mean, then according to Theorem 3.3.3 Donsker's scaling limit is a skew Brownian motion. To prove this, we apply a martingale characterization of a Walsh Brownian motion together with a thorough study of the random walk considered at entrance times to and exit times from the membrane. Our results cover most of the existing results on this topic. Moreover, our methods enable us to prove limit theorems for the number of crossings of 0 and related functionals.

(ii) Assume that the membrane consists of one point, say 0, and the distribution of the jumps from $x \neq 0$ belongs to the domain of attraction of a symmetric α-stable distribution with $\alpha \in (1, 2)$, whereas the distribution of the jumps from 0 belongs to the domain of attraction of a β-stable law with $\beta \in (0, 1)$. If $\beta < \alpha - 1$ then the scaling limit of the random walk with membrane $\{0\}$ is a skew α-stable Lévy process with parameter β. If $\beta > \alpha - 1$, then the perturbations have no effect and the scaling limit is just a symmetric α-stable Lévy process. The latter results that can be found in Theorem 3.4.1 are novel and, to the best of our knowledge, have no analogues in the literature.

Multidimensional locally perturbed random walks are investigated in Sect. 3.5. We only mention our results of two types. Assume that the membrane A consists of a finite number of points. If the distribution tail of the jumps from A is not too heavy, then the scaling limits for the original and the locally perturbed random walks are the same, that is, the membrane plays no role; see Theorem 3.5.1. According to Theorem 3.5.2, if the distribution tail of the jumps from A is too heavy, then the locally perturbed random walk admits no Donsker's scaling limit even though the original random walk does. Another model is a random walk with membrane being a hyperplane. It is discussed in Sect. 3.5.2. If the underlying random walk is a simple random walk and the perturbations have a periodic structure, then the scaling limit is a *Brownian motion with a generalized drift on the hyperplane*, which is a natural generalization of the skew Brownian motion.

To make the book self-contained and accessible for graduate and postgraduate students, we have added an appendix with a collection of important facts and definitions that are frequently used throughout the text. For instance, we recall the definition of the J_1-topology on the space $D([0, \infty), \mathbb{R}^d)$ of càdlàg functions, some basic results on weak convergence such as the continuous mapping theorem, the Skorokhod representation theorem, Donsker's invariance principle, etc. We also give a brief introduction to Itô's excursion theory.

A Short Survey of Some Other Perturbed Random Walks
In this section we provide pointers to articles, which investigate some locally perturbed random walks, not treated in the present book. The list given below is not exhaustive.

Brooks and Chacon in [26] investigate *stretched random walks* and prove that their scaling limits are *stretched Brownian motions*. The stretched random walks include as a special case our random walks with membrane {0}, that is, the Markov chains with transition probabilities given in (1.6).

Perturbed random walks appearing in this book are, for the most part, Markovian. This property streamlines the analysis, granting access to robust analytical tools available in the Markov process theory. Also, it enables us to formulate a reasonably comprehensive asymptotic theory for these processes.

Davis in [36] investigates a random walk perturbed at its local extremes. This is a non-Markovian nearest-neighbor random walk $X = (X(n))_{n \geq 0}$ on $\mathbb{Z}$ which evolves according to the following rules. If the current state $X(n)$ lies strictly between the current running extremes $\min\{X(k) : k \leq n\}$ and $\max\{X(k) : k \leq n\}$, then $X(n+1) = X(n) \pm 1$ with probability $1/2$. If $X(n)$ is equal to the current minimum (respectively, maximum), then $X(n+1) = X(n) + 1$ with probability $1/(2-\alpha)$ (respectively, $1 - 1/(2-\beta)$) for some $\alpha, \beta \in (-\infty, 1)$. If $\alpha = \beta < 0$, then X is a simplest instance of the reinforced random walk; see [35, 161]. It was shown in [36], with the earlier contribution due to Werner [173] in the cases $\alpha = 0$ or $\beta = 0$, that, under Donsker's scaling, X converges in distribution to a unique solution of the equation

$$Y^{\alpha,\beta}(t) = W(t) + \beta \min_{s \in [0,t]} Y^{\alpha,\beta}(s) + \alpha \max_{s \in [0,t]} Y^{\alpha,\beta}(s), \quad t \geq 0$$

provided that $|\alpha\beta/((1-\alpha)(1-\beta))| < 1$, where $(W(t))_{t \geq 0}$ is a standard Brownian motion. Existence and uniqueness of solutions to the latter equation is discussed in [31]. It was proved by Kosygina, Mountford, and Peterson in [89] that, under Donsker's scaling, *excited random walks with Markovian cookie stacks* converge weakly in the Skorokhod space to $Y^{\alpha,\beta}$, with appropriate α and β. An incomplete list of other articles concerned with some perturbed random walks includes [38, 90, 124, 125, 138].

In this book we are concerned, for the most part, with random walks with membranes comprised of a finite number of deterministic points. There are papers, which investigate models with membranes given by countably many deterministic or random points. Andreoletti and Debs consider in [1] a simple random walk on the plane with membrane defined by the integer points of the coordinate axes. Matzavinos et al. in [104] introduce and analyze a *random walk in a sparse random environment*. Further results on this object are obtained by Buraczewski et al. in [27–29] and by Kołodziejska in [84]. This is a simple random walk on the line with membrane given by the successive positions of a two-sided standard random walk. The term "random environment" refers to the fact that probabilities of the unit jumps from the membrane are random. It is shown that as far as one-dimensional distributional convergence is concerned, the properties of this model deviate from those of the corresponding random walk in a non-sparse random environment *weakly* [104], *moderately* [28], or *strongly* [27] depending on whether the distance between the consecutive points of the membrane is a.s. bounded and has a finite or an infinite mean, respectively. Poisat and Simenhaus in [133, 134] deal with a *simple random walk among power-law renewal obstacles* which is defined as follows. A simple random walk moves over real line, dies with a given probability $p \in (0, 1)$ whenever it hits an obstacle, and dies with probability 1 on entering into negative halfline. Obstacles are given by a standard random walk with nonnegative jumps having a distribution μ satisfying $\mu\{n\} \sim \text{const} \cdot n^{-1-\beta}$ as $n \to \infty$ for some $\beta > 0$. The standard random walk is assumed independent of the simple random walk.

Lévy-Type Processes with Singularities

2

> *The singularity is the point where our old models must be discarded, and a new reality rules.*
>
> Vernor Vinge, The Coming Technological Singularity (1993)

Abstract This chapter is devoted to construction of "Lévy-type processes with singularities," which is a class of stochastic processes that serve as scaling limits for locally perturbed random walks and particularly for random walks with membranes. We start by discussing reflected processes derived from the classic and generalized Skorokhod reflection mappings. Then we review both old and new facts about a skew Brownian motion and present the first rigorous definition of a skew stable Lévy process.

2.1 Reflected Processes

2.1.1 The Skorokhod Reflection Problem

Consider the following discrete-time single server queuing system. The working time of the server is divided into discrete cycles indexed by $n = 1, 2, \ldots$. A cycle $n \in \mathbb{N}$ is further split into two parts. During the first part a_n customers arrive at the system and join the queue. During the second part d_n customers are served if the present queue length is larger than d_n, whereas all the customers are served otherwise. Denote by $X(n)$ the number of customers in the queue at the end of the nth cycle. Put

$$b_n := a_n - d_n, \quad n \in \mathbb{N}$$

and note that $X(n+1) = X(n) + b_{n+1}$ provided that $X(n) + b_{n+1} \geq 0$, and $X(n+1) = 0$, otherwise. This yields the recursion

$$X(0) = x_0, \quad X(n+1) = (X(n) + b_{n+1})_+, \quad n \in \mathbb{N}_0 \qquad (2.1)$$

known as Lindley's model. Here, $x_0 \in \mathbb{N}_0$ represents the number of customers initially present in the system and $x_+ = \max(x, 0)$ for $x \in \mathbb{R}$. In this chapter we discuss continuous-time counterparts of the sequence $X = (X(n))_{n \geq 0}$. A natural way to look at X is to regard it as a reflected at zero standard random walk $S_b = (S_b(n))_{n \in \mathbb{N}_0}$ defined by

$$S_b(0) := x_0 \quad \text{and} \quad S_b(n) := x_0 + b_1 + \ldots + b_n, \quad n \in \mathbb{N}$$

(here, x_0 is the same as in (2.1)). This means that whenever S_b hits a negative halfline it is forced to stay at zero until a positive increment of S_b appears. Exploiting a similar idea, a continuous-time counterpart $X = (X(t))_{t \geq 0}$ derived from a sufficiently smooth continuous-time process $Y = (Y(t))_{t \geq 0}$ may be constructed as follows. Assume that Y possesses smooth trajectories and a locally finite number of local extrema. Then the increments of a continuous reflected process X should coincide with the increments of Y while X is positive, whereas after hitting 0 the process X stays at 0 until the derivative of Y becomes positive. *Although this intuition seems to be a good starting point,* it breaks down if trajectories of Y are not smooth enough, for example, if Y is a Brownian motion. To avoid this difficulty, we take another approach, which is based on the fact that $(X(n))_{n \geq 0}$ in (2.1) admits the following representation:

$$X(n) = S_b(n) - \min_{1 \leq k \leq n} (S_b(k) \wedge 0) = S_b(n) + \max_{1 \leq k \leq n} ((S_b(k))_-), \quad n \in \mathbb{N} \qquad (2.2)$$

where $x_- = \max(-x, 0)$ for $x \in \mathbb{R}$. Formula (2.2), which can be checked by induction, suggests the following definition of a continuous-time counterpart of discrete Lindley's model.

Definition 2.1.1 Let $Y = (Y(t))_{t \geq 0}$ be a nonrandom càdlàg function with $Y(0) \geq 0$. The mapping $\Gamma : D([0, \infty)) \to D([0, \infty))$ given by

$$X(t) := \Gamma(Y)(t) = Y(t) - \inf_{s \in [0, t]} (Y(s) \wedge 0) = Y(t) + \sup_{s \in [0, t]} (-(Y(s) \wedge 0))$$

$$= Y(t) + \sup_{s \in [0, t]} ((Y(s))_-), \quad t \geq 0 \qquad (2.3)$$

is called the *Skorokhod reflection map*.

2.1 Reflected Processes

Remark 2.1.1 If Y is only defined on a finite interval $[0, T]$, then so is X, the corresponding Skorokhod reflection map being given by the same formula as before.

It can be checked that the Skorokhod reflection map enjoys the following properties:

(1) If Y is a continuous, respectively, càdlàg function, then so is $\Gamma(Y)$.
(2) The map Γ is Lipschitz continuous in the supremum norm on any finite interval. More precisely, for every $T > 0$ and $Y_1, Y_2 \in D([0, T])$,

$$\sup_{t \in [0, T]} |\Gamma(Y_1)(t) - \Gamma(Y_2)(t)| \leq 2 \sup_{t \in [0, T]} |Y_1(t) - Y_2(t)|. \qquad (2.4)$$

(3) The mapping Γ is continuous as a mapping $D([0, \infty)) \mapsto D([0, \infty))$ and also $C([0, \infty)) \mapsto C([0, \infty))$.
(4) If $\mathcal{F}$ is a filtration on some probability space and Y is an $\mathcal{F}$-adapted càdlàg stochastic process on this space, then

$$\Gamma(Y) \text{ is an } \mathcal{F}\text{-adapted càdlàg stochastic process.} \qquad (2.5)$$

Now we give the original Skorokhod's definition that explains the term "reflection map."

Definition 2.1.2 Assume that y is a càdlàg function on $[0, \infty)$ with $y(0) \geq 0$. A pair (x, l) of càdlàg functions is called a solution to the Skorokhod reflection problem for y if

$$x(t) \geq 0, \quad x(t) = y(t) + l(t), \quad t \geq 0, \qquad (2.6)$$

the function l is nondecreasing with $l(0) = 0$, and

$$\int_{[0, \infty)} \mathbb{1}_{\{x(s) > 0\}} dl(s) = 0. \qquad (2.7)$$

If y is càdlàg on $[0, T]$ for $T > 0$, then a solution to the Skorokhod reflection problem is defined analogously.

We interpret y as a driving noise, x as a position of a particle that reflects at 0, and l as a compensation that does not allow the particle to get through 0. Increments of x coincide with those of y whenever x takes values in $(0, \infty)$. Equation (2.7) tells us that the function l may only increase on the set $\{t \geq 0 : x(t) = 0\}$. Note also that the particle reflects "continuously" from 0, that is, if $x(t-) = 0$ and $x(t) > 0$, then $l(t) - l(t-) = 0$, and the jump-type exit from 0 may only occur due to a jump of y at t. By the monotone convergence theorem equality (2.7) is equivalent to

$$\int_{[0,\infty)} \mathbb{1}_{\{x(s)>\delta\}} dl(s) = 0 \quad \text{for every fixed } \delta > 0.$$

Furthermore, both are equivalent to

$$\int_{[0,\infty)} x(s) dl(s) = 0. \tag{2.8}$$

Indeed, (2.8) obviously implies (2.7), whereas (2.7) entails

$$0 \le \int_{[0,T]} x(s) dl(s) \le \left(\sup_{s \in [0,T]} x(s) \right) \int_{[0,T]} \mathbb{1}_{\{x(s)>0\}} dl(s) = 0$$

for each $T > 0$. Sending $T \to \infty$ gives (2.8).

We denote by $\mathcal{SR}(y)$ the Skorokhod reflection problem for a function y.

Theorem 2.1.1 *For any $y \in D([0, \infty))$ with $y(0) \ge 0$ there is a unique solution to the Skorokhod reflection problem. The solution is given by*

$$x(t) = (\Gamma(y))(t) = x(t) + \sup_{s \in [0,t]} ((y(s))_-), \quad l(t) = \sup_{s \in [0,t]} ((y(s))_-), \quad t \ge 0. \tag{2.9}$$

Remark 2.1.2 In the literature, $\sup_{s \in [0,t]}((y(s))_-)$ is often written in equivalent forms $- \inf_{s \in [0,t]}(y(s) \wedge 0)$ or $-((\inf_{s \in [0,t]}(y(s))) \wedge 0)$.

Proof The fact that the pair (x, l) given in (2.9) is a solution to the Skorokhod reflection problem follows by a routine check. To prove uniqueness, assume that (x_1, l_1) and (x_2, l_2) are two solutions. Put $z(t) := l_1(t) - l_2(t) = x_1(t) - x_2(t)$ and note that the variation of z is locally bounded. For every $t \ge 0$,

$$0 \le (z(t))^2 + \sum_{0 < s \le t} (z(s) - z(s-))^2 = 2 \int_{(0,t]} z(s) dz(s)$$

$$= 2 \int_{(0,t]} (x_1(s) - x_2(s)) d(l_1(s) - l_2(s))$$

$$= -2 \int_{(0,t]} x_1(s) dl_2(s) - 2 \int_{(0,t]} x_2(s) dl_1(s) \le 0.$$

Here, the last equality is a consequence of (2.8). Thus, $z(t) = 0$ for every $t \ge 0$. □

2.1 Reflected Processes

The next corollary will be used in our proofs of functional limit theorems for random walks with reflection at 0.

Corollary 2.1.1 *Assume that a sequence of pairs of càdlàg processes $((Y_n, L_n))_{n\geq 1}$ satisfies the following conditions:*

(I) $Y_n \Longrightarrow Y$ *as* $n \to \infty$ *on* $D([0, \infty))$, *where Y is a.s. continuous with* $Y(0) \geq 0$.
(II) *The sequence of distributions* $(\mathbb{P}\{L_n \in \cdot\})_{n\geq 1}$ *is tight on* $D([0, \infty))$.
(III) *Every distributional limit point (Y, L) of the sequence $((Y_n, L_n))_{n\geq 1}$ in the product topology on $(D([0, \infty)))^2$ is such that:*
  (a) *L is a.s. nondecreasing with $L(0) = 0$.*
  (b) *The process $(X(t))_{t\geq 0}$ given by $X(t) := Y(t) + L(t)$ for $t \geq 0$ is a.s. nonnegative.*
  (c) $\int_{[0,\infty)} \mathbb{1}_{\{X(s)>0\}} \mathrm{d}L(s) = 0$ *a.s..*

Then, with $X_n := Y_n + L_n$,

$$(X_n, Y_n, L_n) \Longrightarrow (X, Y, L), \quad n \to \infty \tag{2.10}$$

in the J_1-topology on $D([0, \infty), \mathbb{R}^3)$, where the pair $(X, L) := (Y + L, L)$ is a solution to the Skorokhod reflection problem $\mathcal{SR}(Y)$.

Proof It follows from conditions (I) and (II) that the sequence of distributions of $((Y_n, L_n))_{n\geq 1}$ is tight in the product topology on $(D([0, \infty), \mathbb{R}))^2$. Then there exists a sequence $(n_k)_{k\geq 1}$ such that

$$(Y_{n_k}, L_{n_k}) \Longrightarrow (Y, L), \quad k \to \infty$$

in the product topology on $(D([0, \infty), \mathbb{R}))^2$.

According to the Skorokhod representation theorem (Theorem 4.1.2) there exist $((\widetilde{Y}_{n_k}, \widetilde{L}_{n_k}))_{k\geq 1}$ versions of $((Y_{n_k}, L_{n_k}))_{k\geq 1}$ and $(\widetilde{Y}, \widetilde{L})$ a version of (Y, L) (all the versions being defined on a common probability space) such that

$$\lim_{k\to\infty} (\widetilde{Y}_{n_k}, \widetilde{L}_{n_k}) = (\widetilde{Y}, \widetilde{L}), \quad \text{a.s.}$$

in the product topology on $(D([0, \infty), \mathbb{R}))^2$. Put $\widetilde{X}_{n_k} := \widetilde{Y}_{n_k} + \widetilde{L}_{n_k}$ and $\widetilde{X} := \widetilde{Y} + \widetilde{L}$ for $k \in \mathbb{N}$. Since $\widetilde{Y}$ is a.s. continuous, we conclude with the help of the first part of Proposition 4.3.6 that $\lim_{k\to\infty} \widetilde{X}_{n_k} = \widetilde{X}$ a.s. in the J_1-topology on $D([0, \infty), \mathbb{R})$ and thereupon

$$\lim_{k\to\infty} (\widetilde{X}_{n_k}, \widetilde{Y}_{n_k}, \widetilde{L}_{n_k}) = (\widetilde{X}, \widetilde{Y}, \widetilde{L}) \quad \text{a.s.} \tag{2.11}$$

in the product topology on $(D([0, \infty), \mathbb{R}))^3$. Condition (III) ensures that $(\widetilde{X}, \widetilde{L})$ is a solution to the Skorokhod problem $\mathcal{SR}(\widetilde{Y})$ which particularly implies that $\widetilde{L}(t) = \max_{s \in [0, t]}((\widetilde{Y}(s))_-)$. This proves a.s. continuity of the limit process $(\widetilde{X}, \widetilde{Y}, \widetilde{L})$ which together with Proposition 4.3.4 guarantees that relation (2.11) holds a.s. in the J_1-topology on $D([0, \infty), \mathbb{R}^3)$. As a consequence,

$$(X_{n_k}, Y_{n_k}, L_{n_k}) \implies (X, Y, L), \quad k \to \infty$$

in the J_1-topology on $D([0, \infty), \mathbb{R}^3)$.

Hence, the distribution of any limit point (X, Y, L) is determined uniquely. This implies that for any subsequence of $((X_n, Y_n, L_n))_{n \geq 1}$ there exists a subsubsequence such that the corresponding triples converge in distribution to the triple (X, Y, L) described above. This secures convergence (2.10) of the whole sequence. □

Remark 2.1.3 The assumption of tightness of the sequence $(\mathbb{P}\{L_n \in \cdot\})_{n \geq 1}$ can be replaced by an equivalent assumption of tightness of the sequence $(\mathbb{P}\{X_n \in \cdot\})_{n \geq 1}$.

We proceed by giving a short introduction to *stochastic differential equations with reflection*, in which the noise is a Brownian motion, and associate the reflecting process L with a two-sided semimartingale local time. Let $\mathcal{F} = (\mathcal{F}_t)_{t \geq 0}$ be a filtration on the underlying probability space and W a standard one-dimensional $\mathcal{F}$-adapted Brownian motion. Let X_0 be a nonnegative $\mathcal{F}_0$-measurable random variable and $a, b : [0, \infty) \to \mathbb{R}$ measurable functions.

Definition 2.1.3 A pair of continuous $\mathcal{F}$-adapted processes $(X(t), L(t))_{t \geq 0}$ is called a solution to the stochastic differential equation with reflection at 0

$$\begin{cases} dX(t) = a(X(t))dt + b(X(t))dW(t) + dL(t), & t \geq 0, \\ X(0) = X_0 \end{cases} \quad (2.12)$$

provided that:

(i) $X(t) \geq 0$ for $t \geq 0$.
(ii) The process L is a.s. nondecreasing with $L(0) = 0$.
(iii) $\int_{[0, \infty)} \mathbb{1}_{\{X(s) > 0\}} dL(s) = 0$ a.s.
(iv) a.s. for every $t \geq 0$

$$X(t) = X_0 + \int_0^t a(X(s))ds + \int_0^t b(X(s))dW(s) + L(t), \quad (2.13)$$

and all the integrals are well-defined.

2.1 Reflected Processes

Observe that (X, L) is a solution to (2.12) if, and only if, (X, L) is a solution to the Skorokhod problem

$$\mathcal{SR}\left(X_0 + \int_0^{(\cdot)} a(X(s))\mathrm{d}s + \int_0^{(\cdot)} b(X(s))\mathrm{d}W(s)\right).$$

In particular,

$$X = \Gamma\left(X_0 + \int_0^{(\cdot)} a(X(s))\mathrm{d}s + \int_0^{(\cdot)} b(X(s))\mathrm{d}W(s)\right) \quad \text{a.s.} \tag{2.14}$$

If a and b are Lipschitz continuous functions, then existence and uniqueness of a solution to (2.14) and consequently to (2.12) follow from Lipschitz property (2.4) and property (2.5) with the help of standard arguments (Gronwall's lemma, Picard's iterations, etc.).

Now we intend to associate L with a local time of X at 0. A solution to (2.12) is a continuous semimartingale. We define the symmetric local time L_0^X as the a.s. limit

$$L_0^X(t) = \lim_{\varepsilon \to 0+} \frac{1}{2\varepsilon} \int_{[0,t]} \mathbb{1}_{\{X(s) \in (-\varepsilon, \varepsilon)\}} \mathrm{d}\langle X, X \rangle(s), \tag{2.15}$$

which is known to exist for every $t \geq 0$; see Corollary 1.9 on p. 227 in [140]. If b is continuous with $b(0) \neq 0$, formula (2.15) can be simplified to

$$L_0^X(t) = \lim_{\varepsilon \to 0+} \frac{(b(0))^2}{2\varepsilon} \int_0^t \mathbb{1}_{\{X(s) \in (-\varepsilon, \varepsilon)\}} \mathrm{d}s \quad \text{a.s.}$$

It follows from Tanaka's formula (see p. 222 in [140]) applied to $(X(t))_+ = X(t)$ that, for each $t \geq 0$,

$$L_0^X(t) = (X(t))_+ - (X(0))_+ - \int_{[0,t]} \mathbb{1}_{\{X(s)>0\}} \mathrm{d}X(s)$$

$$= X(t) - X(0) - \int_0^t a(X(s))\mathbb{1}_{\{X(s)>0\}}\mathrm{d}s - \int_0^t b(X(s))\mathbb{1}_{\{X(s)>0\}}\mathrm{d}W(s)$$

$$\qquad - \int_{[0,t]} \mathbb{1}_{\{X(s)>0\}}\mathrm{d}L(s)$$

$$= X(t) - X(0) - \int_0^t a(X(s))\mathbb{1}_{\{X(s)>0\}}\mathrm{d}s - \int_0^t b(X(s))\mathbb{1}_{\{X(s)>0\}}\mathrm{d}W(s)$$

$$= L(t) + \int_0^t \mathbb{1}_{\{X(s)=0\}}\left(a(X(s))\mathrm{d}s + b(X(s))\mathrm{d}W(s)\right)$$

$$= L(t) + a(0)\int_0^t \mathbb{1}_{\{X(s)=0\}}\mathrm{d}s + b(0)\int_0^t \mathbb{1}_{\{X(s)=0\}}\mathrm{d}W(s).$$

It is known that if $b(0) \neq 0$, then the process X spends zero time at 0 a.s., that is, $\int_0^\infty \mathbb{1}_{\{X(s)=0\}} ds = 0$ a.s. This particularly entails $\int_0^t \mathbb{1}_{\{X(s)=0\}} dW(s) = 0$ a.s. for each $t > 0$. Summarizing, we have checked that the process X defined in (2.12) satisfies

$$\max_{s \in [0,t]} ((X(s))_-) = L(t) = L_0^X(t), \quad t \geq 0. \tag{2.16}$$

If $a(x) = 0$, $b(x) = 1$ for $x \geq 0$ and $X_0 = x \geq 0$, then X is the reflected Brownian motion $(W_x^{\text{refl}}(t))_{t \geq 0}$ which starts at x:

$$W_x^{\text{refl}}(t) = x + W(t) + \max_{s \in [0,t]} ((x + W(s))_-) = x + W(t) + L_0^{W_x^{\text{refl}}}(t), \quad t \geq 0. \tag{2.17}$$

In particular, (2.16) entails a formula for the two-sided local time of W_x^{refl}:

$$L_0^{W_x^{\text{refl}}}(t) = \max_{s \in [0,t]} ((x + W(s))_-), \quad t \geq 0. \tag{2.18}$$

Quite often the term "reflected Brownian motion" is used for the absolute value of a Brownian motion. Here is an explanation of why this is not a coincidence. By Tanaka's formula

$$|x + W(t)| = x + \int_0^t \text{sgn}(x + W(s)) dW(s) + L_0^{x+W}(t)$$

$$= x + \int_0^t \text{sgn}(x + W(s)) dW(s) + L_0^{|x+W|}(t), \quad t \geq 0.$$

The process $(B(t))_{t \geq 0}$ defined by

$$B(t) := \int_0^t \text{sgn}(x + W(s)) dW(s), \quad t \geq 0$$

is a continuous martingale with quadratic variation $\langle B, B \rangle(t) = t$ a.s. for all $t \geq 0$. As a consequence, $(B(t))_{t \geq 0}$ has the same distribution as $(W(t))_{t \geq 0}$. Thus,

$$|x + W(t)| = x + B(t) + L_0^{|x+W|}(t), \quad t \geq 0,$$

which means that

$$(|x + W(t)|)_{t \geq 0} = (\Gamma(x + B(\cdot))(t))_{t \geq 0} \stackrel{d}{=} (\Gamma(x + W(\cdot))(t))_{t \geq 0} = (W_x^{\text{refl}}(t))_{t \geq 0}.$$

2.1 Reflected Processes

Moreover,

$$\left(W_x^{\text{refl}}(t), \max_{s\in[0,t]}((x+W(s))_-)\right)_{t\geq 0} \stackrel{\text{d}}{=} \left(|x+W(t)|, L_0^{x+W}(t)\right)_{t\geq 0}. \qquad (2.19)$$

We stress that the right-hand side of (2.19) is equal to $\left(|x+W(t)|, L_0^{|x+W|}(t)\right)_{t\geq 0}$.

2.1.2 The Generalized Skorokhod Reflection Problem

The following definition is a generalization of Definition 2.1.2 to the case of a jump-type reflection at 0. Throughout this section we write that a function is càdlàg whenever it belongs to $D([0,\infty))$.

Definition 2.1.4 Let y be a càdlàg function with $y(0) \geq 0$ and h an increasing càdlàg function with $h(0) = 0$. A pair (x, ℓ) of càdlàg functions is called a solution to the generalized Skorokhod problem for the noise y and reflection governed by h if

$$x(t) \geq 0, \quad x(t) = y(t) + h(\ell(t)), \quad t \geq 0,$$

the function ℓ is nondecreasing and continuous with $\ell(0) = 0$, and

$$\int_{[0,\infty)} \mathbb{1}_{\{x(s)>0\}} d\ell(s) = 0.$$

We denote the so defined generalized Skorokhod problem by $\mathcal{GSR}(y, h)$.

Theorem 2.1.2 *Assume that h is a strictly increasing càdlàg function with $h(0) = 0$ and*

$$\lim_{t\to\infty} h(t) = +\infty.$$

Then for any càdlàg function y with $y(0) \geq 0$, which does not have negative jumps, there exists a unique solution to the generalized Skorokhod problem $\mathcal{GSR}(y, h)$. With $h^{\leftarrow}(t) := \inf\{s \geq 0 : h(s) > t\}$ denoting the generalized inverse of h, the solution is given by

$$\ell(t) = h^{\leftarrow}(m(t)), \quad t \geq 0, \qquad (2.20)$$

where $m(t) := (m(y))(t) := \sup_{s\in[0,t]}((y(s))_-)$ for $t \geq 0$ and

$$x(t) = y(t) + h(h^{\leftarrow}(m(t))), \quad t \geq 0. \qquad (2.21)$$

Remark 2.1.4 Some properties of generalized inverse functions are discussed in Sect. 4.3.4 of the Appendix.

Proof We first prove that (x, ℓ) is a solution to $\mathcal{GSR}(y, h)$. Since h is strictly increasing, $h^{\leftarrow}$ is continuous and nondecreasing; see Proposition 4.3.13(vii). Moreover, the absence of negative jumps of y implies that m is also continuous. Indeed, since y_- does not have positive jumps and $y(0) \geq 0$ we infer

$$(m(y))(t-) = \sup_{s \in [0,t)} ((y(s))_-) = \Big(\sup_{s \in (0,t)} ((y(s))_-) \Big)_+ = \Big(\sup_{s \in (0,t]} ((y(s-))_-) \Big)_+$$

$$= \Big(\sup_{s \in (0,t]} ((y(s))_-) \Big)_+ = \sup_{s \in [0,t]} ((y(s))_-) = (m(y))(t), \quad t > 0.$$

Therefore, m is continuous and nondecreasing, and so is ℓ. The equality $\ell(0) = 0$ is a trivial consequence of $h(0) = 0 = m(0)$.

In view of $h(h^{\leftarrow}(t)) \geq t$,

$$x(t) = y(t) + h(h^{\leftarrow}(m(t))) \geq y(t) + m(t) =: \widetilde{x}(t), \quad t \geq 0.$$

The pair $(\widetilde{x}, m)$ is a solution to the Skorokhod reflection problem $\mathcal{SR}(y)$. In particular, $x(t) \geq \widetilde{x}(t) \geq 0$. It remains to check that

$$\int_{[0,\infty)} \mathbb{1}_{\{x(s)>0\}} d\ell(s) = 0. \tag{2.22}$$

Let $\{q_k\}$ be an at most countable set of discontinuities of h. Put $a_k := h(q_k-)$, $b_k := h(q_k)$ and observe that

$$h(h^{\leftarrow}(t)) = \begin{cases} t, & t \notin \cup_k [a_k, b_k), \\ b_k, & t \in [a_k, b_k) \text{ for some } k; \end{cases}$$

see Fig. 4.2 in Sect. 4.3.4. Note that $m(s) \notin \cup_k [a_k, b_k)$ entails $x(s) = \widetilde{x}(s)$ and thereupon

$$\int_{[0,\infty)} \mathbb{1}_{\{x(s)>0, m(s) \notin \cup_k [a_k,b_k)\}} d\ell(s) = \int_{[0,\infty)} \mathbb{1}_{\{\widetilde{x}(s)>0, m(s) \notin \cup_k [a_k,b_k)\}} dh^{\leftarrow}(m(s)).$$

Since $(\widetilde{x}, m)$ is a solution to the Skorokhod reflection problem $\mathcal{SR}(y)$, the function m does not increase on the set $\{s \geq 0 : \widetilde{x}(s) > 0\}$. As a consequence, neither does $s \mapsto h^{\leftarrow}(m(s))$, and the right-hand side of the last centered formula vanishes. If $m(s) \in [a_k, b_k)$ for some k, then

$$\int_{[0,\infty)} \mathbb{1}_{\{x(s)>0, m(s) \in [a_k,b_k)\}} d\ell(s) \leq \int_{[0,\infty)} \mathbb{1}_{\{m(s) \in [a_k,b_k)\}} d\ell(s) = 0$$

2.1 Reflected Processes

because ℓ is continuous and $h^{\leftarrow}$ takes a constant value on $[a_k, b_k)$. The proof of (2.22) is complete.

When proving uniqueness we argue as in the proof of Theorem 2.1.1. Assume that (x_1, ℓ_1) and (x_2, ℓ_2) are two solutions. First, we show that

$$\int_{(0,t]} x_1(s) dh(\ell_1(s)) = \int_{(0,t]} x_2(s) dh(\ell_2(s)) = 0. \tag{2.23}$$

Indeed, in view of

$$\int_{(0,t]} \mathbb{1}_{\{x_1(s)>0\}} d\ell_1(s) = \int_{(0,t]} \mathbb{1}_{\{x_2(s)>0\}} d\ell_2(s) = 0,$$

the function ℓ_j does not increase on the set $\{s \geq 0 : x_j(s) > 0\}$ for $j = 1, 2$, and neither does $h(\ell_j)$. As a consequence,

$$\int_{(0,t]} \mathbb{1}_{\{x_1(s)>0\}} dh(\ell_1(s)) = \int_{(0,t]} \mathbb{1}_{\{x_2(s)>0\}} dh(\ell_2(s)) = 0. \tag{2.24}$$

Put $z(t) := x_1(t) - x_2(t) = h(\ell_1(t)) - h(\ell_2(t))$ and note that for each $t \geq 0$

$$0 \leq (z(t))^2 + \sum_{0<s\leq t} (z(s) - z(s-))^2 = 2 \int_{(0,t]} z(s) dz(s)$$

$$= 2 \int_{(0,t]} (x_1(s) - x_2(s)) d(h(\ell_1(s)) - h(\ell_2(s)))$$

$$= -2 \int_{(0,t]} x_1(s) dh(\ell_2(s)) - 2 \int_{(0,t]} x_2(s) dh(\ell_1(s)) \leq 0$$

having utilized (2.23) for the last equality. Thus, $x_1(t) = x_2(t)$ and $h(\ell_1(t)) = h(\ell_2(t))$ for each $t \geq 0$. Since h is strictly increasing, we infer $\ell_1(t) = \ell_2(t)$ for each $t \geq 0$. □

Remark 2.1.5 The problem $\mathcal{GSR}(y, h)$ may have no solutions if y has negative jumps. However, the proof of uniqueness in Theorem 2.1.2 shows that if a solution exists, then it is unique.

Definition 2.1.5 Let $D_{0,\uparrow}^u([0, \infty))$ be the set of nondecreasing unbounded càdlàg functions h with $h(0) = 0$ and $D_+([0, \infty))$ the set of càdlàg functions h with $h(0) \geq 0$. The mapping $\widetilde{\Gamma} : D_+([0, \infty)) \times D_{0,\uparrow}^u([0, \infty)) \to D_+([0, \infty))$ defined by

$$(\widetilde{\Gamma}(y, h))(t) := y(t) + h(h^{\leftarrow}(m(t))), \quad t \geq 0,$$

where $m(t) = \sup_{s \in [0,t]} ((y(s))_-)$, is called the *generalized Skorokhod map*.

Remark 2.1.6 The pair $(\widetilde{\Gamma}(y,h), h^{\leftarrow}(m))$ is a solution to the problem $\mathcal{GSR}(y,h)$ provided that h is strictly increasing. Indeed, the problem $\mathcal{GSR}(y,h)$ is not defined if h is a nondecreasing function which is not strictly increasing. In the sequel, we use the generalized Skorokhod map irrespective of whether h is strictly increasing or nondecreasing.

Remark 2.1.7 If $y = W$ is a Brownian motion and $\mathcal{F}^W$ is the filtration generated by W, then a solution to the Skorokhod reflection problem $\mathcal{SR}(y)$ is $\mathcal{F}^W$-adapted. In contrast, the process x in a solution (x, ℓ) to the generalized Skorokhod problem is not necessarily $\mathcal{F}^W$-adapted. For instance, this is the case when h is an a.s. increasing subordinator with a nonvanishing Lévy measure which is independent of W. Nevertheless, it can be shown that x is a strong Markov process with respect to its own filtration $\mathcal{F}^x$; see Sect. 2.1.3.

To facilitate understanding we now consider three important examples of the generalized Skorokhod problem.

Example 2.1.1 Let h be a continuous strictly increasing function with $h(0) = 0$ and $\lim_{t \to +\infty} h(t) = +\infty$. Then $h^{\leftarrow}$ is the usual inverse function of h and

$$x(t) = y(t) + m(t) = (\Gamma(y))(t), \quad t \geq 0,$$

that is, (x, m) is a solution to the Skorokhod reflection problem.

Denote by $C_+([0, \infty))$ the set of continuous functions f defined on $[0, \infty)$ with $f(0) \geq 0$.

Example 2.1.2 Assume that $y \in C_+([0, \infty))$ and $h(t) = \delta t + \Pi(t)$ for $t \geq 0$. Here, $\delta > 0$ is a constant, and Π is a nondecreasing step function given by either

$$\Pi(t) = \sum_{k=1}^{n} a_k \mathbb{1}_{(b_k, b_{k+1}]}(t), \quad t \geq 0$$

with $0 < a_1 < \ldots < a_n$ and $0 < b_1 < \ldots < b_n < b_{n+1} := +\infty$ or

$$\Pi(t) = \sum_{k \geq 1} a_k \mathbb{1}_{(b_k, b_{k+1}]}(t), \quad t \geq 0$$

with $0 < a_1 < \ldots < a_n < \ldots, 0 < b_1 < \ldots < b_n < \ldots$ and $\lim_{n \to \infty} b_n = +\infty$.

In this case a solution (x, ℓ) to $\mathcal{GSR}(y,h)$ can be obtained as follows. Let (x_0, ℓ_0) be a solution to the Skorokhod reflection problem $\mathcal{SR}(y)$. Denote by t_1 the first time when ℓ_0 hits the level δb_1. Put $x(t) := x_0(t)$ for $t \in [0, t_1)$ and $x(t_1) := a_1$. Next, we solve the Skorokhod reflection problem $\mathcal{SR}(a_1 + y(\cdot) - y(t_1))$ for $t \geq t_1$. Let (x_1, ℓ_1)

2.1 Reflected Processes

be the corresponding solution and t_2 the first time when ℓ_1 hits the level $\delta(b_2 - b_1)$. Put $x(t) := x_1(t)$ for $t \in [t_1, t_2)$ and $x(t_2) := a_2 - a_1$. Proceeding in this way we construct a solution at any $t \geq 0$. Notice that $x(t_1-) = x(t_2-) = \ldots = 0$ and $x(t_1) - x(t_1-) = a_1$, $x(t_2) - x(t_2-) = a_2 - a_1, x(t_3) - x(t_3-) = a_3 - a_2, \ldots$ We call $a_1, a_2 - a_1, a_3 - a_2, \ldots$ the jump-type exits of x from 0.

If $\delta = 0$ and $h(t) = \Pi(t)$, then the function $x(t) := y(t) + \Pi(\Pi^{\leftarrow}(m(t)))$ for $t \geq 0$ can be constructed via the following recurrent procedure (see Fig. 4.2 in Sect. 4.3.4). Put $t_1 := \inf\{t \geq 0 : y(t) = 0\}$ and then $x(t) := y(t)$ for $t \in [0, t_1)$. Subsequently, for $n \in \mathbb{N}$, put $t_{n+1} := \inf\{t \geq t_n : b_n + y(t) - y(t_n) = 0\}$ and then $x(t) := b_n + y(t) - y(t_n)$ for $t \in [t_n, t_{n+1})$. Note that $x(t_n-) = 0$ if $t_n < \infty$ and

$$x(t) = y(t) + \sum_{t_n \leq t} b_n. \tag{2.25}$$

Example 2.1.3 Let $W = (W(t))_{t \geq 0}$ be a standard Brownian motion and $\Pi = (\Pi(t))_{t \geq 0}$ a nondecreasing compound Poisson process which is independent of W. Denote the (finite and nonzero) Lévy measure of Π by μ. Then

$$\Pi(t) = \sum_{n=1}^{N(t)} \varepsilon_n, \tag{2.26}$$

where $N = (N(t))_{t \geq 0}$ is a homogeneous Poisson process with intensity $\alpha = \mu((0, \infty))$ and $(\varepsilon_n)_{n \geq 1}$ are independent random variables which are independent of N and have the distribution function

$$\mathbb{P}\{\varepsilon_n \leq x\} = \mu((0, x])/\alpha, \quad x \geq 0.$$

The process Π a.s. satisfies the assumptions of Example 2.1.2. Hence, a solution (X, L) to the generalized Skorokhod problem $\mathcal{GSR}(W, h)$ with $h(t) = \delta t + \Pi(t)$ can be obtained by successively solving the Skorokhod reflection problems and adding jump-type exits from 0. It can be checked that the process L is a two-sided local time of X at zero defined by (2.15).

If Π is a subordinator with an infinite Lévy measure, then Π a.s. has an infinite number of jumps in any open interval and the construction of Example 2.1.2 is not applicable. It can be shown in this case that L is a continuous additive functional of a Markov process X that may increase only at times when X is equal to 0. Hence, L is a Blumenthal-Getoor local time up to a multiplicative constant. This characterization of L as a local time will be discussed in the next section.

In contrast to the Skorokhod reflection map, the generalized Skorokhod map $\widetilde{\Gamma}$ is not continuous on $D_+([0, \infty)) \times D_{0,\uparrow}^u([0, \infty))$. Even worse, it is not continuous on

$C_+([0,\infty)) \times D_{0,\uparrow\uparrow}^u([0,\infty))$, where $D_{0,\uparrow\uparrow}^u([0,\infty))$ is a subset of $D_{0,\uparrow}^u([0,\infty))$ comprised of strictly increasing functions. The next theorem provides sufficient conditions for a point $(y,h) \in D_+([0,\infty)) \times D_{0,\uparrow\uparrow}^u([0,\infty))$ to be a continuity point of $\widetilde{\Gamma}$.

Recall that we denote $\sup_{s \in [0,t]}((y(s))_-)$ by $m(t)$ or $(m(y))(t)$. A proof of the theorem stated next follows from Propositions 4.3.8 and 4.3.9.

Theorem 2.1.3 *Assume that $(h_n)_{n \geq 0}$ and $(y_n)_{n \geq 0}$ are sequences of deterministic functions living in the spaces $D_{0,\uparrow}^u([0,\infty))$ and $D_+([0,\infty))$, respectively, and satisfying the following conditions:*

(I) $\lim_{n \to \infty} h_n = h_0$ *on* $D([0,\infty))$ *and h_0 is strictly increasing.*
(II) $\lim_{n \to \infty} y_n = y_0$ *on* $D([0,\infty))$ *and y_0 does not have negative jumps.*
(III) *Let $\{t_1, t_2, \ldots\}$ be the set of discontinuities of h_0. Assume that, for any $k \in \mathbb{N}$, the equation $m_0(t) = h_0(t_k-)$ has at most one solution, where $m_0 = m(y_0)$.*

Then

$$x_n := \widetilde{\Gamma}(y_n, h_n) = y_n + h_n(h_n^{\leftarrow}(m_n))$$
$$\to y_0 + h_0(h_0^{\leftarrow}(m_0)) = \widetilde{\Gamma}(y_0, h_0) =: x_0, \quad n \to \infty \quad (2.27)$$

on $D([0,\infty))$, where $m_n = m(y_n)$.

Applying the Skorokhod representation theorem (Theorem 4.1.2) we obtain the following corollaries:

Corollary 2.1.2 *Assume that $(y_n, h_n)_{n \geq 0}$ is a sequence of stochastic processes taking values in $D_+([0,\infty)) \times D_{0,\uparrow}^u([0,\infty))$ and satisfying the following conditions:*

(I) *For each $n \in \mathbb{N}_0$, y_n and h_n are independent.*
(II) $y_n \Longrightarrow y_0$ *as $n \to \infty$ on $D([0,\infty))$, where y_0 does not have negative jumps and $y_0(0) \geq 0$ a.s.*
(III) $h_n \Longrightarrow h_0$ *as $n \to \infty$ on $D([0,\infty))$, where h_0 is an a.s. strictly increasing process.*
(IV) *For $u > 0$ and $t \geq 0$, $\mathbb{P}\{(m_0(y_0))(t) = u\} = 0$.*

Then $x_n \Longrightarrow x_0$ as $n \to \infty$ on $D([0,\infty))$, where the sequence $(x_n)_{n \geq 0}$ is as defined in (2.27).

2.1 Reflected Processes

Corollary 2.1.3 *Assume that $(y_n, h_n)_{n \geq 1}$ is a sequence of càdlàg stochastic processes satisfying the following conditions:*

(I) *For each $n \in \mathbb{N}$, y_n and h_n are independent; h_n is a.s. nondecreasing with $h_n(0) = 0$ a.s.*
(II) *$y_n \Longrightarrow W$ as $n \to \infty$ on $D([0, \infty))$, where W is a Brownian motion.*
(III) *$h_n \Longrightarrow h_0$ as $n \to \infty$ on $D([0, \infty))$, where h_0 is an a.s. increasing subordinator which is independent of W.*

Then $x_n \Longrightarrow x_0$ as $n \to \infty$ on $D([0, \infty))$, where the sequence $(x_n)_{n \geq 0}$ is as defined in (2.27).

For fixed $n \in \mathbb{N}$, $(x_n, h_n^{\leftarrow}(m_n))$ is not necessarily a solution to a generalized Skorokhod reflection problem. Indeed, Corollary 2.1.3 applies to $h_n = \Pi_n$ a compound Poisson process with positive jumps defined by (2.26). Thus, h_n is a subordinator which is not a.s. strictly increasing and, as discussed in Remark 2.1.6, the generalized Skorokhod problem $\mathcal{GSR}(y_n, h_n)$ is not defined.

2.1.3 Properties of a Reflected Brownian Motion with Jump-Type Exit from 0

Let W be a standard Brownian motion, H an a.s. strictly increasing subordinator, and ζ a nonnegative random variable. Assume that W, H, and ζ are independent. Let $X = \widetilde{\Gamma}(\zeta + W(\cdot), H(\cdot))$ be a solution to the generalized Skorokhod problem $\mathcal{GSR}(\zeta + W, H)$ given by

$$X(t) = \zeta + W(t) + H(L(t)), \quad L(t) = H^{\leftarrow}(M(t)),$$

$$M(t) := \sup_{s \in [0, t]} ((\zeta + W(s))_-), \quad t \geq 0. \quad (2.28)$$

Definition 2.1.6 *The process $X = (X(t))_{t \geq 0}$ defined in (2.28) is called Brownian motion with jump-type exit from 0 and reflection governed by H.*

Theorem 2.1.4 *The process $X = (X(t))_{t \geq 0}$ is a Markov process.*

Proof Observe that in view of (2.20) for each $s, t \geq 0$

$$X(t + s) = X(t) + W(t + s) - W(t) + H(L(t + s)) - H(L(t))$$
$$= X(t) + W_t(s) + H_t(L_t(s)),$$

where

$$W_t(s) := W(t+s) - W(t), \quad L_t(s) := L(t+s) - L(t), \quad H_t(z) := H(L(t)+z) - H(L(t)).$$

The equality

$$\int_{(0,\infty)} \mathbb{1}_{\{X(t+s)>0\}} dL_t(s) = \int_{(t,\infty)} \mathbb{1}_{\{X(s)>0\}} dL(s) = 0$$

demonstrates that the pair $(X(t + \cdot), L_t(\cdot))$ is a solution to the generalized Skorokhod reflection problem $\mathcal{GSR}(X(t) + W_t(\cdot), H_t(\cdot))$. Since the solution must be unique by Theorem 2.1.2, we conclude that

$$X(t + \cdot) = \widetilde{\Gamma}(X(t) + W_t(\cdot), H_t(\cdot)).$$

We denote by $\mathcal{F}^Z := (\mathcal{F}^Z_t)_{t \geq 0}$ a natural filtration[1] of a stochastic process $Z = (Z(t))_{t \geq 0}$. The claim of the theorem follows if we can show that for any $t \geq 0$

$$(W_t(s), H_t(s))_{s \geq 0} \overset{d}{=} (W(s), H(s))_{s \geq 0}$$

and that the process $(W_t(s), H_t(s))_{s \geq 0}$ is independent of $\mathcal{F}^X_t$. To do this, it suffices to verify that

$$(W_t(s))_{s \geq 0} \overset{d}{=} (W(s))_{s \geq 0}, \tag{2.29}$$

$$(H_t(s))_{s \geq 0} \overset{d}{=} (H(s))_{s \geq 0}, \tag{2.30}$$

$(W_t(s))_{s \geq 0}$ is independent of $\sigma(\mathcal{F}^X_t \cup \mathcal{F}^H_\infty)$, $\qquad(2.31)$

$(H_t(s))_{s \geq 0}$ is independent of $\sigma(\mathcal{F}^X_t)$. $\qquad(2.32)$

Formula (2.29) is obvious. Claim (2.31) follows from independence of W and H, and independence of $(W_t(s))_{s \geq 0}$ and $\sigma(\zeta, \mathcal{F}^W_t \cup \mathcal{F}^H_\infty)$, the σ-algebra which is larger than $\sigma(\mathcal{F}^X_t \cup \mathcal{F}^H_\infty)$. Define the filtration $\mathcal{G}^H := (\mathcal{G}^H_t)_{t \geq 0}$ by $\mathcal{G}^H_t := \sigma(\zeta, \mathcal{F}^H_t \cup \mathcal{F}^W_\infty)$ for $t \geq 0$. Notice that the process H is a Lévy process (and a strong Markov process) with respect to this filtration. Thus, (2.30) and (2.32) follow from the strong Markov property of H and

---

[1] Throughout this section we always assume that all σ-algebras are completed by subsets of zero-probability sets.

2.1 Reflected Processes

the fact that $L(t)$ is a stopping time with respect to the filtration $\mathcal{G}^H$. The latter is justified by Remark 2.1.2 and the equalities

$$\{L(t) \leq x\} = \{H^{\leftarrow}(\max_{s \in [0,t]} ((\zeta + W(s))_-)) \leq x\}$$

$$= \{\max_{s \in [0,t]} ((\zeta + W(s))_-) \leq H(x)\} \in \mathcal{G}_x^H, \quad x \geq 0.$$

Here, the last equality is secured by Proposition 4.3.13 (i). Observe that $H^{\leftarrow} = H^-$ a.s. as a consequence of a.s. strict monotonicity of H. □

Denote by $C_b([0, \infty))$ and $C_0([0, \infty))$ the spaces of bounded continuous functions defined on $[0, \infty)$ and continuous functions defined on $[0, \infty)$ which vanish at infinity, respectively. Actually, the process X is strong Markov. According to Theorem 5.10 in [45], this follows from the facts that X is càdlàg and that for each $t > 0$ and each $f \in C_b([0, \infty))$ the function

$$P_t f(x) := \mathbb{E}^x f(X(t)) = \mathbb{E} f(\widetilde{\Gamma}(x + W(\cdot), H(\cdot)))(t), \quad x \geq 0$$

is continuous in x. Now we justify the latter claim. By Theorem 2.1.3, the map $x \to \widetilde{\Gamma}(x + W(\cdot), H(\cdot))(t)$ is continuous in probability and, therefore, $x \to \mathbb{E} f(\widetilde{\Gamma}(x + W(\cdot), H(\cdot)))(t)$ is a continuous function by the Lebesgue dominated convergence theorem. Moreover, a stronger result can be proved, which says that $(P_t)_{t \geq 0}$ is a Feller semigroup, that is, $(P_t)_{t \geq 0}$ is strongly continuous on $C_0([0, \infty))$; see Definition 4.5.1.

Theorem 2.1.5 *The processes $W = (W(t))_{t \geq 0}$ and $L = (L(t))_{t \geq 0}$ are $\mathcal{F}^X$-adapted. Moreover, for each $s, t \geq 0$ the random variables $W(t+s) - W(t)$ and $L(t+s) - L(t)$ are $\mathcal{F}_s^{X(t+\cdot)-X(t)}$-measurable.*

Proof Observe that $X(t)$ may only be equal to 0 if the minimum of W on $[0, t]$ is attained at the point t. This is an event of probability zero. Hence, X has no sojourn at 0 a.s., that is,

$$\mathbb{P}\left\{\int_0^\infty \mathbb{1}_{\{X(t)=0\}} dt = 0\right\} = 1. \tag{2.33}$$

Fix $\varepsilon > 0$ and introduce the stopping times:

$$\tau_0^{(\varepsilon)} := \inf\{t \geq 0 : X(t) \geq \varepsilon\},$$

$$\sigma_k^{(\varepsilon)} := \inf\{t \geq \tau_k^{(\varepsilon)} : X(t) = 0\} \quad \text{and} \quad \tau_{k+1}^{(\varepsilon)} := \inf\{t \geq \sigma_k^{(\varepsilon)} : X(t) \geq \varepsilon\}, \quad k \in \mathbb{N}_0.$$

The increments of X and W coincide on $[\tau_k^{(\varepsilon)}, \sigma_k^{(\varepsilon)}]$ for all $\varepsilon > 0$ and all $k \in \mathbb{N}_0$. Hence, the process $W^{(\varepsilon)} := (W^{(\varepsilon)}(t))_{t \geq 0}$ defined by

$$W^{(\varepsilon)}(t) := \sum_{k \geq 0} \left(X(\sigma_k^{(\varepsilon)} \wedge t) - X(\tau_k^{(\varepsilon)} \wedge t) \right) = \sum_{k \geq 0} \left(W(\sigma_k^{(\varepsilon)} \wedge t) - W(\tau_k^{(\varepsilon)} \wedge t) \right), \quad t \geq 0$$

is $\mathcal{F}^X$-adapted. Since W is a $\sigma(\zeta, \mathcal{F}^W \cup \mathcal{F}_\infty^H)$-Brownian motion, and $\tau_k^{(\varepsilon)}$ and $\sigma_k^{(\varepsilon)}$ are $\sigma(\zeta, \mathcal{F}^W \cup \mathcal{F}_\infty^H)$-stopping times, we conclude that

$$W^{(\varepsilon)}(t) := \int_0^t \mathbb{1}_{\{s \in \cup_{k \geq 0}[\tau_k^{(\varepsilon)}, \sigma_k^{(\varepsilon)}]\}} dW(s), \quad t \geq 0.$$

It follows from (2.33) that $\lim_{\varepsilon \to 0+} \int_0^t \left(1 - \mathbb{1}_{\{s \in \cup_{k \geq 0}[\tau_k^{(\varepsilon)}, \sigma_k^{(\varepsilon)}]\}} \right)^2 ds = 0$ a.s. As a consequence, $W^{(\varepsilon)}$ converges to W uniformly in probability on each finite interval as $\varepsilon \to 0+$. This proves that the process W is $\mathcal{F}^X$-adapted.

The subordinator H can be represented by

$$F(t) = at + \int_{[0,t]} \int_{[0,\infty)} u N(ds, du), \tag{2.34}$$

where $a \geq 0$ is a drift, N is a Poisson random measure with intensity LEB $\otimes \nu$, and the Lévy measure ν satisfies

$$\int_{(0,\infty)} (1 \wedge u) \nu(du) < \infty.$$

Since H is a.s. strictly increasing, we conclude that $a > 0$ or $\nu((0,\infty)) = \infty$. Observe that $X(z) - X(z-) = H(L(z)) - H(L(z)-)$ a.s. for $z \geq 0$.

CASE $a > 0$. Since

$$H(t) - \sum_{s \leq t} (H(s) - H(s-)) = at \quad \text{for all } t \geq 0 \quad \text{a.s.,}$$

we conclude that

$$H(L(t)) - \sum_{s \leq L(t)} (H(s) - H(s-)) = aL(t) \quad \text{for all } t \geq 0 \quad \text{a.s.}$$

and, using continuity of L, that

$$X(t) - \zeta - W(t) - \sum_{z \leq t} (X(z) - X(z-)) = aL(t) \quad \text{for all } t \geq 0 \quad \text{a.s.}$$

2.1 Reflected Processes

Since W is $\mathcal{F}^X$-adapted and $\zeta = X(0)$, the last formula implies that L is $\mathcal{F}^X$-adapted as well.

CASE $\nu((0, \infty)) = \infty$. Observe that

$$\sum_{s \leq t} \mathbb{1}_{\{H(s)-H(s-) \geq \varepsilon\}} = N([0, t] \times [\varepsilon, \infty))$$

and that for fixed $t > 0$ $(N([0, t] \times [u^{-1}, \infty)))_{u>0}$ is an inhomogeneous Poisson process of intensity $\nu([u^{-1}, \infty))$. This entails

$$\lim_{\varepsilon \to 0+} \frac{\sum_{s \leq t} \mathbb{1}_{\{H(s)-H(s-) \geq \varepsilon\}}}{\nu([\varepsilon, \infty))} = \lim_{u \to \infty} \frac{N([0, t] \times [u^{-1}, \infty))}{\nu([u^{-1}, \infty))} = t \quad \text{a.s.} \quad (2.35)$$

for each fixed $t \geq 0$ and thereupon for all rational $t \geq 0$. For each fixed $\varepsilon > 0$

$$t \mapsto \frac{\sum_{s \leq t} \mathbb{1}_{\{F(s)-F(s-) \geq \varepsilon\}}}{\nu([\varepsilon, \infty))}$$

is a.s. nondecreasing. This together with the fact that the limit in (2.35) is continuous enables us to conclude that for all $T > 0$

$$\lim_{\varepsilon \to 0+} \sup_{t \in [0, T]} \left| \frac{\sum_{s \leq t} \mathbb{1}_{\{H(s)-H(s-) \geq \varepsilon\}}}{\nu([\varepsilon, \infty))} - t \right| = 0 \quad \text{a.s.}$$

Thus,

$$\lim_{\varepsilon \to 0+} \frac{\sum_{z \leq t} \mathbb{1}_{\{X(z)-X(z-) \geq \varepsilon\}}}{\nu([\varepsilon, \infty))} = \lim_{\varepsilon \to 0+} \frac{\sum_{s \leq L(t)} \mathbb{1}_{\{H(s)-H(s-) \geq \varepsilon\}}}{\nu([\varepsilon, \infty))}$$

$$= L(t) \quad \text{for all } t \geq 0 \quad \text{a.s.}$$

This proves that L is $\mathcal{F}^X$-adapted.

Fix any $t, s \geq 0$. Measurability of the random variables $W(t+s) - W(t)$ and $L(t+s) - L(t)$ with respect to $\mathcal{F}_s^{X(t+\cdot)-X(t)}$ can be proved similarly. □

An inspection of the proof of Theorem 2.1.5 reveals that the process L is a continuous additive functional of the strong Markov process X; see, for instance, [14] for the definition. Moreover, L can only increase on the set $\{t \geq 0 : X(t) = 0\}$. Hence, L is a Blumenthal-Getoor local time of X at 0 up to a multiplicative constant. We shall investigate the properties of L in Sect. 2.3 in a more general case.

Remark 2.1.8 Although we assume that all σ-algebras are completed by events of zero probability, one has to be careful when working with σ-algebras generated by Markov processes. Further technical details are left to an interested reader; see also Section II.3(c) in [14].

To close the section, we calculate the resolvent of X

$$R_\lambda^X f(x) := \int_0^\infty e^{-\lambda t} \mathbb{E}^x f(X(t)) dt, \quad \lambda > 0, \quad x \geq 0, \quad f \in C_b([0, \infty))$$

in the situation when representation (2.34) holds with $a = 0$ and $\nu((0, \infty)) = \infty$. Without loss of generality we assume that $X = \widetilde{\Gamma}(W, H)$, where W is a Brownian motion with $W(0) = x$ under $\mathbb{P}^x$.

Put $\sigma(X) := \inf\{t \geq 0 : X(t) = 0\}$. According to formula (4.18) in Lemma 4.5.1 from the Appendix

$$R_\lambda^X f(x) = V_\lambda^X f(x) + \mathbb{E}^x e^{-\lambda \sigma(X)} R_\lambda^X f(0),$$

where

$$V_\lambda^X f(x) = \mathbb{E}^x \int_0^{\sigma(X)} e^{-\lambda t} f(X(t)) dt$$

is the resolvent of X killed at $\sigma(X)$. Since $X(t) = W(t)$ for $t \in [0, \sigma(X)]$, $V_\lambda^X f$ is equal to $V_\lambda^W f$, the resolvent of a killed Brownian motion, namely,

$$V_\lambda^W f(x) = \mathbb{E}^x \int_0^{\sigma(W)} e^{-\lambda t} f(W(t)) dt = \int_0^\infty v^{(\lambda)}(x, y) f(y) dy, \quad (2.36)$$

where $\sigma(W) := \inf\{t \geq 0 : W(t) = 0\}$ and

$$v^{(\lambda)}(x, y) := \frac{1}{\sqrt{2\lambda}} \left(e^{-\sqrt{2\lambda}|x-y|} - e^{-\sqrt{2\lambda}(x+y)} \right), \quad x, y > 0; \quad (2.37)$$

see p. 56 in [14]. Putting in (2.36) $f(x) \equiv 1$ for $x \geq 0$ and calculating the resulting integral, we infer $\mathbb{E}^x e^{-\lambda \sigma(W)} = e^{-\sqrt{2\lambda}x}$ for $x > 0$. Since $-W$ has the same distribution as W, we conclude that $\mathbb{E}^{-x} e^{-\lambda \sigma(W)} = \mathbb{E}^x e^{-\lambda \sigma(W)}$ for $x > 0$. This yields a known formula

$$\mathbb{E}^x e^{-\lambda \sigma(W)} = e^{-\sqrt{2\lambda}|x|}, \quad x \in \mathbb{R}; \quad (2.38)$$

2.1 Reflected Processes

see, for instance, formula (5) on p. 26 in [75]. Obviously, (2.38) ensures that

$$\mathbb{E}^x e^{-\lambda \sigma(X)} = e^{-\sqrt{2\lambda}|x|}, \quad x \in \mathbb{R}.$$

It remains to calculate

$$R_\lambda^X f(0) = \mathbb{E}^0 \int_0^\infty e^{-\lambda t} f(W(t) + H(H^\leftarrow(\max_{s \in [0,t]}((W(s))_-)))) dt.$$

Put

$$H_n(t) = \int_{[0,t]} \int_{[1/n,\infty)} u N(ds, du) \quad \text{and} \quad X_n(t) := \widetilde{\Gamma}(W, H_n)(t), \quad n \in \mathbb{N}, \ t \geq 0.$$

By Corollary 2.1.3 and the Lebesgue dominated convergence theorem

$$R_\lambda^X f(0) = \mathbb{E}^0 \int_0^\infty e^{-\lambda t} f(X(t)) dt = \lim_{n \to \infty} \mathbb{E}^0 \int_0^\infty e^{-\lambda t} f(X_n(t)) dt. \tag{2.39}$$

Under $\mathbb{E}^0$, $W(0) = 0$, whence

$$X_n(0) = W(0) + H_n(H_n^\leftarrow((W(0))_-)) = H_n(H_n^\leftarrow(0)) = \varepsilon_{n,1}$$

(see representation (2.25)), where $\varepsilon_{n,k}$ is the size of the kth jump of H_n. Further

$$X(\sigma(\varepsilon_{n,1} + W)) = \varepsilon_{n,2}, \tag{2.40}$$

where $\sigma(\varepsilon_{n,1} + W) := \inf\{t \geq 0 : \varepsilon_{n,1} + W(t) = 0\}$. The strong Markov property of the Brownian motion together with independence of W and N guarantees that the expectation under the limit in the right-hand side of (2.39) is equal to

$$\mathbb{E} R_\lambda^{X_n} f(\varepsilon_{n,1}) = \mathbb{E}^0 \left(\int_0^{\sigma(\varepsilon_{n,1}+W)} + \int_{\sigma(\varepsilon_{n,1}+W)}^\infty \right) e^{-\lambda t} f(\varepsilon_{n,1} + X_n(t)) dt$$

$$= \mathbb{E}^0 \int_0^{\sigma(\varepsilon_{n,1}+W)} e^{-\lambda t} f(\varepsilon_{n,1} + W(t)) dt + \mathbb{E}^0 e^{-\lambda \sigma(\varepsilon_{n,1}+W)} \mathbb{E} R_\lambda^{X_n} f(\varepsilon_{n,2})$$

$$= \mathbb{E}^0 \int_0^{\sigma(\varepsilon_{n,1}+W)} e^{-\lambda t} f(\varepsilon_{n,1} + W(t)) dt + \mathbb{E}^0 e^{-\lambda \sigma(\varepsilon_{n,1}+W)} \mathbb{E} R_\lambda^{X_n} f(\varepsilon_{n,1}), \tag{2.41}$$

where the second equality is a consequence of (2.40) and the third equality follows from $\varepsilon_{n,1} \stackrel{d}{=} \varepsilon_{n,2}$. Therefore, with $1(x) \equiv 1$ denoting the function that is equal to one everywhere,

$$\mathbb{E} R_\lambda^{X_n} f(\varepsilon_{n,1}) = \frac{\mathbb{E}^0 \int_0^{\sigma(\varepsilon_{n,1}+W)} e^{-\lambda t} f(\varepsilon_{n,1} + W(t)) dt}{1 - \mathbb{E}^0 e^{-\lambda \sigma(\varepsilon_{n,1}+W)}}$$

$$= \frac{\mathbb{E}^0 \int_0^{\sigma(\varepsilon_{n,1}+W)} e^{-\lambda t} f(\varepsilon_{n,1} + W(t)) dt}{\lambda \mathbb{E}^0 \int_0^{\sigma(\varepsilon_{n,1}+W)} e^{-\lambda t} dt} = \frac{\int_{[1/n,\infty)} V_\lambda^W f(x) \nu(dx)}{\lambda \int_{[1/n,\infty)} V_\lambda^W 1(x) \nu(dx)}.$$

In view of (2.38)

$$|V_\lambda^W f(x)| \leq \|f\|_\infty V_\lambda^W 1(x) = \|f\|_\infty \lambda^{-1}(1 - \mathbb{E}^x e^{-\lambda \sigma(W)}) = \|f\|_\infty \lambda^{-1}(1 - e^{-\sqrt{2\lambda}|x|}),$$

where $\|f\|_\infty = \sup_{x \geq 0} |f(x)|$. This together with the fact that ν is a Lévy measure ensures that the integrals

$$\int_{(0,\infty)} V_\lambda^W f(x) \nu(dx) \quad \text{and} \quad \int_{(0,\infty)} V_\lambda^W 1(x) \nu(dx)$$

converge. With this at hand, invoking (2.39) yields

$$R_\lambda^X f(0) = \lim_{n \to \infty} \frac{\int_{[1/n,\infty)} V_\lambda^W f(x) \nu(dx)}{\lambda \int_{[1/n,\infty)} V_\lambda^W 1(x) \nu(dx)} = \frac{\int_{(0,\infty)} V_\lambda^W f(x) \nu(dx)}{\lambda \int_{(0,\infty)} V_\lambda^W 1(x) \nu(dx)}.$$

Putting things together we obtain a final formula: for $\lambda > 0$, $x \in \mathbb{R}$ and $f \in C_b([0,\infty))$,

$$R_\lambda^X f(x) = V_\lambda^W f(x) + e^{-\sqrt{2\lambda}|x|} \frac{\int_{(0,\infty)} V_\lambda^W f(x) \nu(dx)}{\lambda \int_{(0,\infty)} V_\lambda^W 1(x) \nu(dx)},$$

where $V_\lambda^W f$ is defined by (2.36) and (2.37).

Similar formulas appear in Sect. 2.3, where we analyze stable Lévy processes that admit jump-type exits from 0. We believe that the arguments based on the Skorokhod reflection problem that we developed in this section are easier for description and analysis of reflected processes in comparison to the resolvent theory or Itô's excursion theory exploited in Sect. 2.3. We stress that in the situation when exits from zero occur by jumps taking values of both signs, the generalized Skorokhod map is not well-defined, and we have to resort to other methods.

Bibliographic Comments

Skorokhod introduced in 1961 a reflection problem for continuous functions in [149, 150]; see also [152]. Nowadays the problem and its extension for càdlàg functions are commonly known as the Skorokhod reflection problem. In the aforecited papers existence and uniqueness of solutions in the deterministic setting were proved and existence and pathwise uniqueness of the weak solutions to the stochastic differential equations with reflection were settled. The process l was associated with some additive functional of the solution. Today it is known that the Yamada-Watanabe theorem entails existence of the strong solutions to the stochastic differential equation with reflection. However, at that time the Yamada-Watanabe theorem and the Tanaka formula had yet to be discovered. A detailed survey of the stochastic differential equations with reflection can be found in [123]. To the best of our knowledge, formula (2.3) for continuous-time processes appeared for the first time on p. 190 of Watanabe's paper [169]. Watanabe gave credit to Itô for this discovery. The formula was used in [169] to construct stable processes with boundary conditions.

Since the Skorokhod map is continuous, many functional limit theorems for reflected random walks, storage processes, and objects in a heavy traffic regime are a simple consequence of the continuous mapping theorem; see, for instance [24, 92, 175].

A generalization of the reflection problem for the boundary of a multidimensional domain is not as simple as the original one-dimensional problem; see [32, 33, 43, 64, 100, 139, 143, 147, 160]. The cited articles investigated existence, uniqueness, continuity, Lipschitz property, etc. of the multidimensional reflection map. The construction of a multidimensional diffusion with reflection at the boundary of a multidimensional domain is quite standard, if the coefficients of the corresponding stochastic differential equation are smooth and the corresponding Skorokhod map is Lipschitz continuous. For instance, if the boundary of a domain is smooth, then an appropriate space transformation in combination with localization reduce the problem to the case where the domain is a half-space and the reflection is normal; see [170]. By now this approach has become classical; see [67, 123].

Another approach to constructing and analyzing processes with reflection is via martingale and submartingale problems. It has proved useful in cases where the coefficients of stochastic differential equations are not smooth enough [156, 157]. The approach also applies if the boundary of a domain is not smooth. In this situation the corresponding (deterministic) Skorokhod map may not exist [37, 93, 94, 163]. In the latter setting the reflection term may have a strange nature and a reflected Brownian motion in a wedge, say, may not be a semimartingale; see [176, 177].

The Skorokhod map and its generalizations discussed above describe a "continuous" reflection from the boundary of a domain in the following sense: (a) the resulting process is continuous if so is the driving process, and (b) if the driving process is càdlàg, then a jump-type exit from the boundary inwards the domain can only happen because of a jump of the driving process, rather than the reflection term. Feller and Wentzell [51, 164, 165] described all the possible exits, including jump-type exits, from the boundary of diffusion processes. These authors investigated stochastic processes, for the most part, from the viewpoint of semigroup theory and found the most general boundary conditions for the generators of the corresponding diffusions. Itô and McKean [74, 75] constructed diffusions on a halfline as derived processes of the reflected Brownian motion and its local time. Informally, the most general behavior at the boundary can be thought of as a "mixture" of killing at the boundary, Skorokhod reflection, jump-type exit, and a delay proportional to a local time at the boundary. It is relevant to look at the corresponding theory from the perspective of Itô's excursion theory; see [73] or, for a modern exposition, [14].

Constructing diffusion processes with jump-type exit from the boundary is a much more complicated task in comparison to those with "continuous" reflection; see [2, 3, 112, 113] for a martingale problem and stochastic differential equations methods, [171, 172] for methods related to Itô's excursion theory, and [56, 72, 86, 87, 159] for partial differential equations and semigroup methods. We stress that there is no general "meta-theorem" that justifies the equivalence of stochastic and analytic approaches for diffusions with jump-type exits from the boundary.

As far as we know, representation (2.21) made its first appearance in formula (3) of Section 12 in [74] in a particular case where y is a standard Brownian motion and h is an a.s. strictly increasing subordinator. With the same y and h, formula (2.21) was also used in Theorem 3.11 on p. 62 in [14] for constructing a Brownian motion with jump-type exit from 0 as well as some other regular extensions of a Brownian motion on a halfline. With y being a spectrally positive Lévy process and h a subordinator which is independent of y, formula (2.21) can be found in Theorem 3 of [95]. The formulation of the generalized Skorokhod problem as given in Definition 2.1.4 appeared in [122]. The results of Sect. 2.1.3 are taken from [71, 122].

2.2 The Skew Brownian Motion and Its Generalizations

2.2.1 The Skew Brownian Motion

Consider a diffusion in a medium with semipermeable membrane, which is a thin slice of a material whose properties are other than the properties of the environment. Such a diffusion arises naturally in biological tissues, geophysics, porous and composite materials, ecology models, and many other applied models; see the review paper [97]. In this section we introduce and describe properties of a skew Brownian motion, which is one of the simplest continuous-time model for a diffusion of the aforementioned type. Roughly speaking, a skew Brownian motion is a time-homogeneous continuous strong Markov process that behaves like a Brownian motion until hitting 0 and has some asymmetry (a membrane) at 0.

To provide an intuitive definition of a skew Brownian motion, we first recall an interpretation of the Brownian motion as a scaling limit of random walks. Let $S = (S(n))_{n \geq 0}$ be a simple symmetric random walk on $\mathbb{Z}$, that is,

$$p_{i,i\pm 1} := \mathbb{P}\{S_{n+1} = i \pm 1 | S_n = i\} = 1/2, \quad i \in \mathbb{Z}.$$

For $n \in \mathbb{N}$, the corresponding n-step transition probabilities are given by

$$p_{i_1,i_2}^{(n)} := \mathbb{P}\{S_{n+k} = i_2 | S_k = i_1\} = \phi_n(i_2 - i_1),$$

2.2 The Skew Brownian Motion and Its Generalizations

where

$$\phi_n(i) = \begin{cases} 2^{-n} \binom{n}{\frac{n+i}{2}}, & \text{if } |i| \leq n \text{ and } n+i \text{ is even,} \\ 0, & \text{otherwise.} \end{cases} \quad (2.42)$$

Donsker's theorem implies that

$$\left(\frac{S(\lfloor nt \rfloor)}{\sqrt{n}}\right)_{t \geq 0} \Longrightarrow (W(t))_{t \geq 0}, \quad n \to \infty$$

on $D([0, \infty))$, where W is a standard Brownian motion.

Let $\widetilde{S} = (\widetilde{S}(n))_{n \geq 0}$ be a perturbation of S at 0 having transition probabilities

$$\widetilde{p}_{i, i \pm 1} = 1/2, \quad i \neq 0; \quad \widetilde{p}_{0,1} = p \quad \text{and} \quad \widetilde{p}_{0,-1} = 1 - p,$$

where $p \in [0, 1]$. If $S(0) \stackrel{d}{=} \widetilde{S}(0)$, then the sequences $(|S(n)|)_{n \geq 0}$ and $(|\widetilde{S}(n)|)_{n \geq 0}$ have the same distribution, and either of these is a simple random walk reflected at 0. If $\widetilde{S}(0) = S(0) = 0$, then for $n \in \mathbb{N}$ the n-step transition probabilities of $\widetilde{S}$ are given by

$$\widetilde{p}_{0,j}^{(n)} = \begin{cases} p\mathbb{P}\{|S(n)| = j\}, & \text{if } j > 0, \\ (1-p)\mathbb{P}\{|S(n)| = |j|\}, & \text{if } j < 0, \\ \mathbb{P}\{|S(n)| = 0\}, & \text{if } j = 0. \end{cases}$$

This formula follows from $(|S(n)|)_{n \geq 0} \stackrel{d}{=} (|\widetilde{S}(n)|)_{n \geq 0}$ upon noticing that for every $j \neq 0$ and all $n \in \mathbb{N}$

$$\mathbb{P}\{\widetilde{S}(n) > 0 \mid |\widetilde{S}(n)| = |j|\} = p, \quad \mathbb{P}\{\widetilde{S}(n) < 0 \mid |\widetilde{S}(n)| = |j|\} = 1 - p.$$

More generally, by using classical Andre's reflection principle, it can be checked that

$$\widetilde{p}_{i,j}^{(n)} = p_{i,j}^{(n)} + \gamma \, \text{sgn}(j) p_{|i|, -|j|}^{(n)}, \quad i, j \in \mathbb{Z}, \quad n \in \mathbb{N},$$

where $\gamma := 2p - 1$. By the de Moivre-Laplace theorem the following convergence of transition probabilities (2.42) holds true:

$$\lim_{n \to \infty} p_{\lfloor \sqrt{n} x \rfloor, \lfloor \sqrt{n} y \rfloor}^{(\lfloor nt \rfloor)} = \varphi_t(x - y), \quad x, y \in \mathbb{R}, \quad t > 0.$$

Here, for each $t > 0$ the function φ_t defined by $\varphi_t(x) = (2\pi t)^{-1/2}e^{-x^2/(2t)}$ for $x \in \mathbb{R}$ is a density of the normal distribution with mean 0 and variance t. Hence,

$$\lim_{n \to \infty} \sqrt{n}\, \widetilde{p}^{(\lfloor nt \rfloor)}_{\lfloor \sqrt{n}x \rfloor, \lfloor \sqrt{n}y \rfloor}$$
$$= \varphi_t(x - y) + \gamma\, \text{sgn}(y)\varphi_t(|x| + |y|) =: \widetilde{p}^{(t)}_\gamma(x, y), \quad x, y \in \mathbb{R}, \quad t > 0.$$

Remark 2.2.1 One can show that for each $\gamma \in \mathbb{R}$ and each $x \in \mathbb{R}$ the function $y \mapsto \widetilde{p}^{(t)}_\gamma(x, y)$ satisfies the Chapman-Kolmogorov equation and $\int_\mathbb{R} \widetilde{p}^{(t)}_\gamma(x, y)dy = 1$. Furthermore, $\widetilde{p}^{(t)}_\gamma$ is nonnegative if, and only if, $|\gamma| \leq 1$.

The discussion just presented suggests the following definition:

Definition 2.2.1 A skew Brownian motion with permeability parameter $\gamma \in [-1, 1]$ (to be denoted by W_γ or W^{skew}_γ) is a Markov process on $\mathbb{R}$ with transition probability density function

$$\widetilde{p}^{(t)}_\gamma(x, y) = \varphi_t(x - y) + \gamma\, \text{sgn}(y)\varphi_t(|x| + |y|), \quad x, y \in \mathbb{R}, \quad t > 0,$$

where $\varphi_t(x) = (2\pi t)^{-1/2}e^{-x^2/(2t)}$ for $x \in \mathbb{R}$ and $t > 0$.

It was stated by Harrison and Shepp in Section 4 of [65] that

$$\left(\frac{\widetilde{S}(\lfloor nt \rfloor)}{\sqrt{n}}\right)_{t \geq 0} \Longrightarrow (W_\gamma(t))_{t \geq 0}, \quad n \to \infty \qquad (2.43)$$

on $D([0, \infty))$, where $\gamma = 2p - 1$. The skew Brownian motion W_γ inherits many properties of the perturbed random walk $\widetilde{S}$, for instance,

$$(|W_\gamma(t)|)_{t \geq 0} \stackrel{d}{=} (|W(t)|)_{t \geq 0}, \quad (W_0(t))_{t \geq 0} \stackrel{d}{=} (W(t))_{t \geq 0}, \quad (W_1(t))_{t \geq 0} \stackrel{d}{=} (|W(t)|)_{t \geq 0}.$$

In Problem 1 of Section 4.2 in [75] Itô and McKean obtained a skew Brownian motion from a reflected Brownian motion (both started from 0, for simplicity) by enumerating excursions of the reflected Brownian motion in any measurable way and changing the sign of each excursion independently with probability $1 - p$ or, equivalently, keeping it positive with probability p. It was shown by Walsh in Proposition 1 of [168] that a diffusion X on $\mathbb{R}$ is a skew Brownian motion if, and only if, $(|X(t)|)_{t \geq 0}$ is a reflected Brownian motion. Hence, any continuous strong Markov process, which behaves like a Brownian motion outside 0 and spends zero time at 0, is a skew Brownian motion with some parameter $\gamma \in [-1, 1]$.

2.2 The Skew Brownian Motion and Its Generalizations

Harrison and Shepp in Section 3 of [65] also characterized the skew Brownian motion via an equation with a local time. Here is their result.

Theorem 2.2.1 *Let $(W(t))_{t \geq 0}$ be an $(\mathcal{F}_t)_{t \geq 0}$-Brownian motion.*

(I) *Let $\gamma \in [-1, 1]$. For any $\mathcal{F}_0$-measurable random variable $X(0)$ there exists a unique strong solution to the stochastic differential equation*

$$dX(t) = dW(t) + \gamma dL_0^X(t), \quad t \geq 0, \tag{2.44}$$

where L_0^X is a local time at 0 of the solution X given by

$$L_0^X(t) = \lim_{\varepsilon \to 0+} \frac{1}{2\varepsilon} \int_0^t \mathbb{1}_{\{|X(s)| \leq \varepsilon\}} ds.$$

Moreover, X is a skew Brownian motion with permeability parameter γ.
(II) *If $|\gamma| > 1$, then there is no solution (neither strong nor weak) to Eq. (2.44).*

The family of functions $x \mapsto (2\varepsilon)^{-1} \mathbb{1}_{[-\varepsilon,\varepsilon]}(x)$ converges in the sense of distributions to Dirac's delta function $x \mapsto \delta_0(x)$ as $\varepsilon \to 0+$. This observation advocates using the term *stochastic differential equation with generalized drift $\gamma \delta_0$* for (2.44). However, one should be careful when attempting to approximate a solution to (2.44) by solutions to stochastic differential equations whose drifts converge to $\gamma \delta_0$. The following surprising result is due to Portenko [137].

Theorem 2.2.2 *Assume that a family $(a_\varepsilon)_{\varepsilon > 0}$ of nonnegative integrable functions converges in the sense of distributions to $\alpha \delta_0$, that is,*

$$\lim_{\varepsilon \to 0} \int_{\mathbb{R}} a_\varepsilon(x) \varphi(x) dx = \alpha \varphi(0)$$

for any bounded continuous function φ, where α is a constant. For each $\varepsilon > 0$, let $X_\varepsilon = (X_\varepsilon(t))_{t \geq 0}$ be a solution to

$$dX_\varepsilon(t) = a_\varepsilon(X_\varepsilon(t))dt + dW(t), \quad t \geq 0, \quad X_\varepsilon(0) = x \in \mathbb{R},$$

where W is a standard Brownian motion. Then $X_\varepsilon \Longrightarrow W_\gamma$ as $\varepsilon \to 0+$ on $C([0, \infty))$, where $\gamma = \frac{e^\alpha - e^{-\alpha}}{e^\alpha + e^{-\alpha}} = \tanh(\alpha)$ and $W_\gamma(0) = x$.

Remark 2.2.2 The assumption of nonnegativity of a_ε can be replaced by

$$\sup_{\varepsilon>0} \int_\mathbb{R} |a_\varepsilon(x)| \mathrm{d}x < \infty.$$

A proof in this case repeats verbatim the one given in [137].

Here is a typical example of the family (a_ε) satisfying the assumptions of Theorem 2.2.2. For an integrable function a, put

$$a_\varepsilon(x) = \varepsilon^{-1} a(\varepsilon^{-1} x), \quad \varepsilon > 0.$$

Then $\alpha = \int_\mathbb{R} a(x) \mathrm{d}x$.

In the remainder of this section we show that a skew Brownian motion can be obtained as a scaling limit of a sequence of Brownian motions exiting 0 by independent identically distributed jumps with magnitudes tending to 0. To this end, we need some additional notation. Let $(\eta_k)_{k\geq 1}$ be a sequence of independent copies of a random variable η with $\eta \neq 0$ a.s. and $\mathbb{E}|\eta| < \infty$, $W = (W(t))_{t\geq 0}$ a standard Brownian motion, and $(X_\varepsilon(0))_{\varepsilon>0}$ a family of random variables taking values in $\mathbb{R} \setminus \{0\}$. Assume that $(\eta_k)_{k\geq 1}$, W and $(X_\varepsilon(0))_{\varepsilon>0}$ are mutually independent. We define a family of strong Markov processes $((X_\varepsilon(t))_{t\geq 0})_{\varepsilon>0}$ as follows. Introduce the stopping times

$$\sigma_1^{(\varepsilon)} := \inf\{t \geq 0 : X_\varepsilon(0) + W(t) = 0\},$$

$$\sigma_{k+1}^{(\varepsilon)} := \inf\{t > \sigma_k^{(\varepsilon)} : \varepsilon\eta_k + W(t) - W(\sigma_k^{(\varepsilon)}) = 0\}, \quad k \in \mathbb{N}$$

and put

$$X_\varepsilon(t) := \begin{cases} X_\varepsilon(0) + W(t), & t \in [0, \sigma_1^{(\varepsilon)}), \\ \varepsilon\eta_k + W(t) - W(\sigma_k^{(\varepsilon)}), & t \in [\sigma_k^{(\varepsilon)}, \sigma_{k+1}^{(\varepsilon)}), \quad k \in \mathbb{N}. \end{cases} \quad (2.45)$$

In other words,

(I) $\sigma_k^{(\varepsilon)}$ is the kth epoch for which $X_\varepsilon(\sigma_k^{(\varepsilon)}-) = 0$.
(II) The jump of X_ε at $\sigma_k^{(\varepsilon)}$ is $\varepsilon\eta_k$.
(III) The increments of X_ε and W coincide on each interval $[\sigma_k^{(\varepsilon)}, \sigma_{k+1}^{(\varepsilon)}), k \in \mathbb{N}_0$.

By definition $X_\varepsilon(t) \neq 0$ for each $t \geq 0$.

In what follows, $\overset{\mathrm{d}}{\to}$ denotes convergence in distribution of real-valued random variables.

2.2 The Skew Brownian Motion and Its Generalizations

Theorem 2.2.3 *Assume that $X_\varepsilon(0) \xrightarrow{d} \kappa$ as $\varepsilon \to 0+$ for some real-valued random variable κ. Then*

$$(X_\varepsilon(t))_{t \geq 0} \implies (W_\gamma^{\text{skew}}(t))_{t \geq 0}, \quad \varepsilon \to 0+$$

on $D([0, \infty))$, where $\gamma = \mathbb{E}\eta/\mathbb{E}|\eta|$ and $W_\gamma^{\text{skew}}(0) \stackrel{d}{=} \kappa$.

Proof Let $\lambda > 0$, $x \neq 0$, and $f \in C_b(\mathbb{R})$. For each $\varepsilon > 0$, we shall work with the resolvents of X_ε and X_ε killed at 0 defined by

$$R_\lambda^\varepsilon f(x) := \mathbb{E}^x \int_0^\infty e^{-\lambda t} f(X_\varepsilon(t)) dt$$

and

$$V_\lambda^\varepsilon f(x) := \mathbb{E}^x \int_0^{\sigma_1^{(\varepsilon)}} e^{-\lambda t} f(X_\varepsilon(t)) dt,$$

respectively. Put $\sigma(x + W) := \inf\{t \geq 0 : x + W(t) = 0\}$ and note that

$$\mathbb{P}^x\{\sigma_1^{(\varepsilon)} \in \cdot\} = \mathbb{P}\{\sigma_1^{(\varepsilon)} \in \cdot | X_\varepsilon(0) = x\} = \mathbb{P}\{\sigma(x + W) \in \cdot\}.$$

Therefore,

$$V_\lambda^\varepsilon f(x) := \mathbb{E}^x \int_0^{\sigma_1^{(\varepsilon)}} e^{-\lambda t} f(X_\varepsilon(t)) dt = \mathbb{E} \int_0^{\sigma(x+W)} e^{-\lambda t} f(x + W(t)) dt =: V_\lambda f(x).$$

In particular, $V_\lambda^\varepsilon f$ does not depend on ε. Here and hereafter, we write $\mathbb{E}$ for $\mathbb{E}^0$. Similarly to Lemma 4.5.1 (see also the derivation of (2.41)),

$$R_\lambda^\varepsilon f(x) = V_\lambda f(x) + \mathbb{E}e^{-\lambda \sigma(x+W)} \mathbb{E} R_\lambda^\varepsilon f(\varepsilon \eta). \tag{2.46}$$

According to Lemma 4.5.1 the resolvent $R_\lambda^{\text{skew}} f$ of a skew Brownian motion satisfies

$$R_\lambda^{\text{skew}} f(x) = V_\lambda f(x) + \mathbb{E}e^{-\lambda \sigma(x+W)} R_\lambda^{\text{skew}} f(0).$$

By Theorem 4.5.2, the claim of Theorem 2.2.3 follows if we can show that

$$\lim_{\varepsilon \to 0+} \mathbb{E} R_\lambda^\varepsilon f(\varepsilon \eta) = R_\lambda^{\text{skew}} f(0). \tag{2.47}$$

Note that Theorem 4.5.1 is not applicable because X_ε is a Markov process on $\mathbb{R} \setminus \{0\}$, whereas W_γ^{skew} is a Markov process on $\mathbb{R}$. The process X_ε cannot hit 0, nor start from 0 by definition. In order to prove (2.47) we integrate (2.46) with respect to the distribution of $\varepsilon\eta$. This yields

$$\mathbb{E} R_\lambda^\varepsilon f(\varepsilon\eta) = \mathbb{E} V_\lambda f(\varepsilon\eta) + \mathbb{E} e^{-\lambda \sigma(\varepsilon\eta + W)} \mathbb{E} R_\lambda^\varepsilon f(\varepsilon\eta)$$

or equivalently

$$\mathbb{E} R_\lambda^\varepsilon f(\varepsilon\eta) = \frac{\mathbb{E} V_\lambda f(\varepsilon\eta)}{1 - \mathbb{E} e^{-\lambda \sigma(\varepsilon\eta + W)}}. \tag{2.48}$$

Recalling the form of the transition density function of a skew Brownian motion given in Definition 2.2.1 and using the known formula

$$\int_0^\infty e^{-\lambda t} \frac{e^{-z^2/(2t)}}{\sqrt{2\pi t}} dt = \frac{1}{\sqrt{2\lambda}} e^{-\sqrt{2\lambda}|z|}, \quad z \in \mathbb{R}$$

(see, for instance, p. 17 in [75]), we conclude that the resolvent of a skew Brownian motion is defined by

$$R_\lambda^{\text{skew}} f(x) = \int_\mathbb{R} r_\lambda^{\text{skew}}(x, y) f(y) dy,$$

where

$$r_\lambda^{\text{skew}}(x, y) = \frac{1}{\sqrt{2\lambda}} (e^{-\sqrt{2\lambda}|x-y|} + \gamma \operatorname{sgn}(y) e^{-\sqrt{2\lambda}(|x|+|y|)}), \quad x, y \in \mathbb{R}.$$

In particular,

$$r_\lambda^{\text{skew}}(0, y) = \sqrt{\frac{2}{\lambda}} (p \mathbb{1}_{(0,\infty)}(y) + (1-p) \mathbb{1}_{(-\infty,0)}(y)) e^{-\sqrt{2\lambda}|y|}, \quad y \in \mathbb{R} \setminus \{0\},$$

where $p := (1+\gamma)/2 = \mathbb{E}\eta_+/\mathbb{E}|\eta|$ and $1 - p = (1-\gamma)/2 = \mathbb{E}\eta_-/\mathbb{E}|\eta|$. Here, as usual, $x_+ = x \vee 0$ and $x_- = (-x) \vee 0$ for $x \in \mathbb{R}$.

Thus, we have to prove that

$$\lim_{\varepsilon \to 0+} \mathbb{E} R_\lambda^\varepsilon f(\varepsilon\eta) = \sqrt{\frac{2}{\lambda}} \left(\frac{\mathbb{E}\eta_+}{\mathbb{E}|\eta|} \int_0^\infty e^{-\sqrt{2\lambda} y} f(y) dy + \frac{\mathbb{E}\eta_-}{\mathbb{E}|\eta|} \int_{-\infty}^0 e^{-\sqrt{2\lambda}|y|} f(y) dy \right).$$

2.2 The Skew Brownian Motion and Its Generalizations

Recall from (2.38) that $\mathbb{E}^x e^{-\lambda \sigma(W)} = e^{-\sqrt{2\lambda}|x|}$ for $x \in \mathbb{R}$. With this at hand, we obtain

$$1 - \mathbb{E}e^{-\lambda\sigma(\varepsilon\eta + W)} = 1 - \mathbb{E}e^{-\sqrt{2\lambda}\varepsilon|\eta|} \sim \sqrt{2\lambda}\mathbb{E}|\eta|\varepsilon, \quad \varepsilon \to 0+.$$

Summarizing, in view of (2.48) our task boils down to showing that

$$\lim_{\varepsilon \to 0+} \varepsilon^{-1} \mathbb{E} V_\lambda f(\varepsilon\eta) \mathbb{1}_{\{\eta > 0\}} = 2\mathbb{E}\eta_+ \int_0^\infty e^{-\sqrt{2\lambda}y} f(y) \mathrm{d}y$$

and

$$\lim_{\varepsilon \to 0+} \varepsilon^{-1} \mathbb{E} V_\lambda f(\varepsilon\eta) \mathbb{1}_{\{\eta < 0\}} = 2\mathbb{E}\eta_- \int_{-\infty}^0 e^{-\sqrt{2\lambda}|y|} f(y) \mathrm{d}y.$$

Observe that

$$|V_\lambda f(x)| \leq \|f\|_\infty V_\lambda 1(x) = \|f\|_\infty \lambda^{-1}(1 - \mathbb{E}e^{-\lambda\sigma(x+W)})$$
$$= \|f\|_\infty \lambda^{-1}(1 - \mathbb{E}^x e^{-\lambda\sigma(W)}) = \|f\|_\infty \lambda^{-1}(1 - e^{-\sqrt{2\lambda}|x|})$$
$$\leq \|f\|_\infty \lambda^{-1}\sqrt{2\lambda}|x| = \|f\|_\infty \sqrt{2/\lambda}|x|, \quad x \in \mathbb{R},$$

where $\|f\|_\infty = \sup_{x \in \mathbb{R}} |f(x)|$. In particular,

$$|V_\lambda f(\varepsilon\eta)| \leq \|f\|_\infty \sqrt{2/\lambda}\,\varepsilon|\eta|.$$

It follows from the previous reasoning and the Lebesgue dominated convergence theorem that it suffices to prove that

$$\lim_{\varepsilon \to 0+} \varepsilon^{-1} V_\lambda f(\varepsilon x) \mathbb{1}_{(0,\infty)}(\pm x) = 2|x| \int_0^\infty e^{-\sqrt{2\lambda}y} f(\pm y) \mathrm{d}y \qquad (2.49)$$

for $\lambda > 0$ and $f \in C_b([0, \infty))$. Assume that $x > 0$. A combination of (2.36) and (2.37) yields

$$V_\lambda f(x) = \frac{1}{\sqrt{2\lambda}} \int_0^\infty \left(e^{-\sqrt{2\lambda}|x-y|} - e^{-\sqrt{2\lambda}(x+y)}\right) f(y) \mathrm{d}y$$
$$= \frac{1}{\sqrt{2\lambda}} \int_0^x \left(e^{-\sqrt{2\lambda}(x-y)} - e^{-\sqrt{2\lambda}(x+y)}\right) f(y) \mathrm{d}y$$
$$+ \int_x^\infty \left(e^{-\sqrt{2\lambda}(y-x)} - e^{-\sqrt{2\lambda}(x+y)}\right) f(y) \mathrm{d}y.$$

Hence, for each $x > 0$,

$$\begin{aligned}
\varepsilon^{-1} V_\lambda f(\varepsilon x) &= \frac{1}{\sqrt{2\lambda}\varepsilon} \int_0^{\varepsilon x} \left(e^{-\sqrt{2\lambda}(\varepsilon x - y)} - e^{-\sqrt{2\lambda}(\varepsilon x + y)}\right) f(y) dy \\
&\quad + \frac{1}{\sqrt{2\lambda}\varepsilon} \int_{\varepsilon x}^{\infty} \left(e^{-\sqrt{2\lambda}(y - \varepsilon x)} - e^{-\sqrt{2\lambda}(\varepsilon x + y)}\right) f(y) dy \\
&= \frac{e^{-\sqrt{2\lambda}\varepsilon x}}{\sqrt{2\lambda}\varepsilon} \int_0^{\varepsilon x} \left(e^{\sqrt{2\lambda} y} - e^{-\sqrt{2\lambda} y}\right) f(y) dy \\
&\quad + \frac{e^{\sqrt{2\lambda}\varepsilon x} - e^{-\sqrt{2\lambda}\varepsilon x}}{\sqrt{2\lambda}\varepsilon} \int_{\varepsilon x}^{\infty} e^{-\sqrt{2\lambda} y} f(y) dy \\
&\to 2x \int_0^{\infty} e^{-\sqrt{2\lambda} y} f(y) dy, \quad \varepsilon \to 0+,
\end{aligned}$$

which proves the case $x > 0$ of (2.49). The case $x < 0$ can be treated similarly. The proof[2] of Theorem 2.2.3 is complete. □

Theorem 2.2.3 could have been proved in several ways. We decided to provide an analytic proof based on the analysis of resolvents. The reason is that a similar argument will be used in Sect. 2.3 when investigating perturbations of a symmetric α-stable Lévy process with $\alpha \in (1, 2)$. Other possible approaches are discussed in "Bibliographic Comments."

If the distribution tail of η is regularly varying at ∞ of index in $(-1, 0)$, then $\mathbb{E} R_\lambda^\varepsilon f(\varepsilon \eta)$ does not converge to $R_\lambda^{\text{skew}} f(0)$. The corresponding limit will be found in Sect. 2.3, where perturbations that are more general than those in Theorem 2.2.3 are allowed, and the underlying process is a symmetric α-stable Lévy process with $\alpha \in (1, 2)$.

2.2.2 The Walsh Brownian Motion

The Walsh Brownian motion is a natural multidimensional generalization of the skew Brownian motion. While the state space of a skew Brownian motion is $\mathbb{R}$, the state space of a Walsh Brownian motion consists of d rays with a common endpoint [168]. On each ray the Walsh Brownian motion is a standard one-dimensional Brownian motion, and the probability that "an excursion selects" the kth ray upon hitting the origin is equal to $p_k > 0$, where $p_1 + \ldots + p_d = 1$. Usually it is assumed that the rays belong to a plane.

---

[2] We are grateful to Adam Bobrowski for giving us a valuable advice, which simplified our original proof.

2.2 The Skew Brownian Motion and Its Generalizations

However, for our needs it is convenient to assume that the phase space E_d for a Walsh Brownian motion is the union of nonnegative coordinate semi-axes in $\mathbb{R}^d$, that is,

$$E_d := \{x \in \mathbb{R}^d \ : \ x_i \geq 0 \text{ and } x_i x_j = 0, \ i \neq j, \ i, j = 1, \ldots, d\}.$$

Here is a formal definition of the Walsh Brownian motion in terms of transition probability density function.

Definition 2.2.2 The Walsh Brownian motion with parameters $p_k > 0$ for $1 \leq k \leq d$ satisfying $\sum_{k=1}^{d} p_k = 1$ is a Markov process on E_d with the transition probability density function $p^{(t)}$ given by

$$p^{(t)}((0, \ldots, 0, x_i, 0, \ldots, 0), (0, \ldots, 0, y_j, 0, \ldots, 0))$$
$$= \begin{cases} \varphi_t(x_i - y_j) + (2p_i - 1)\varphi_t(x_i + y_j), & i = j \\ 2p_j \varphi_t(x_i + y_j), & i \neq j \end{cases}$$

for nonnegative x_i and y_j, where $\varphi_t(z) = (2\pi t)^{-1/2} e^{-z^2/(2t)}$ for $z \in \mathbb{R}$ and $t > 0$ is the same as in Definition 2.2.1.

Remark 2.2.3 It can be checked that, with $x_i \geq 0$ and $t > 0$ fixed, the function given in Definition 2.2.2 is indeed a probability density function in y_j and that it satisfies the Chapman-Kolmogorov equation.

We proceed by giving a Markovian characterization of the Walsh Brownian motion which can be derived from the discussion given in Section 2 of [6].

Proposition 2.2.1 *A continuous strong Markov process*

$$(W_{\mathbf{p}}(t))_{t \geq 0} = ((X_1(t), \ldots, X_d(t)))_{t \geq 0}$$

on E_d is a Walsh Brownian motion with parameter $\mathbf{p} = (p_1, \ldots, p_d)$, where $p_k > 0$ and $p_1 + \ldots + p_d = 1$, if, and only if,

(a) For each $x = (0, \ldots, 0, x_k, 0, \ldots, 0) \in E_d$

$$\mathbb{P}\{(X_k(t)\mathbb{1}_{\{t \leq \sigma(X)\}})_{t \geq 0} \in \cdot \big| X(0) = x\}$$
$$= \mathbb{P}\{(W(t)\mathbb{1}_{\{t \leq \sigma(W)\}})_{t \geq 0} \in \cdot \big| W(0) = x_k\},$$

where W is a one-dimensional standard Brownian motion, $\sigma(W_\mathbf{p}) = \inf\{t \geq 0 : W_\mathbf{p}(t) = 0\}$ and $\sigma(W) = \inf\{t \geq 0 : W(t) = 0\}$.
(b) For each $t > 0$, $k = 1, \ldots, d$ and a Borel set $A \subset (0, \infty)$

$$\mathbb{P}\{X_k(t) \in A \mid X(0) = 0\} = p_k \mathbb{P}\{|W(t)| \in A \mid W(0) = 0\}.$$

The following result follows from Definition 2.2.2.

Theorem 2.2.4 *Let $W_\mathbf{p}(t) = (X_1, \ldots, X_d)$ be a Walsh Brownian motion with parameter $\mathbf{p} = (p_1, \ldots, p_d)$. Then, for any set $I \subseteq \{1, \ldots, d\}$, the process $(Y(t))_{t \geq 0}$ defined by*

$$Y(t) := \sum_{i \in I} X_i(t) - \sum_{j \in I^c} X_j(t), \quad t \geq 0 \tag{2.50}$$

is a skew Brownian motion with permeability parameter

$$\gamma = \sum_{i \in I} p_i - \sum_{j \in I^c} p_j = 2 \sum_{i \in I} p_i - 1 \in [-1, 1].$$

In particular, the process $\left(\sum_{k=1}^d X_k(t)\right)_{t \geq 0}$ is a reflected Brownian motion, and there exists a local time of $W_\mathbf{p}$ at 0 defined by

$$\nu(t) := L_0^{W_\mathbf{p}}(t) := \lim_{\varepsilon \to 0+} \frac{1}{2\varepsilon} \int_0^t \mathbb{1}_{\{\max_{1 \leq i \leq d} |X_i(s)| \leq \varepsilon\}} ds. \tag{2.51}$$

Remark 2.2.4 We stress that for each $t > 0$ there is at most one nonzero summand in (2.50) and $\max_{1 \leq i \leq d} |X_i(s)| = \sum_{i=1}^d X_i(s)$.

Given next is a martingale characterization of a Walsh Brownian motion, which is a powerful tool for proving limit theorems for locally perturbed random walks.

Theorem 2.2.5 *Let $X = (X_1(t), \ldots, X_d(t))_{t \geq 0}$ and $\nu = (\nu(t))_{t \geq 0}$ be continuous processes, and $\mathcal{F}^X$ a filtration generated by X. Then X is a Walsh Brownian motion with parameter $\mathbf{p} = (p_1, \ldots, p_d)$, and ν is the local time of X at 0 if, and only if,*

(I) $X_i(t) \geq 0$ for $1 \leq i \leq d$ and $X_i(t)X_j(t) = 0$ for $1 \leq i \neq j \leq d$ and $t \geq 0$.
(II) ν is a.s. nondecreasing $\mathcal{F}^X$-adapted process with $\nu(0) = 0$ and

$$\int_{[0, \infty)} \mathbb{1}_{\{X(s) \neq 0\}} d\nu(s) = 0 \quad a.s.$$

(III) The processes $M_1, \ldots, M_d$ defined by

$$M_i(t) := X_i(t) - p_i v(t), \quad t \geq 0$$

are continuous square integrable martingales with respect to $\mathcal{F}^X$ having the predictable quadratic variations

$$\langle M_i, M_i \rangle(t) = \int_0^t \mathbb{1}_{\{X_i(s)>0\}} \, ds.$$

(IV) $\int_0^\infty \mathbb{1}_{\{X(s)=0\}} \, ds = 0$ a.s.

In the remainder of this section we state a result on a distributional approximation of a Walsh Brownian motion, which bears a close resemblance to Theorem 2.2.3. We start by setting the scene. Let $(\eta_k)_{k \geq 1}$ be a sequence of independent copies of an E_d-valued random variable $\eta = (\eta^{(1)}, \ldots, \eta^{(d)})$ with $\eta \neq 0$ a.s. and $\mathbb{E}\|\eta\| < \infty$; $W = (W(t))_{t \geq 0}$ a standard one-dimensional Brownian motion, and $(X_\varepsilon(0))_{\varepsilon > 0}$ a family of random variables taking values in $E_d \setminus \{0\}$. Assume that $(\eta_k)_{k \geq 1}$, W and $(X_\varepsilon(0))_{\varepsilon > 0}$ are mutually independent. We intend to define a family of strong Markov processes $((X_\varepsilon(t))_{t \geq 0})_{\varepsilon > 0}$. To this end, put

$$l(x) := \sum_{k=1}^d \mathbb{1}_{\{x^{(k)}>0\}} e_k \quad x = (x^{(1)}, \ldots, x^{(d)})$$

where $(e_i)_{1 \leq i \leq d}$ is the standard orthonormal basis of $\mathbb{R}^d$, and define the stopping times $\sigma_0^{(\varepsilon)} = 0$,

$$\sigma_1^{(\varepsilon)} := \inf\{t \geq 0 : \|X_\varepsilon(0)\| + W(t) = 0\},$$

$$\sigma_{k+1}^{(\varepsilon)} := \inf\{t > \sigma_k^{(\varepsilon)} : \varepsilon\|\eta_k\| + (W(t) - W(\sigma_k^{(\varepsilon)})) = 0\}, \quad k \in \mathbb{N}.$$

With these at hand,

$$X_\varepsilon(t) := \begin{cases} X_\varepsilon(0) + W(t) l(X_\varepsilon(0)), & t \in [0, \sigma_1^{(\varepsilon)}), \\ \varepsilon \eta_k + (W(t) - W(\sigma_k^{(\varepsilon)})) l(\eta_k), & t \in [\sigma_k^{(\varepsilon)}, \sigma_{k+1}^{(\varepsilon)}), \quad k \in \mathbb{N}. \end{cases}$$

In other words,

(I) For $k \in \mathbb{N}$, $\sigma_k^{(\varepsilon)}$ is the kth epoch for which $X_\varepsilon(\sigma_k^{(\varepsilon)}-) = 0$.
(II) For $k \in \mathbb{N}$, the jump of X_ε at $\sigma_k^{(\varepsilon)}$ is $\varepsilon \eta_k$.
(III) On each interval $[\sigma_k^{(\varepsilon)}, \sigma_{k+1}^{(\varepsilon)})$, $k \in \mathbb{N}_0$ only one coordinate of X_ε is nonzero, and the increments of X_ε along this coordinate coincide with the increments of W.

By definition $X_\varepsilon(t) \neq 0$ for each $t \geq 0$.

Theorem 2.2.6 *Assume that* $X_\varepsilon(0) \overset{d}{\to} \kappa$ *as* $\varepsilon \to 0+$ *for some* E_d-*valued random variable* κ. *Then*

$$(X_\varepsilon(t))_{t\geq 0} \implies (W_\mathbf{p}(t))_{t\geq 0}, \quad \varepsilon \to 0+$$

on $D([0, \infty))$, *where* $p_i = \mathbb{E}\eta^{(i)}/\mathbb{E}\|\eta\|$ *for* $i = 1, \ldots, d$ *and* $W_\mathbf{p}(0) \overset{d}{=} \kappa$.

Theorem 2.2.6 can be proved similarly to Theorem 2.2.3. We omit details.

Bibliographic Comments

As has already been mentioned in Sect. 2.2.1 Itô and McKean [75] constructed a skew Brownian motion by randomly assigning signs to excursions of a reflected Brownian motion. Portenko developed the theory of diffusions with generalized drifts [136, 137]. In particular, a skew Brownian motion is a particular instance of a one-dimensional diffusion whose drift is proportional to Dirac's delta function concentrated at 0. We stress that, with the exception of the values $\gamma \in \{-1, 0, 1\}$, the skew Brownian motion W_γ^{skew} cannot be obtained as an image under the action of a reasonably simple continuous Skorokhod-type map on the underlying Brownian motion. As a consequence, most of the results concerned with a skew Brownian motion and its generalizations rely on more advanced approaches. Various properties of a skew Brownian motion were exhibited in the paper by Walsh [168]. The proof of existence and uniqueness of a solution to the stochastic differential equation with a local time (2.44) for a skew Brownian motion was given in [65]. We also mention the detailed review paper [97] on a skew Brownian motion.

Now we discuss briefly two articles in which a counterpart of (2.43) on the space $C([0, 1])$ was derived. Brooks and Chacon proved in Theorem 2 of [26] that Donsker's scaling of the sequence of random polygonal functions obtained by the linear interpolation of $\widetilde{S}$ converges in distribution on $C([0, 1])$ to a skew Brownian motion. These authors exploited a standard approach for proving functional limit theorems, that is, they showed weak convergence of finite-dimensional distributions and then checked tightness. Bhattacharya and Waymire in Theorem 25.4 on p. 454 of their recent book [11] gave an alternative proof based on the Skorokhod embedding for a skew Brownian motion.

The proof of Theorem 2.2.3 is new. We state this result and prove it as a preliminary step and a warm-up argument for construction of a skew stable Lévy process in Sect. 2.3. When proving Theorem 2.2.3, instead of the resolvent technique we could have exploited an approach based on the martingale characterization of the Walsh Brownian motion as given in Theorem 2.2.5. Alternatively, one could have argued along the lines of the article [129] in which a method based on cutting and merging trajectories was used.

2.3 The Skew Stable Lévy Process

The Walsh Brownian motion, which is a natural generalization of the skew Brownian motion from the viewpoint of excursions, was introduced in the Epilogue of the paper [168]. The articles [6, 7, 53–55, 61, 62, 78, 88, 141, 144], as well as the papers cited therein, discuss in depth many properties of the Walsh Brownian motion including representations for the resolvent and the generator, a martingale characterization, Ito's formula, and some others. The proof of Theorem 2.2.5 can be found in [6]. The present formulation of the theorem is taken from [22].

The skew Brownian motion and the Walsh Brownian motion can also be constructed as a concatenation of reflected Brownian motions taken at certain random times. Such an approach comes from the theory of multiarmed bandits; see, for instance, [5, 8, 80, 102]. We stress that questions/problems related to filtrations are rather delicate in this setting. On the one hand, the Walsh Brownian motion (in particular, the skew Brownian motion) can be represented as a measurable function of several Brownian motions. On the other hand, the filtration generated by a Walsh Brownian motion on a graph with more than three rays is not a Brownian filtration [162].

In some sense, point 0 is an instantaneous penetration point for a skew Brownian motion X (or a Walsh Brownian motion). If a stopping time τ satisfies $X(\tau) = 0$, then, for each $\varepsilon > 0$, the process $(X(t))_{t \in [\tau, \tau+\varepsilon]}$ visits any halfline infinitely often with probability one. A snapping out Brownian motion X^{SNOB} introduced in the papers [98, 103] can be thought of as a Brownian motion in a media with "non-instantaneous" penetration. Informally, its behavior can be described as follows. Assume that $X^{\text{SNOB}}(0) > 0$ and fix positive α_+ and α_-. The process moves like a reflected Brownian motion until the time σ_1^+, when its local time at 0 process reaches a random variable ξ_1^+ which is independent of the past and has an exponential distribution with mean $1/\alpha_+$. Then X^{SNOB} snaps to the negative halfline and moves like a reflected down Brownian motion until the time σ_1^- ($\sigma_1^- > \sigma_1^+$), when its local time process reaches a variable $\xi_1^+ + \xi_1^-$, where ξ_1^- is independent of the past and has an exponential distribution with mean $1/\alpha_-$. Afterwards X^{SNOB} snaps up and behaves like a reflected up Brownian motion until the time σ_2^+ ($\sigma_2^+ > \sigma_1^-$), when its local time process reaches a variable $\xi_1^+ + \xi_1^- + \xi_2^+$, where ξ_2^+ is independent of the past and has an exponential distribution with mean $1/\alpha_+$, and so on. Although X^{SNOB} is not a strong Markov process on $\mathbb{R}$, it becomes a Feller process on a modified state space $(-\infty, 0-] \cup [0+, +\infty)$ obtained by splitting the point 0 into two points $0-$ and $0+$. Approximations of a skew Brownian motion and a Walsh Brownian motion by snapping out processes with increasing intensities can be found in [18, 19].

2.3 The Skew Stable Lévy Process

2.3.1 Definition

In this section we address the following question: what is a natural analogue of the skew Brownian motion in the case when an underlying Brownian motion is replaced with a stable Lévy process? The corresponding problem is highly nontrivial. We shall discuss a few approaches for defining a counterpart of the skew Brownian motion. We suggest that the reader decides whether the term "skew stable Lévy process" is suitable for the process that we are constructing.

Let $U_\alpha := (U_\alpha(t))_{t \geq 0}$ be a symmetric α-stable Lévy process with $\alpha \in (1, 2)$ and the characteristic function

$$\mathbb{E} \exp(i z U_\alpha(t)) = \exp(-t|z|^\alpha), \quad z \in \mathbb{R}, \quad t \geq 0.$$

It is known that U_α hits any point with probability 1; see, for instance, p. 63 in [10] or Example 43.22 on p. 325 in [146]. We are going to construct a strong Markov process that behaves like U_α on the set $\mathbb{R} \setminus \{0\}$ and has some "natural asymmetry" at 0.

Towards that aim we first attempt to follow ideas which have proved useful in constructing and investigating the skew Brownian motion. We start by writing a formal analogue of the transition probability density function of the skew Brownian motion given in Definition 2.2.1

$$p^{(t,\alpha)}(x, y) := \varphi_t^{(\alpha)}(x - y) + (2p - 1)\mathrm{sgn}(y)\varphi_t^{(\alpha)}(|x| + |y|), \quad x, y \in \mathbb{R}, \quad t \geq 0, \tag{2.52}$$

where $\varphi_t^{(\alpha)}$ is a density of $U_\alpha(t)$ and $p \in [0, 1]$. The function $p^{(t,\alpha)}$ is a transition probability density function of a Markov process $Z^{(\alpha)}$; see [135] for more details. If $p = 1/2$, then $Z^{(\alpha)}$ has the same distribution as U_α. It is known that if $p = 1$, then $Z^{(\alpha)}$ has the same distribution as $|U_\alpha|$. A martingale characterization of $Z^{(\alpha)}$ given in [135] implies that the intensity of jumps of $Z^{(\alpha)}$ is other than that of U_α.

The approach of Itô and McKean, based on a random assignment of signs to excursions of a (reflected) Brownian motion, does not apply in the present situation because U_α does not hit 0 by a single jump. It crosses 0 infinitely often before hitting 0 (see Theorem 6.4 in [169] and Proposition 8 on p. 226 in [10]) and also changes sign infinitely often on exiting 0. The latter can be seen with the help of a time-inversion argument.

To the best of our knowledge, at the moment there are no results on solutions to stochastic equations involving a non-Gaussian Lévy noise and a local time. The proof of existence and uniqueness of solutions to (2.44) given in [65] relies heavily upon Nakao's result [115], which, in turn, is based on the assumption that the noise is a Brownian motion. In view of our arguments presented later in the text, it is unlikely that there is a solution to a counterpart of (2.44), with the process U_α replacing the Brownian motion. On the other hand, some other equations involving a local time will appear in our subsequent presentation.

To define the asymmetry at 0, we shall use the approach of Harrison and Shepp [65]. Recall that these authors obtained a skew Brownian motion as a scaling limit of a sequence of simple symmetric random walks perturbed at 0. We shall prove a limit theorem in the spirit of Theorem 2.2.3, in which a Brownian motion experiences small jump-type exits from 0 upon every hitting 0.

We proceed by introducing the model. Let $(\zeta_k)_{k \geq 1}$ be independent copies of a random variable ζ, $\zeta \neq 0$ a.s. Assume that the sequence $(\zeta_k)_{k \geq 1}$ and the process U_α are

2.3 The Skew Stable Lévy Process

jointly independent and also independent of a family $(X_\varepsilon(0))_{\varepsilon>0}$ of random variables taking values in $\mathbb{R} \setminus \{0\}$. Similarly to (2.45), we intend to define a family of processes $((X_\varepsilon(t))_{t\geq 0})_{\varepsilon>0}$. To this end, we define the stopping times

$$\sigma_1^{(\varepsilon)} := \inf\{t > 0 : X_\varepsilon(0) + U_\alpha(t) = 0\}$$

and

$$\sigma_{k+1}^{(\varepsilon)} := \inf\{t > \sigma_k^{(\varepsilon)} : \varepsilon\zeta_k + U_\alpha(t) - U_\alpha(\sigma_k^{(\varepsilon)}) = 0\}, \quad k \in \mathbb{N}.$$

Finally, we put

$$X_\varepsilon(t) := \begin{cases} X_\varepsilon(0) + U_\alpha(t), & t \in [0, \sigma_1^{(\varepsilon)}), \\ \varepsilon\zeta_k + U_\alpha(t) - U_\alpha(\sigma_k^{(\varepsilon)}), & t \in [\sigma_k^{(\varepsilon)}, \sigma_{k+1}^{(\varepsilon)}), \quad k \in \mathbb{N}. \end{cases} \quad (2.53)$$

As has already been mentioned, the process X_ε cannot hit 0 by a single jump, so that $X_\varepsilon(\sigma_k^{(\varepsilon)}-) = 0$ for $k \in \mathbb{N}$. Although $X_\varepsilon(\sigma_k^{(\varepsilon)}) \neq 0$ a.s. by definition, we call $\sigma_k^{(\varepsilon)}$ the time of the kth visit of X_ε to 0. Let $N_\varepsilon(t) := \#\{k \in \mathbb{N} : \sigma_k^{(\varepsilon)} \leq t\}$ be the number of visits to 0 up to time $t \geq 0$.

Observe that

$$X_\varepsilon(t) = X_\varepsilon(0) + U_\alpha(t) + \varepsilon \sum_{k=1}^{N_\varepsilon(t)} \zeta_k =: X_\varepsilon(0) + U_\alpha(t) + \varepsilon S_\zeta(N_\varepsilon(t)), \quad t \geq 0, \quad (2.54)$$

where $S_\zeta(0) := 0$ and $S_\zeta(n) := \zeta_1 + \ldots + \zeta_n$ for $n \in \mathbb{N}$. Thus, to ensure a distributional convergence of X_ε as $\varepsilon \to 0+$ it is natural to assume that either $\mathbb{E}|\zeta| < \infty$ or the distribution of ζ belongs to the domain of attraction of a β-stable distribution with $\beta \in (0, 1)$. In the former case, $S_\zeta(n)$ satisfies a law of large numbers. In the latter case, the variables $S_\zeta(n)$, properly normalized, converge in distribution to a random variable with a β-stable distribution. It is known (see Theorem 4.4.1) that this happens if, and only if, the function $x \mapsto \mathbb{P}\{|\zeta| > x\}$ is regularly varying at $+\infty$ of index $-\beta$ and the limits

$$c_\pm := \lim_{x \to +\infty} \frac{\mathbb{P}\{\pm\zeta > x\}}{\mathbb{P}\{|\zeta| > x\}} \quad (2.55)$$

exist. Observe that if these exist, then necessarily $c_- + c_+ = 1$.

Recall that, for $\lambda > 0$, the resolvent V_λ^Y of a strong Markov process Y killed at 0 is given by

$$V_\lambda^Y f(x) := \mathbb{E}^x \int_0^{\sigma(Y)} e^{-\lambda s} f(Y(s)) ds, \quad x \in \mathbb{R}$$

for bounded measurable $f : \mathbb{R} \to \mathbb{R}$. Here, $\sigma(Y)$ is the first hitting time of 0 by Y, that is, $\sigma(Y) = \inf\{t \geq 0 : Y(t) = 0\}$. The notation $V_\lambda 1$ is understood as $V_\lambda f$ with $f(x) = 1$ for $x \in \mathbb{R}$. Given a measure ν on $\mathbb{R}$ we write $\langle \nu, f \rangle$ for $\int_\mathbb{R} f(x)\nu(dx)$ provided that the integral is well-defined.

Theorem 2.3.1 *Let $\alpha \in (1,2)$. Assume that $(\zeta_k)_{k \geq 1}$, $(X_\varepsilon(0))_{\varepsilon > 0}$ and U_α are jointly independent, and that*

$$X_\varepsilon(0) \xrightarrow{d} \kappa, \quad \varepsilon \to 0+$$

for some real-valued random variable κ.

(a) Assume that the distribution of ζ belongs to the domain of attraction of a β-stable distribution with $\beta < \alpha - 1$ (in particular, relation (2.55) holds). Then the processes X_ε converge in distribution on $D([0, \infty))$ as $\varepsilon \to 0+$ to a Feller process X, with $X(0)$ having the same distribution as κ. The resolvent of X is given by

$$R_\lambda^X f(x) = V_\lambda^{U_\alpha} f(x) + \lambda^{-1} \mathbb{E}^x e^{-\lambda \sigma(U_\alpha)} \frac{\langle \eta^*, V_\lambda^{U_\alpha} f \rangle}{\langle \eta^*, V_\lambda^{U_\alpha} 1 \rangle} \qquad (2.56)$$

for $\lambda > 0$, $x \in \mathbb{R}$ and bounded measurable functions $f : \mathbb{R} \to \mathbb{R}$, where η^* is the measure defined by

$$\eta^*(dx) = (c_- \mathbb{1}_{(-\infty, 0)}(x) + c_+ \mathbb{1}_{(0, \infty)}(x))|x|^{-(1+\beta)} dx, \quad x \in \mathbb{R}, \qquad (2.57)$$

and the constants $c_\pm$ are given in (2.55).

(b) Assume that either $\mathbb{E}|\zeta| < \infty$ or the distribution of ζ belongs to the domain of attraction of a β-stable distribution with $\beta \in (\alpha - 1, 1)$. Then the processes X_ε converge in distribution on $D([0, \infty))$ as $\varepsilon \to 0+$ to a process X defined by $X(t) = \kappa + U_\alpha(t)$ for $t \geq 0$.

Definition 2.3.1 For $\alpha \in (1, 2)$ and $\beta \in (0, \alpha - 1)$, we call *skew α-stable Lévy process with parameter β* a Markov process with the resolvent given in (2.56) and denote it by $U_{\alpha, \beta} = (U_{\alpha, \beta}(t))_{t \geq 0}$. Occasionally, we write $U_{\alpha, \beta}(x, t)$ to bring out the dependence on the initial state x.

Remark 2.3.1 We attract the reader's attention to an essential difference between Theorem 2.2.3 and part (b) of Theorem 2.3.1. If the perturbations have a finite nonzero mean, then the limit process in the latter result is U_α for $\alpha \in (1, 2)$, whereas it is a skew Brownian motion in Theorem 2.2.3 rather than a standard Brownian motion. The origin of this distinction is rather deep. On the one hand, there is a unique Feller process that

2.3 The Skew Stable Lévy Process

behaves like a symmetric α-stable Lévy process outside 0, has a zero sojourn at 0, and "exits continuously" from 0. The distribution of this Feller process coincides with the distribution of U_α. On the other hand, a skew Brownian motion and some other processes behave like a standard Brownian motion outside 0 and "exit continuously" from 0. We shall discuss excursion properties of the skew stable Lévy processes in Sect. 2.3.3.

2.3.2 Proof of Theorem 2.3.1

PART (A) IN WHICH $\beta < \alpha - 1$. It follows from Corollary 18 on p. 64 in [10] that

$$\mathbb{E}^x e^{-\lambda \sigma(U_\alpha)} = \frac{v_\lambda(-x)}{v_\lambda(0)}, \quad x \in \mathbb{R}, \quad \lambda > 0, \tag{2.58}$$

where v_λ is the density of the resolvent kernel of U_α. According to Theorem 19(iii) on p. 65 in [10], applied with $\Psi(z) = |z|^\alpha$,

$$v_\lambda(x) = \frac{1}{\pi} \int_0^\infty \frac{\cos(x\theta)}{\lambda + \theta^\alpha} d\theta, \quad x \in \mathbb{R}, \quad \lambda > 0. \tag{2.59}$$

Alternatively, the latter formula can be derived as follows. Recall that $\varphi_t^{(\alpha)}$ denotes a density of $U_\alpha(t)$. Passing to the Fourier transform in

$$v_\lambda(x) = \int_0^\infty e^{-\lambda t} \varphi_t^{(\alpha)}(x) dt, \quad x \in \mathbb{R}$$

and using Fubini's theorem we obtain

$$\int_\mathbb{R} e^{i\theta x} v_\lambda(x) dx = \int_\mathbb{R} e^{i\theta x} \left(\int_0^\infty e^{-\lambda t} \varphi_t^{(\alpha)}(x) dt \right) dx$$

$$= \int_0^\infty e^{-\lambda t} \int_\mathbb{R} \left(e^{i\theta x} \varphi_t^{(\alpha)}(x) dx \right) dt = \int_0^\infty e^{-\lambda t} \mathbb{E} e^{i\theta U_\alpha(t)} dt$$

$$= \int_0^\infty e^{-\lambda t} e^{-|\theta|^\alpha t} dt = \frac{1}{\lambda + |\theta|^\alpha}.$$

Formula (2.59) now follows by Fourier inversion. According to the aforementioned Corollary 18 in [10], formula (2.58) is valid for any Lévy process, whose resolvent kernel is absolutely continuous with a bounded density.

Lemma 2.3.1 collects a couple of formulas to be used in what follows.

Lemma 2.3.1 *Let $\alpha \in (1, 2)$.*

(a) For $\gamma \in [0, \alpha - 1)$ and $\lambda > 0$,

$$\int_0^\infty \frac{\theta^\gamma}{\lambda + \theta^\alpha} d\theta = \frac{\Gamma(1 - \frac{\gamma+1}{\alpha})\Gamma(\frac{\gamma+1}{\alpha})}{\alpha \lambda^{1-\frac{\gamma+1}{\alpha}}} = \frac{\pi}{\alpha \sin \frac{\pi(\gamma+1)}{\alpha}} \frac{1}{\lambda^{1-\frac{\gamma+1}{\alpha}}},$$

where Γ is the Euler gamma function. In particular,

$$v_\lambda(0) = \frac{1}{\alpha \sin \frac{\pi}{\alpha}} \frac{1}{\lambda^{1-\frac{1}{\alpha}}}. \tag{2.60}$$

(b) For $x \in \mathbb{R}$,

$$\int_0^\infty \frac{1 - \cos(xy)}{y^\alpha} dy = \frac{\Gamma(2-\alpha) \sin \frac{\pi\alpha}{2}}{\alpha - 1} |x|^{\alpha-1}. \tag{2.61}$$

Proof While the first equality in the first formula of part (a) is a consequence of formula (3.241)(2) in [59], the second equality follows from Euler's reflection formula $\Gamma(1-z)\Gamma(z) = \frac{\pi}{\sin(\pi z)}$ which holds true for any noninteger z. Equality (2.60) is implied by the first formula with $\gamma = 0$ and equality (2.59) with $x = 0$. Part (b) follows from formula (14.18) in [146] or from equation (3.823) in [59]. □

We proceed by showing that formula (2.56) does define the resolvent of a Feller process. To this end, according to Lemma 4.5.1 with $x^* = 0$, the discussion in Example 4.6.4 and particularly formula (4.31), it is enough to check that

$$\langle \eta^*, V_\lambda^{U_\alpha} 1 \rangle = \lambda^{-1} \int_\mathbb{R} \mathbb{E}^x (1 - e^{-\lambda \sigma(U_\alpha)}) \eta^*(dx) < \infty. \tag{2.62}$$

Formula (2.62) is secured by the result given next.

Lemma 2.3.2 *For $\alpha \in (1, 2)$ and $\lambda > 0$,*

$$\lambda V_\lambda 1(x) = \mathbb{E}^x (1 - e^{-\lambda \sigma(U_\alpha)}) \sim A_{\lambda,\alpha} |x|^{\alpha-1}, \quad x \to 0, \tag{2.63}$$

where

$$A_{\lambda,\alpha} := \frac{\alpha \Gamma(2-\alpha) \sin \frac{\pi}{\alpha} \sin \frac{\pi\alpha}{2}}{\pi(\alpha - 1)} \lambda^{1-\frac{1}{\alpha}},$$

2.3 The Skew Stable Lévy Process

and

$$\mathbb{P}^1\{\sigma(U_\alpha) > y\} \sim B_\alpha y^{-1+\frac{1}{\alpha}}, \quad y \to \infty, \tag{2.64}$$

where $B_\alpha := A_{1,\alpha}/\Gamma(1/\alpha) = \frac{\Gamma(2-\alpha)\sin\frac{\pi}{\alpha}\sin\frac{\pi\alpha}{2}}{\pi(\alpha-1)\Gamma(1+\frac{1}{\alpha})}$.

Proof We start with proving (2.63). In view of (2.58) and (2.59)

$$\mathbb{E}^x(1 - e^{-\lambda\sigma(U_\alpha)}) = \frac{1}{\pi v_\lambda(0)} \int_0^\infty \frac{1 - \cos(x\theta)}{\lambda + \theta^\alpha} d\theta$$

$$= \frac{|x|^{\alpha-1}}{\pi v_\lambda(0)} \int_0^\infty \frac{1 - \cos(y)}{\lambda|x|^\alpha + y^\alpha} dy \sim \frac{|x|^{\alpha-1}}{\pi v_\lambda(0)} \int_0^\infty \frac{1 - \cos(y)}{y^\alpha} dy, \quad x \to 0. \tag{2.65}$$

Invoking Lemma 2.3.1 we arrive at (2.63).

Using the first equality in (2.65) and Lemma 2.3.1 we infer

$$\mathbb{E}^1(1 - e^{-\lambda\sigma(U_\alpha)}) = \frac{\alpha\sin\frac{\pi}{\alpha}}{\pi} \lambda^{1-\frac{1}{\alpha}} \int_0^\infty \frac{1 - \cos\theta}{\lambda + \theta^\alpha} d\theta$$

$$\sim \frac{\alpha\sin\frac{\pi}{\alpha}}{\pi} \lambda^{1-\frac{1}{\alpha}} \int_0^\infty \frac{1 - \cos\theta}{\theta^\alpha} d\theta$$

$$= \frac{\alpha\Gamma(2-\alpha)\sin\frac{\pi}{\alpha}\sin\frac{\pi\alpha}{2}}{\pi(\alpha-1)} \lambda^{1-\frac{1}{\alpha}} = A_{\lambda,\alpha}, \quad \lambda \to 0+.$$

An application of Corollary 8.1.7 in [13] yields (2.64). □

Lemma 2.3.3(a) is the principal ingredient of the proof of Theorem 2.3.1(a). Part (b) of Lemma 2.3.3 will be used in Sect. 3.4.2.

Lemma 2.3.3 *Assume that the function* $x \mapsto \mathbb{P}\{|\zeta| > x\}$ *is regularly varying at* $+\infty$ *of index* $-\beta \in (-1, 0)$, *and relation (2.55) holds.*

(a) *For a bounded continuous function* $g : \mathbb{R} \to \mathbb{R}$ *satisfying for some* $\gamma > 0$ $g(x) = O(|x|^{\beta+\gamma})$ *as* $x \to 0$,

$$\lim_{u \to \infty} \frac{\mathbb{E}g(\zeta/u)}{\mathbb{P}\{|\zeta| > u\}} = \int_\mathbb{R} g(x)\eta^*(dx) = \langle \eta^*, g \rangle < \infty, \tag{2.66}$$

where η^* *is the measure defined in (2.57).*

(b) *Let* $(g_u)_{u>0}$ *be a family of uniformly bounded measurable functions which satisfy the conditions:*

(i) *For a continuous function g*

$$\lim_{u\to\infty} \sup_{x\in\mathbb{R}} |g_u(x) - g(x)| = 0.$$

(ii) *For some positive constants u_0 and c, a constant $\gamma \in (0, 1-\beta)$, and a nonnegative constant r which do not depend on x nor u*

$$|g_u(x)| \le c(|x| + r/u)^{\beta+\gamma}, \quad x \in \mathbb{R}, \quad u \ge u_0. \tag{2.67}$$

Then

$$\lim_{u\to\infty} \frac{\mathbb{E} g_u(\zeta/u)}{\mathbb{P}\{|\zeta|>u\}} = \int_{\mathbb{R}} g(x) \eta^*(\mathrm{d}x) \in \mathbb{R}.$$

Remark 2.3.2 By uniform boundedness of $(g_u)_{u>0}$, if inequality (2.67) holds for all x in some vicinity of 0, then it holds for all $x \in \mathbb{R}$. Formulating (2.67) in the present form makes the subsequent proof notationally simpler.

Proof Part (a). The functions $g_+ := \max(g, 0)$ and $g_- := \max(-g, 0)$ are nonnegative, bounded, and continuous and satisfy $g_\pm(x) = O(|x|^{\beta+\gamma})$ as $x \to 0$. Thus, without loss of generality we can and do assume that g is nonnegative.
Put $G(x) := \mathbb{P}\{\zeta \le x\}$ for $x \in \mathbb{R}$. We shall show that

$$\lim_{u\to\infty} \frac{\int_{[0,\infty)} g(x) \mathrm{d}_x G(ux)}{\mathbb{P}\{|\zeta|>u\}} = c_+ \beta \int_0^\infty g(x) x^{-\beta-1} \mathrm{d}x. \tag{2.68}$$

Fix any $\varepsilon \in (0, 1)$. Then

$$\lim_{u\to\infty} \frac{\int_{(\varepsilon,\infty)} g(x) \mathrm{d}_x G(ux)}{\mathbb{P}\{|\zeta|>u\}} = c_+ \beta \int_\varepsilon^\infty g(x) x^{-\beta-1} \mathrm{d}x$$

follows from

$$\lim_{u\to\infty} \frac{\mathbb{P}\{\zeta > ux\}}{\mathbb{P}\{|\zeta|>u\}} = c_+ x^{-\beta}, \quad x > 0.$$

There is a constant $c > 0$ such that $g(x) \le c x^{\beta+\gamma}$ whenever $x \in (0, 1]$. With this at hand we conclude that

$$\int_{[0,\varepsilon]} g(x) \mathrm{d}_x G(ux) \le \int_{[0,\varepsilon]} c x^{\beta+\gamma} \mathrm{d}_x G(ux) = \frac{c}{u^{\beta+\gamma}} \int_{[0,u\varepsilon]} x^{\beta+\gamma} \mathrm{d}G(x).$$

2.3 The Skew Stable Lévy Process

Further, as $u \to +\infty$,

$$\int_{[0,u\varepsilon]} x^{\beta+\gamma} dG(x) \sim \frac{\beta}{\gamma}(u\varepsilon)^{\beta+\gamma}(1-G(u\varepsilon))$$

$$\sim \frac{\beta}{\gamma}\varepsilon^{\gamma} u^{\beta+\gamma}(1-G(u)) \leq \frac{\beta}{\gamma}\varepsilon^{\gamma} u^{\beta+\gamma}\mathbb{P}\{|\zeta|>u\},$$

where the first asymptotic relation follows from Karamata's theorem (Theorem 1.6.4 in [13]). We infer

$$\limsup_{u\to\infty} \int_{[0,\varepsilon]} g(x) \frac{d_x G(ux)}{\mathbb{P}\{|\zeta|>u\}} \leq \frac{c\beta}{\gamma}\varepsilon^{\gamma}$$

and

$$\limsup_{u\to\infty} \int_{[0,\infty)} g(x) \frac{d_x G(ux)}{\mathbb{P}\{|\zeta|>u\}} \leq \frac{c\beta}{\gamma}\varepsilon^{\gamma} + c_{+}\beta \int_{\varepsilon}^{\infty} g(x) x^{-\beta-1} dx.$$

Sending $\varepsilon \to 0+$ we arrive at

$$\limsup_{u\to\infty} \int_{[0,\infty)} g(x) \frac{d_x G(ux)}{\mathbb{P}\{|\zeta|>u\}} \leq c_{+}\beta \int_{0}^{\infty} g(x) x^{-\beta-1} dx.$$

For the lower bound, write, for any $\varepsilon > 0$,

$$\liminf_{u\to\infty} \int_{[0,\infty)} g(x) \frac{d_x G(ux)}{\mathbb{P}\{|\zeta|>u\}} \geq \liminf_{u\to\infty} \int_{(\varepsilon,\infty)} g(x) \frac{d_x G(ux)}{\mathbb{P}\{|\zeta|>u\}}$$

$$= c_{+}\beta \int_{\varepsilon}^{\infty} g(x) x^{-\beta-1} dx, \quad u \to \infty.$$

Sending $\varepsilon \to 0+$ we obtain

$$\liminf_{u\to\infty} \int_{[0,\infty)} g(x) \frac{d_x G(ux)}{\mathbb{P}\{|\zeta|>u\}} \geq c_{+}\beta \int_{0}^{\infty} g(x) x^{-\beta-1} dx,$$

and (2.68) follows.
Starting with

$$\lim_{u\to\infty} \frac{\mathbb{P}\{-\zeta>ux\}}{\mathbb{P}\{|\zeta|>u\}} = c_{-} x^{-\beta}, \quad x > 0$$

and arguing analogously we also conclude that

$$\int_{(-\infty,0)} g(x) \frac{d_x G(ux)}{\mathbb{P}\{|\zeta| > u\}} = c_- \beta \int_{-\infty}^{0} g(x)|x|^{-\beta-1} dx.$$

Combining this with (2.68) completes the proof of part (a) of the lemma.
Part (b). Write

$$\mathbb{E} g_u(\zeta/u) = \int_{\mathbb{R}} g_u(x) d_x G(ux), \quad u > 0.$$

Given $\delta > 0$ there exists a u_δ such that

$$|g_u(x) - g(x)| \leq 2c((|x| + r/u) \wedge \delta)^{\beta+\gamma}, \quad x \in \mathbb{R}$$

whenever $u \geq u_\delta$. Now we prove that

$$\lim_{u \to \infty} \frac{\int_{\mathbb{R}} \left(((|x| + r/u) \wedge \delta)^{\beta+\gamma} - (|x| \wedge \delta)^{\beta+\gamma}\right) d_x G(ux)}{\mathbb{P}\{|\zeta| > u\}} = 0. \quad (2.69)$$

Indeed, let u satisfy $u \geq r/\delta$. Then using

$$((|x| + r/u) \wedge \delta)^{\beta+\gamma} - (|x| \wedge \delta)^{\beta+\gamma} = (\delta^{\beta+\gamma} - |x|^{\beta+\gamma}) \mathbb{1}_{(\delta-r/u, \delta]}(|x|)$$
$$+ \left((|x| + r/u)^{\beta+\gamma} - |x|^{\beta+\gamma}\right) \mathbb{1}_{[0, \delta-r/u]}(|x|)$$

we conclude that the contribution of the first summand on the right-hand side to the expression under the limit in (2.69) does not exceed

$$\frac{\delta^{\beta+\gamma} \int_{\delta-r/u < |x| \leq \delta} d_x G(ux)}{\mathbb{P}\{|\zeta| > u\}} = \frac{\delta^{\beta+\gamma} \left(\mathbb{P}\{|\zeta| > u\delta - r\} - \mathbb{P}\{|\zeta| > u\delta\}\right)}{\mathbb{P}\{|\zeta| > u\}}.$$

This converges to 0 as $u \to \infty$. The remaining piece of the expression under the limit in (2.69) is equal to

$$\frac{\int_{|x| \leq \delta-r/u} \left((|x| + r/u)^{\beta+\gamma} - |x|^{\beta+\gamma}\right) d_x G(ux)}{\mathbb{P}\{|\zeta| > u\}} \leq \frac{r^{\beta+\delta}}{u^{\beta+\delta} \mathbb{P}\{|\zeta| > u\}}.$$

The right-hand side also converges to 0 as $u \to \infty$. We have used subadditivity of $z \mapsto z^{\beta+\gamma}$ on $[0, \infty)$. The proof of (2.69) is complete.

2.3 The Skew Stable Lévy Process

According to (2.66) and (2.69)

$$\limsup_{u\to\infty} \left| \frac{\int_{\mathbb{R}} g_u(x) \mathrm{d}_x G(ux)}{\mathbb{P}\{|\zeta| > u\}} - \frac{\int_{\mathbb{R}} g(x) \mathrm{d}_x G(ux)}{\mathbb{P}\{|\zeta| > u\}} \right|$$

$$\leq \limsup_{u\to\infty} \frac{2c \int_{\mathbb{R}} (|x| \wedge \delta)^{\beta+\gamma} \mathrm{d}_x G(ux)}{\mathbb{P}\{|\zeta| > u\}} = 2c \int_{\mathbb{R}} (|x| \wedge \delta)^{\beta+\gamma} \eta^*(\mathrm{d}x).$$

Sending $\delta \to 0+$ and invoking the Lebesgue dominated convergence theorem, we conclude that the right-hand side vanishes. The proof closes with an application of part (a). We omit details. □

Let $(\varepsilon_n)_{n\geq 1}$ be any sequence of positive numbers which converges to 0 as $n \to \infty$. Noting that for each $n \in \mathbb{N}$ the process X_{ε_n} is strong Markov, we intend to prove part (a) by an application of Theorem 4.5.2 with $E_n = \mathbb{R} \setminus \{0\}$ and $Z_n = X_{\varepsilon_n}$ for $n \in \mathbb{N}$. Note that $X_\varepsilon(0)$ converges in distribution to $X(0) = \kappa$ as $\varepsilon \to 0+$ by assumption. Mimicking the proof of Lemma 4.5.1 (see also (2.41) and (2.46)), we infer

$$R_\lambda^{X_{\varepsilon_n}} f(x) = V_\lambda^{U_\alpha} f(x) + \mathbb{E}^x \mathrm{e}^{-\lambda \sigma(U_\alpha)} \mathbb{E} R_\lambda^{X_{\varepsilon_n}} f(\varepsilon_n \zeta), \quad n \in \mathbb{N} \qquad (2.70)$$

for $\lambda > 0$ and bounded measurable $f : \mathbb{R} \to \mathbb{R}$. Integrating both sides of (2.70) with respect to the distribution of $\varepsilon_n \zeta$ (see also derivation of (2.48)), we conclude that

$$\lambda \mathbb{E} R_\lambda^{X_{\varepsilon_n}} f(\varepsilon_n \zeta) = \frac{\langle P_{\varepsilon_n \zeta}, V_\lambda^{U_\alpha} f \rangle}{\langle P_{\varepsilon_n \zeta}, V_\lambda^{U_\alpha} 1 \rangle}, \quad n \in \mathbb{N} \qquad (2.71)$$

for $\lambda > 0$ and bounded measurable $f : \mathbb{R} \to \mathbb{R}$, where P_Z denotes the distribution (probability measure) of a random variable Z.

We already know that (2.56) defines the resolvent of a Feller process. Thus, according to Lemma 4.5.1 and Theorem 4.5.2 it remains to prove that for $\lambda > 0$ and bounded measurable $f : \mathbb{R} \to \mathbb{R}$

$$\lim_{n\to\infty} \frac{\langle P_{\varepsilon_n \zeta}, V_\lambda^{U_\alpha} f \rangle}{\langle P_{\varepsilon_n \zeta}, V_\lambda^{U_\alpha} 1 \rangle} = \frac{\langle \eta^*, V_\lambda^{U_\alpha} f \rangle}{\langle \eta^*, V_\lambda^{U_\alpha} 1 \rangle}. \qquad (2.72)$$

Now we check that $g = V_\lambda^{U_\alpha} 1$ and $g = V_\lambda^{U_\alpha} f$ satisfy the assumptions of part (a) of Lemma 2.3.3. The function $V_\lambda^{U_\alpha} 1$ is nonnegative and bounded (by λ^{-1}). Continuity of $V_\lambda^{U_\alpha} 1$ follows from (2.58) and (2.59). According to (2.63), $V_\lambda^{U_\alpha} 1(x) = O(|x|^{\alpha-1})$ as $x \to 0$. By virtue of

$$|V_\lambda^{U_\alpha} f(x)| \leq \|f\|_\infty V_\lambda^{U_\alpha} 1(x), \quad x \in \mathbb{R}, \qquad (2.73)$$

and the fact that f is bounded by assumption, we conclude that $V_\lambda^{U_\alpha} f$ is a bounded function satisfying $V_\lambda^{U_\alpha} f(x) = O(|x|^{\alpha-1})$ as $x \to 0$.

The function $x \mapsto R_\lambda^{U_\alpha} f(x)$ is continuous whenever $f : \mathbb{R} \to \mathbb{R}$ is bounded and measurable. Even though this follows from Proposition 10 on p. 25 in [10], we give a simple argument. Using $R_\lambda^{U_\alpha} f(x) = \int_\mathbb{R} f(y) v_\lambda(y - x) \mathrm{d}y$ for $x \in \mathbb{R}$ and $\lambda > 0$, where v_λ is as given in (2.59), we infer for any fixed $h \in \mathbb{R}$

$$\left| R_\lambda^{U_\alpha} f(x + h) - R_\lambda^{U_\alpha} f(x) \right| = \left| \int_\mathbb{R} f(y)(v_\lambda(y - x - h) - v_\lambda(y - x)) \mathrm{d}y \right|$$

$$\leq \|f\|_\infty \int_\mathbb{R} |v_\lambda(y - h) - v_\lambda(y)| \mathrm{d}y.$$

Since v_λ is nonnegative and continuous and $\int_\mathbb{R} v_\lambda(y - h) \mathrm{d}y = \int_\mathbb{R} v_\lambda(y) \mathrm{d}y = 1/\lambda$, we conclude that

$$\int_\mathbb{R} |v_\lambda(y - h) - v_\lambda(y)| \mathrm{d}y = 2 \int_\mathbb{R} (v_\lambda(y - h) - v_\lambda(y))_+ \mathrm{d}y \to 0, \quad h \to 0,$$

with the help of the Lebesgue dominated convergence theorem (actually, the latter argument is known as Scheffé's lemma).

Continuity of $x \mapsto V_\lambda^{U_\alpha} f(x)$ is now secured by continuity of $x \mapsto R_\lambda^{U_\alpha} f(x)$, formulas (2.58) and (2.59), and also the formula

$$V_\lambda^{U_\alpha} f(x) = R_\lambda^{U_\alpha} f(x) - \mathbb{E}^x e^{-\lambda \sigma(U_\alpha)} R_\lambda^{U_\alpha} f(0); \quad (2.74)$$

see Lemma 4.5.1. Thus, relation (2.72) follows by an application of Lemma 2.3.3(a) with $\gamma = \alpha - 1 - \beta$.

PART (B) IN WHICH $\beta > \alpha - 1$ OR $\mathbb{E}|\zeta| < \infty$. We only treat the case $\beta > \alpha - 1$, the analysis of the case $\mathbb{E}|\zeta| < \infty$ being similar.

In view of (2.54) it suffices to show that for each $t > 0$

$$\varepsilon S_{|\zeta|}(N_\varepsilon(t)) \overset{\mathbb{P}}{\to} 0, \quad \varepsilon \to 0+. \quad (2.75)$$

This ensures that $(\varepsilon S_{|\zeta|}(N_\varepsilon(t)))_{t \geq 0}$ converges in distribution on $D([0, \infty))$ as $\varepsilon \to 0+$ to a zero process, for $t \mapsto S_{|\zeta|}(N_\varepsilon(t))$ is a.s. nondecreasing.

Let $\theta_1, \theta_2, \ldots$ be independent copies of $\theta := \inf\{t > 0 : U_\alpha(t) = -1\}$, which are also independent of $\zeta_1, \zeta_2, \ldots$. Self-similarity of U_α entails

$$(\sigma_{k+1}^{(\varepsilon)} - \sigma_1^{(\varepsilon)})_{k \geq 1} \overset{\mathrm{d}}{=} (\varepsilon^\alpha S_{|\zeta|^\alpha \theta}(k))_{k \geq 1}.$$

2.3 The Skew Stable Lévy Process

Observe that for $k \in \mathbb{N}_0$

$$\mathbb{P}\{N_\varepsilon(t) > k\} = \mathbb{P}\{\sigma_{k+1}^{(\varepsilon)} \leq t\} \leq \mathbb{P}\{\sigma_{k+1}^{(\varepsilon)} - \sigma_1^{(\varepsilon)} \leq t\}$$
$$= \mathbb{P}\{\varepsilon^\alpha S_{|\zeta|^\alpha \theta}(k) \leq t\} \leq \mathbb{P}\{\varepsilon^\alpha S_{\hat{\theta}}(k) \leq t\},$$

where $\hat{\theta} := \mathbb{1}_{\{|\zeta|>1\}}\theta$.

Pick $\delta \in (0, \beta)$ satisfying $\beta - \delta > \alpha - 1$ (its existence is secured by the assumption $\beta > \alpha - 1$). For each $t > 0$ and each $\rho > 0$,

$$\mathbb{P}\{\varepsilon S_{|\zeta|}(N_\varepsilon(t)) > \rho\} \leq \mathbb{P}\{\varepsilon S_{|\zeta|}(\lfloor \varepsilon^{-\beta+\delta} \rfloor) > \rho\} + \mathbb{P}\{N_\varepsilon(t) > \lfloor \varepsilon^{-\beta+\delta} \rfloor\}$$
$$\leq \mathbb{P}\{\varepsilon S_{|\zeta|}(\lfloor \varepsilon^{-\beta+\delta} \rfloor) > \rho\} + \mathbb{P}\{\varepsilon^\alpha S_{\hat{\theta}}(\lfloor \varepsilon^{-\beta+\delta} \rfloor) \leq t\}.$$

By Theorem 4.4.1, $S_{|\zeta|}(\lfloor \varepsilon^{-\beta+\delta} \rfloor)/c_1(\lfloor \varepsilon^{-\beta+\delta} \rfloor)$ converges in distribution to a positive random variable having a β-stable distribution. Since, by (2.64), the function $t \mapsto \mathbb{P}\{\theta > t\} = \mathbb{P}^1\{\sigma(U_\alpha) > t\}$ is regularly varying at ∞ of index $1/\alpha - 1 \in (-1, 0)$, so is $t \mapsto \mathbb{P}\{\hat{\theta} > t\}$. Hence, by another appeal to Theorem 4.4.1, $S_{\hat{\theta}}(\lfloor \varepsilon^{-\beta+\delta} \rfloor)/c_2(\lfloor \varepsilon^{-\beta+\delta} \rfloor)$ converges in distribution to a positive random variable having a $(1 - 1/\alpha)$-stable distribution. According to Remark 4.4.1, the normalizing functions c_1 and c_2 are regularly varying at ∞ of indices $1/\beta$ and $\alpha/(\alpha-1)$, respectively. As a consequence, $\lim_{\varepsilon \to 0+} \varepsilon c_1(\lfloor \varepsilon^{-\beta+\delta} \rfloor) = 0$ and, by the choice of δ, $\lim_{\varepsilon \to 0+} \varepsilon^\alpha c_2(\lfloor \varepsilon^{-\beta+\delta} \rfloor) = +\infty$. This proves $\lim_{\varepsilon \to 0+} \mathbb{P}\{\varepsilon S_{|\zeta|}(N_\varepsilon(t)) > \rho\} = 0$ for each $t > 0$ and each $\rho > 0$ which is equivalent to (2.75).

The proof of Theorem 2.3.1 is complete.

2.3.3 An Equation for the Skew Stable Lévy Process

For $\alpha \in (1, 2)$, let U_α be a symmetric α-stable Lévy process as in the previous section. Also, let H be a Lévy process of bounded variation and κ a random variable, U_α, H and κ being mutually independent. In this section we discuss the problem of existence and uniqueness of solutions to the equation

$$X(t) = \kappa + U_\alpha(t) + H(L_0^X(t)), \quad t \geq 0, \tag{2.76}$$

where L_0^X is a local time of X at 0. Among other things, it will be shown that the skew α-stable Lévy process with parameter $\beta \in (0, 1)$ satisfies (2.76) for an appropriate process H.

Before we start our considerations, we have to choose a definition of the local time and a class of appropriate processes X in (2.76). A local time can be defined in numerous ways: via the occupation times formula, as a limit of the number of intersections of certain intervals, as a limit of the number of excursions of certain length, as a semimartingale local time, etc. We shall work with a Blumenthal-Getoor local time L_0^X. As a consequence, the class of processes X satisfying (2.76) has to be restricted to homogeneous Feller processes. Since L_0^X can only increase on the set $\{t \geq 0 : X(t) = 0\}$, the increments of X and U_α coincide, while X is away from 0. Hence, the strong Markov property of U_α implies that X is a Feller extension of the minimal process associated with U_α whenever X is a solution to (2.76).

Our proofs in this section are based on Itô's excursion theory. A brief review of the main results and definitions on this topic can be found in Sect. 4.6. In particular, we refer to that section for the definitions of the Blumenthal-Getoor local time, the minimal extension of a process, and the Feller extension of a process.

For a Lévy process $Z = (Z(t))_{t \geq 0}$ and a filtration $\mathcal{H} = (\mathcal{H}_t)_{t \geq 0}$ we say that Z is an $\mathcal{H}$-Lévy process if the variable $Z(t)$ is $\mathcal{H}_t$-measurable for each $t \geq 0$ and the process $(Z(t+s) - Z(t))_{s \geq 0}$ is independent of $\mathcal{H}_t$ for each $t \geq 0$.

Definition 2.3.2 We say that there is a (weak) solution to (2.76) if there exist a filtered probability space $(\Omega, \mathcal{G} := (\mathcal{G}_t)_{t \geq 0}, \mathcal{F}, \mathbb{P})$, a $\mathcal{G}_0$-measurable random variable κ, a symmetric α-stable $\mathcal{G}$-Lévy process U_α, $\alpha \in (1, 2)$, a Lévy process H of bounded variation, and a $\mathcal{G}$-adapted Feller process X such that κ, U_α, and H are independent, and (2.76) holds with probability one.

Uniqueness of a weak solution is understood in a distributional sense: distinct weak solutions to (2.76) having the same distribution are identified.

It is unlikely that the variable $X(t)$ is $\sigma(U_\alpha(s), H(s), s \in [0, t])$-measurable because the time in the term $H(L_0^X(t))$ depends on the local time of X. In view of this we do not assume that H is a $\mathcal{G}$-Lévy process. Perhaps, it is not quite natural to consider a solution to (2.76) in the situation where a filtration $\mathcal{G}$ and the processes U_α and H are given in advance, whereas a $\mathcal{G}$-adapted X has to be found. Nevertheless, it makes sense to include a filtration into the definition. One reason is that if U_α is a symmetric α-stable Lévy process, then the process X is a $\mathcal{G}$-semimartingale, which is a nice property.

Let X be a Feller extension of the minimal process corresponding to U_α killed at 0, with the resolvent given by

$$R_\lambda f(x) = V_\lambda f(x) + \mathbb{E}^x e^{-\lambda \sigma(X)} \frac{\langle \theta, V_\lambda f \rangle}{\lambda \langle \theta, V_\lambda 1 \rangle}, \quad \lambda > 0, \ x \in \mathbb{R} \tag{2.77}$$

(see (4.18) and (4.31)), where θ is an infinite measure on $\mathbb{R} \setminus \{0\}$ satisfying

$$\mathbb{E}^\theta (1 - e^{-\sigma(X)}) = \int_{\mathbb{R} \setminus \{0\}} \mathbb{E}^x (1 - e^{-\sigma(X)}) \theta(\mathrm{d}x) = 1. \tag{2.78}$$

2.3 The Skew Stable Lévy Process

Remark 2.3.3 The process X with the resolvent given in (2.77) has zero sojourn at 0 and the excursion measure of X is $\bar{\mathbb{P}}^\theta$ defined by

$$\bar{\mathbb{P}}^\theta\{\cdot\} = \int_{\mathbb{R}\setminus\{0\}} \bar{\mathbb{P}}^x\{\cdot\} \theta(\mathrm{d}x), \quad \text{where}$$

$$\bar{\mathbb{P}}^x\{\cdot\} := \mathbb{P}\{(X(t \wedge \sigma(X))_{t\geq 0} \in \cdot | X(0) = x\}, \quad x \in \mathbb{R}; \quad (2.79)$$

see Example 4.6.4 in Sect. 4.6 for further details and a background. Assumption (2.78) is a canonical normalization of θ. Assume that θ is an infinite measure satisfying $\mathbb{E}^\theta(1 - \mathrm{e}^{-\sigma(X)}) < \infty$ rather than (2.78). Then the corresponding resolvent is given by (2.77) with θ replaced by the measure $\hat{\theta}$ defined by

$$\hat{\theta}(\mathrm{d}x) := \frac{\theta(\mathrm{d}x)}{\mathbb{E}^\theta(1 - \mathrm{e}^{-\sigma(X)})}, \quad x \in \mathbb{R} \setminus \{0\}.$$

Observe that $\mathbb{E}^{\hat{\theta}}(1 - \mathrm{e}^{-\sigma(X)}) = 1$.

Theorem 2.3.2 *Assume that the resolvent of a Feller process X is given by (2.77), and an infinite measure θ satisfies (2.78). Then the process X is a (weak) solution to the equation*

$$X(t) = X(0) + U_\alpha(t) + \mathcal{S}^{(\theta)}(L_0^X(t)), \quad t \geq 0, \quad (2.80)$$

where U_α is a symmetric α-stable Lévy process, $\mathcal{S}^{(\theta)}$ is a pure-jump Lévy process of locally bounded variation with the Lévy measure θ, which is independent of U_α, and L_0^X is the Blumenthal-Getoor local time of X at 0.

Remark 2.3.4 It follows from $\sigma(X) = \sigma(U_\alpha)$ a.s., (2.63), and

$$\lim_{x \to \infty} \mathbb{E}^x(1 - \mathrm{e}^{-\sigma(X)}) = 1$$

that $\mathbb{E}^\theta(1 - \mathrm{e}^{-\sigma(X)}) < \infty$ entails $\int_{\mathbb{R}\setminus\{0\}}(|x| \wedge 1)\theta(\mathrm{d}x) < \infty$, thereby proving that the process $\mathcal{S}^{(\theta)}$ is well-defined and has a locally bounded variation; see p. 15 in [10].

As an immediate corollary we conclude that the skew α-stable Lévy process $U_{\alpha,\beta}$ with parameter β is a solution to Eq. (2.76). Recall that the resolvent of $U_{\alpha,\beta}$ is given in (2.56).

Corollary 2.3.1 *For $\alpha \in (1, 2)$, let $X = (U_{\alpha,\beta}(t))_{t \geq 0}$ be a skew α-stable Lévy process with parameter $\beta \in (0, \alpha - 1)$. Also, let $\mathcal{S}_\beta$ be a β-stable Lévy process which is independent of U_α and has the Lévy measure $\hat{\eta}$ given by*

$$\hat{\eta}(\mathrm{d}x) = C\eta^*(\mathrm{d}x). \quad (2.81)$$

Here, the measure η^* is as defined in (2.57),

$$C := (\mathbb{E}^{\eta^*}(1 - e^{-\sigma(U_\alpha)}))^{-1} = \left(\int_{\mathbb{R}\setminus\{0\}} \mathbb{E}^x(1 - e^{-\sigma(U_\alpha)})\eta^*(\mathrm{d}x)\right)^{-1}$$

$$= \frac{\beta \sin \frac{\pi(1+\beta)}{\alpha}}{(c_- + c_+)\Gamma(1-\beta)\cos\frac{\pi\beta}{2}\sin\frac{\pi}{\alpha}}.$$

Then X is a weak solution to the equation

$$X(t) = X(0) + U_\alpha(t) + \mathcal{S}_\beta(L_0^X(t)), \quad t \geq 0, \tag{2.82}$$

where L_0^X is the Blumenthal-Getoor local time of X at 0.

Proof of Corollary 2.3.1 We only have to show that

$$\left(\int_{\mathbb{R}\setminus\{0\}} \mathbb{E}^x(1 - e^{-\sigma(U_\alpha)})\eta^*(\mathrm{d}x)\right)^{-1} = \frac{\beta \sin \frac{\pi(1+\beta)}{\alpha}}{(c_- + c_+)\Gamma(1-\beta)\cos\frac{\pi\beta}{2}\sin\frac{\pi}{\alpha}}.$$

This follows from (2.57), the first equality in (2.65), Lemma 2.3.1(a) with $\gamma = \beta$ and $\lambda = 1$, and (2.61) with $\alpha = 1 + \beta$:

$$\int_{\mathbb{R}\setminus\{0\}} \mathbb{E}^x(1 - e^{-\sigma(U_\alpha)})\eta^*(\mathrm{d}x) = \frac{c_- + c_+}{\pi v_1(0)} \int_0^\infty \int_0^\infty \frac{1-\cos(x\theta)}{1+\theta^\alpha}\mathrm{d}\theta \frac{\mathrm{d}x}{x^{1+\beta}}$$

$$= \frac{c_- + c_+}{\pi v_1(0)} \int_0^\infty \left(\int_0^\infty \frac{1-\cos(x\theta)}{x^{1+\beta}}\mathrm{d}x\right)\frac{\mathrm{d}\theta}{1+\theta^\alpha}$$

$$= \frac{c_- + c_+}{\pi v_1(0)} \frac{\Gamma(1-\beta)\sin\frac{\pi(1+\beta)}{2}}{\beta} \int_0^\infty \frac{\theta^\beta}{1+\theta^\alpha}\mathrm{d}\theta$$

$$= \frac{(c_- + c_+)\Gamma(1-\beta)\cos\frac{\pi\beta}{2}\sin\frac{\pi}{\alpha}}{\beta \sin \frac{\pi(1+\beta)}{\alpha}}.$$

$\square$

Proof of Theorem 2.3.2 We intend to construct processes U_α, $\mathcal{S}^{(\theta)}$, and X satisfying (2.80), with the resolvent of X given by (2.77). While doing so, we exploit certain Poisson random measures and appeal to Itô's excursion theory. In the role of filtration appearing in Definition 2.3.2 we take $\mathcal{F}^X$ the filtration generated by X and completed by subsets of events of probability 0. As has already been mentioned in Remark 2.3.3, formula (2.77) implies that the excursion measure of X is $\bar{\mathbb{P}}^\theta$.

2.3 The Skew Stable Lévy Process

Let $N_\theta := \sum_{k\geq 1} \delta_{(s_k,x_k)}$ be a Poisson random measure on $[0,\infty) \times \mathbb{R}$ with intensity LEB $\otimes\, \theta$, where $\delta_{(s,x)}$ is a Dirac measure at (s,x) and $((s_k, x_k))_{k\geq 1}$ is a (measurable) enumeration of the atoms of N_θ. Let $U_\alpha^{(1)}, U_\alpha^{(2)}, \ldots$ denote independent copies of U_α, which are independent of N_θ. Without loss of generality assume that these copies are càdlàg. Put

$$N := \sum_{k\geq 1} \delta_{(s_k, x_k + U_\alpha^{(k)}(\cdot \wedge \sigma_k))},$$

where

$$\sigma_k := \sigma(x_k + U_\alpha^{(k)}(\cdot)) := \inf\{t \geq 0 : x_k + U_\alpha^{(k)}(t) = 0\}, \quad k \in \mathbb{N}.$$

It can be checked that N is a Poisson random measure on $[0,\infty) \times D([0,\infty))$ with intensity LEB $\otimes\, \bar{\mathbb{P}}^\theta$, where $\bar{\mathbb{P}}^\theta$ is defined by (2.79) with $X = U_\alpha$.

Without loss of generality we can and do assume that $X(0) = 0$ and that the process X is built upon the Poisson point measure N with the help of Itô's procedure, as presented in Sect. 4.6. For the reader's convenience, we now give some details of the construction. Put

$$\tau(s) := \sum_{s_k \leq s} \sigma(x_k + U_\alpha^{(k)}(\cdot)), \quad s \geq 0 \tag{2.83}$$

and

$$\varphi(t) := \inf\{s \geq 0 : \tau(s) > t\}, \quad t \geq 0.$$

The process that we wanted to construct is then given by

$$X(t) := \begin{cases} x_k + U_\alpha^{(k)}(t - \tau(s_k-)), & t \in [\tau(s_k-), \tau(s_k)), \quad k \in \mathbb{N}, \\ 0, & t \notin \cup_{k\geq 1}[\tau(s_k-), \tau(s_k)). \end{cases}$$

According to Theorem 4.6.5, the process φ is a Blumenthal-Getoor local time of X at 0.

A moment's reflection reveals that $X(t)$ is a concatenation of α-stable Lévy processes and a sum of the atoms ordinates x_k of N_θ satisfying $s_k \leq \varphi(t)$. In view of this, it is natural to expect that X admits representation (2.80) with appropriate U_α. The remainder of the proof is concerned with a justification of this claim. It is not easy to work with the process X directly. Indeed, its definition involves the sequence $(s_k)_{k\geq 1}$ which, as a consequence of $\theta(\mathbb{R} \setminus \{0\}) = \infty$, is dense in any nonempty interval of the positive halfline. As a remedy, we are going to use an approximation procedure.

For each $\varepsilon > 0$, we define restrictions of the Poisson random measures N_θ and N by

$$N_\theta^{(\varepsilon)} := \sum_{k \geq 1} \mathbb{1}_{\{|x_k| > \varepsilon\}} \delta_{(s_k, x_k)} \quad \text{and} \quad N^{(\varepsilon)} := \sum_{k \geq 1} \mathbb{1}_{\{|x_k| > \varepsilon\}} \delta_{(s_k, x_k + U_\alpha^{(k)}(\cdot \wedge \sigma_k))}. \quad (2.84)$$

With the measure θ_ε given by $\theta_\varepsilon(dx) := \mathbb{1}_{\{|x| > \varepsilon\}} \theta(dx)$, $N_\theta^{(\varepsilon)}$ is a Poisson random measure on $[0, \infty) \times \mathbb{R}$ with intensity $\text{LEB} \otimes \theta_\varepsilon$ and $N^{(\varepsilon)}$ is a Poisson random measure on $[0, \infty) \times D([0, \infty))$ with intensity $\text{LEB} \otimes \tilde{\mathbb{P}}^{\theta_\varepsilon}$. Since $\theta(\{x \in \mathbb{R} : |x| > \varepsilon\}) < \infty$, $N_\theta^{(\varepsilon)}$ has finitely many atoms in the strip $[0, T] \times \mathbb{R}$ a.s. for each $T > 0$. We denote the atoms of $N_\theta^{(\varepsilon)}$ by $(s_1^{(\varepsilon)}, x_1^{(\varepsilon)})$, $(s_2^{(\varepsilon)}, x_2^{(\varepsilon)})$, ... and assume without loss of generality that $0 < s_1^{(\varepsilon)} < s_2^{(\varepsilon)} < s_3^{(\varepsilon)} < \ldots$. Plainly, the sequence $(s_k^{(\varepsilon)})_{k \geq 1}$ is much easier to deal with in comparison to the original sequence $(s_k)_{k \geq 1}$.

For $k \in \mathbb{N}$, denote by $U_\alpha^{(k,\varepsilon)}$ a process from the collection $(U_\alpha^{(j)})_{j \geq 1}$, which corresponds to $s_k^{(\varepsilon)}$ in the definition of N. With this notation, we obtain representations

$$N_\theta^{(\varepsilon)} := \sum_{k \geq 1} \delta_{(s_k^{(\varepsilon)}, x_k^{(\varepsilon)})} \quad \text{and} \quad N^{(\varepsilon)} := \sum_{k \geq 1} \delta_{(s_k^{(\varepsilon)}, x_k^{(\varepsilon)} + U_\alpha^{(k,\varepsilon)}(\cdot \wedge \sigma_k^{(\varepsilon)}))}.$$

The processes $U_\alpha^{(1,\varepsilon)}$, $U_\alpha^{(2,\varepsilon)}$, ... are still independent copies of U_α, which are independent of $N_\theta^{(\varepsilon)}$.

We now construct a process $X^{(\varepsilon)}$ with the help of the Poisson random measure $N^{(\varepsilon)}$, along the lines of construction of X with the help of N. For each $\varepsilon > 0$, put

$$\tau^{(\varepsilon)}(s) := \sum_{s_j \leq s} \mathbb{1}_{\{|x_j| > \varepsilon\}} \sigma(x_j + U_\alpha^{(j)}(\cdot)) =: \sum_{s_k^{(\varepsilon)} \leq s} \hat{\sigma}_k^{(\varepsilon)}, \quad s \geq 0, \quad (2.85)$$

$$\varphi^{(\varepsilon)}(t) := \inf\{s \geq 0 : \tau^{(\varepsilon)}(s) > t\}, \quad t \geq 0 \quad (2.86)$$

and then, for $t \in [\tau^{(\varepsilon)}(s_k^{(\varepsilon)}-), \tau^{(\varepsilon)}(s_k^{(\varepsilon)}))$, $X^{(\varepsilon)}(t) := x_k^{(\varepsilon)} + U_\alpha^{(k,\varepsilon)}(t - \tau^{(\varepsilon)}(s_k^{(\varepsilon)}-))$. Among other things, formula (2.85) serves a definition of $\hat{\sigma}_1^{(\varepsilon)}$, $\hat{\sigma}_2^{(\varepsilon)}$, ...

We stress that the process $X^{(\varepsilon)}$ is not a holding and jumping process; see Example 4.6.3 for the definition. In contrast to the process X, a.s. there does not exist a t such that $X^{(\varepsilon)}(t) = 0$. The process $X^{(\varepsilon)}$ jumps upon "touching" 0. The distribution of this jump is $\mathbb{P}\{x_k^{(\varepsilon)} \in dx\} = \theta_\varepsilon(dx)/\theta_\varepsilon(\mathbb{R})$. Since

$$\mathbb{P}\{U_\alpha(s-) = 0, \ U_\alpha(s) \neq 0 \text{ for some } s \geq 0\} = 0,$$

we conclude that with probability one the equality $X^{(\varepsilon)}(s-) = 0$ with some (random) s entails $X^{(\varepsilon)}(s) = x_k^{(\varepsilon)}$ for some k. Analogously,

$$\mathbb{P}\{X(s-) = 0 \text{ and } X(s) \notin \{x_k : k \geq 1\} \text{ for some } s \geq 0\} = 0. \quad (2.87)$$

2.3 The Skew Stable Lévy Process

Returning to the proof of Theorem 2.3.2, we define the process $U_\alpha^{(\varepsilon)}$ by $U_\alpha^{(\varepsilon)}(0) := 0$, $U_\alpha^{(\varepsilon)}(t) := U_\alpha^{(1,\varepsilon)}(t)$ for $t \in (0, \hat{\sigma}_1^{(\varepsilon)}]$ and

$$U_\alpha^{(\varepsilon)}(t) := U_\alpha^{(\varepsilon)}(\hat{\sigma}_1^{(\varepsilon)} + \ldots + \hat{\sigma}_k^{(\varepsilon)}) + U_\alpha^{(k+1,\varepsilon)}(t - (\hat{\sigma}_1^{(\varepsilon)} + \ldots + \hat{\sigma}_k^{(\varepsilon)}))$$

for $t \in (\hat{\sigma}_1^{(\varepsilon)} + \ldots + \hat{\sigma}_k^{(\varepsilon)}, \hat{\sigma}_1^{(\varepsilon)} + \ldots + \hat{\sigma}_k^{(\varepsilon)} + \hat{\sigma}_{k+1}^{(\varepsilon)}]$ and $k \in \mathbb{N}$. Observe that $U_\alpha^{(\varepsilon)}$ is a copy of U_α, which is independent of N_θ, hence of

$$\mathcal{S}^{(\theta,\varepsilon)}(s) := \int_{[0,s]} \int_{|x|>\varepsilon} x N_\theta^{(\varepsilon)}(\mathrm{d}z, \mathrm{d}x)$$

$$= \int_{[0,s]} \int_{\mathbb{R}\setminus\{0\}} x \mathbb{1}_{(\varepsilon,\infty)}(|x|) N_\theta(\mathrm{d}z, \mathrm{d}x), \quad s \geq 0. \quad (2.88)$$

According to the construction of $X^{(\varepsilon)}$,

$$X^{(\varepsilon)}(t) = U_\alpha^{(\varepsilon)}(t) + \mathcal{S}^{(\theta,\varepsilon)}(\varphi^{(\varepsilon)}(t)), \quad t \geq 0.$$

To prove the theorem it suffices to show that for each fixed $t > 0$ a.s.

$$\lim_{\varepsilon \to 0+} X^{(\varepsilon)}(t) = X(t), \quad \lim_{\varepsilon \to 0+} U_\alpha^{(\varepsilon)}(t) = U_\alpha(t), \quad \lim_{\varepsilon \to 0+} \mathcal{S}^{(\theta,\varepsilon)}(\varphi^{(\varepsilon)}(t)) = \mathcal{S}^{(\theta)}(\varphi(t)),$$

where X and φ are as defined earlier with the help of Itô's construction, U_α is a symmetric α-stable Lévy process and

$$\mathcal{S}^{(\theta)}(s) := \int_{[0,s]} \int_{\mathbb{R}\setminus\{0\}} x N_\theta(\mathrm{d}z, \mathrm{d}x), \quad s \geq 0.$$

We believe that $\lim_{\varepsilon \to 0+} X^{(\varepsilon)} = X$ a.s. on $D([0,\infty))$. On the one hand, the proof of this fact is beyond our reach at the moment. On the other hand, this stronger convergence is not required for the present proof.

Recall that we assume that $X(0) = 0$. To proceed we need a lemma.

Lemma 2.3.4 *For each $T > 0$, almost surely*

$$\lim_{\varepsilon \to 0+} \sup_{s \in [0,T]} |\tau^{(\varepsilon)}(s) - \tau(s)| = 0, \quad (2.89)$$

$$\lim_{\varepsilon \to 0+} \sup_{t \in [0,T]} |\varphi^{(\varepsilon)}(t) - \varphi(t)| = 0 \quad (2.90)$$

and

$$\lim_{\varepsilon \to 0+} \sup_{s \in [0, T]} |\mathcal{S}^{(\theta,\varepsilon)}(s) - \mathcal{S}^{(\theta)}(s)| = 0. \qquad (2.91)$$

Proof Note that

$$\sup_{s \in [0, T]} |\tau^{(\varepsilon)}(s) - \tau(s)| = \sum_{s_j \leq T} \mathbb{1}_{\{|x_j| \leq \varepsilon\}} \sigma(x_j + U_\alpha^{(j)}(\cdot))$$

and

$$\sup_{s \in [0, T]} |\mathcal{S}^{(\theta,\varepsilon)}(s) - \mathcal{S}^{(\theta)}(s)| \leq \int_{[0, T]} \int_{\mathbb{R} \setminus \{0\}} |x| \mathbb{1}_{[-\varepsilon, \varepsilon]}(x) N_\theta(\mathrm{d}z, \mathrm{d}x).$$

In view of $\int_{\mathbb{R} \setminus \{0\}} (|x| \wedge 1)\theta(\mathrm{d}x) < \infty$ the integral is a.s. finite for each $\varepsilon > 0$. Now we conclude with the help of the monotone convergence theorem that equalities (2.89) and (2.91) hold true. Since $s \mapsto \tau(s)$ is a.s. strictly increasing, relation (2.90) follows from (2.89) according to Proposition 4.3.14(ii). □

Lemma 2.3.5 *For any $t > 0$,*

$$\mathbb{P}\left\{\text{there exists } \varepsilon_0 > 0 \text{ such that } \varphi^{(\varepsilon)}(t) = \varphi(t) \text{ for all } \varepsilon \in (0, \varepsilon_0)\right\} = 1. \qquad (2.92)$$

Proof Fix $t > 0$. It follows from (2.83) that τ is a drift-free subordinator with an infinite Lévy measure. Theorem 4 on p. 77 in [10] implies that with probability one there exists k such that $t \in (\tau(s_k-), \tau(s_k))$. Furthermore, $\varphi(t) = \tau(s_k-)$ a.s., for φ takes a constant value on $(\tau(s_k-), \tau(s_k))$ by definition. According to (2.89), $\lim_{\varepsilon \to 0+} \tau^{(\varepsilon)}(s_k-) = \tau(s_k-)$ and $\lim_{\varepsilon \to 0+} \tau^{(\varepsilon)}(s_k) = \tau(s_k)$ a.s. Hence, there exists an a.s. finite $\varepsilon_0 > 0$ such that $t \in (\tau^{(\varepsilon)}(s_k-), \tau^{(\varepsilon)}(s_k))$ whenever $\varepsilon \in (0, \varepsilon_0)$ and $t \in (\tau(s_k-), \tau(s_k))$. Since $\varphi^{(\varepsilon)}(t) = \tau^{(\varepsilon)}(s_k-)$ a.s., and the intervals $(\tau(s_k-), \tau(s_k))$ and $(\tau^{(\varepsilon)}(s_k-), \tau^{(\varepsilon)}(s_k))$ intersect, we arrive at (2.92). □

A combination of relation (2.91) and Lemma 2.3.5 leads to the following result:

Corollary 2.3.2 *For all $t > 0$,*

$$\lim_{\varepsilon \to 0+} \mathcal{S}^{(\theta,\varepsilon)}(\varphi^{(\varepsilon)}(t)) = \mathcal{S}^{(\theta)}(\varphi(t)) \quad \text{a.s.} \qquad (2.93)$$

Now we check that for each $t > 0$

$$\lim_{\varepsilon \to 0+} X^{(\varepsilon)}(t) = X(t) \quad \text{a.s.} \qquad (2.94)$$

2.3 The Skew Stable Lévy Process

We know that $t \in (\tau(s_{k_0}-), \tau(s_{k_0}))$ for some (random) $k_0 = k_0(t)$ with probability one. Thus,

$$X(t) = x_{k_0} + U_\alpha^{(k_0)}(t - \tau(s_{k_0}-)).$$

Put $\varepsilon_0 := |x_{k_0}| > 0$ and note that, for all $\varepsilon \in (0, \varepsilon_0)$, the atom (s_{k_0}, x_{k_0}) of N_θ belongs to the set of atoms of $N_\theta^{(\varepsilon)}$. Let $k_1(\varepsilon) \in \mathbb{N}$ with $\varepsilon \in (0, \varepsilon_0)$ satisfy

$$(s_{k_0}, x_{k_0}) = (s_{k_1(\varepsilon)}^{(\varepsilon)}, x_{k_1(\varepsilon)}^{(\varepsilon)}).$$

It follows from the proof of Lemma 2.3.5 that $t \in (\tau^{(\varepsilon)}(s_{k_0}-), \tau^{(\varepsilon)}(s_{k_0}))$ for all $\varepsilon \in (0, \varepsilon_1)$, for some sufficiently small (random) $\varepsilon_1 > 0$ with probability one. Thus, with probability one, for all $\varepsilon \in (0, \min(\varepsilon_0, \varepsilon_1))$,

$$t \in (\tau^{(\varepsilon)}(s_{k_1(\varepsilon)}^{(\varepsilon)}-), \tau^{(\varepsilon)}(s_{k_1(\varepsilon)}^{(\varepsilon)}))$$

and thereupon

$$X^{(\varepsilon)}(t) = x_{k_1(\varepsilon)}^{(\varepsilon)} + U_\alpha^{(k_1(\varepsilon),\varepsilon)}(t - \tau^{(\varepsilon)}(s_{k_1(\varepsilon)}^{(\varepsilon)}-)) = x_{k_0} + U_\alpha^{(k_1(\varepsilon),\varepsilon)}(t - \tau^{(\varepsilon)}(s_{k_0}-)).$$

By definition, $U_\alpha^{(k_1(\varepsilon),\varepsilon)}$ is a process from the collection $(U_\alpha^{(j)})_{j \geq 1}$, which corresponds to $s_{k_1(\varepsilon)}^{(\varepsilon)}$ in the definition of N. Since $s_{k_1(\varepsilon)}^{(\varepsilon)} = s_{k_0}$, we conclude that $U_\alpha^{(k_1(\varepsilon),\varepsilon)} = U_\alpha^{(k_0)}$, for $\varepsilon \in (0, \min(\varepsilon_0, \varepsilon_1))$. Summarizing we have shown that, for all sufficiently small $\varepsilon > 0$,

$$X(t) = x_{k_0} + U_\alpha^{(k_0)}(t - \tau(s_{k_0}-)) \quad \text{and} \quad X^{(\varepsilon)}(t) = x_{k_0} + U_\alpha^{(k_0)}(t - \tau^{(\varepsilon)}(s_{k_0}-)).$$

It suffices to check that

$$\lim_{\varepsilon \to 0+} U_\alpha^{(k_0)}(t - \tau^{(\varepsilon)}(s_{k_0}-)) = U_\alpha^{(k_0)}(t - \tau(s_{k_0}-)) \quad \text{a.s.}$$

But this is implied by the fact that $U_\alpha^{(k_0)}$ has right-continuous paths and $t - \tau^{(\varepsilon)}(s_{k_0}-)$ approaches $t - \tau(s_{k_0}-)$ from the right as $\varepsilon \to 0+$. For the latter observe that $\tau^{(\varepsilon)}(s_{k_0}-)$ approaches $\tau(s_{k_0}-)$ from the left as $\varepsilon \to 0+$.

It follows from (2.93) and (2.94) that for all $t > 0$

$$\lim_{\varepsilon \to 0+} U_\alpha^{(\varepsilon)}(t) = \lim_{\varepsilon \to 0+} (X^{(\varepsilon)}(t) - \mathcal{S}^{(\theta,\varepsilon)}(\varphi^{(\varepsilon)}(t))) = X(t) - \mathcal{S}^{(\theta)}(\varphi(t)) = U_\alpha(t) \quad \text{a.s.} \tag{2.95}$$

The process U_α is càdlàg a.s. as the difference of càdlàg processes. Thus, U_α is a symmetric α-stable Lévy process as an a.s. pointwise limit of a sequence of such processes.

Remark 2.3.5 For each $T > 0$, the process $(U_\alpha^{(\varepsilon)}(t))_{t \in [0,T]}$ is constructed by merging finitely (but randomly) many fragments of paths of independent α-stable Lévy processes. The fact that $U_\alpha^{(\varepsilon)}$ is a symmetric α-stable Lévy process follows from the construction and the strong Markov property. The process $(U_\alpha(t))_{t \in [0,T]}$ is built upon a countable number of paths. Although one may anticipate that U_α is a symmetric α-stable Lévy process, this fact is not that immediate as in the case of $U_\alpha^{(\varepsilon)}$ and does require a proof.

To complete the proof of the theorem, it remains to check that the variable $U_\alpha(t)$ is $\mathcal{F}_t^X$-measurable for each $t \geq 0$ and the process $(U_\alpha(t+s) - U_\alpha(t))_{s \geq 0}$ is independent of $\mathcal{F}_t^X$ for each $t \geq 0$. It follows from the construction and the strong Markov property of Lévy processes that $U_\alpha^{(\varepsilon)}$ is a $\mathcal{F}^{X^{(\varepsilon)}}$-Lévy process. As a consequence, for each $t > 0$, each $m, n \in \mathbb{N}$, any bounded continuous functions $f : \mathbb{R}^n \to \mathbb{R}$ and $g : \mathbb{R}^m \to \mathbb{R}$, and any $t_1, \ldots, t_n$ and $v_1, \ldots, v_m$ satisfying $0 \leq t_1 \leq \ldots \leq t_n = t \leq t + v_1, \ldots, t + v_m$ (we call such t_i and v_j admissible)

$$\mathbb{E} f(X^{(\varepsilon)}(t_1), \ldots, X^{(\varepsilon)}(t_n)) g(U_\alpha^{(\varepsilon)}(t+v_1) - U_\alpha^{(\varepsilon)}(t), \ldots, U_\alpha^{(\varepsilon)}(t+v_m) - U_\alpha^{(\varepsilon)}(t))$$
$$= \mathbb{E} f(X^{(\varepsilon)}(t_1), \ldots, X^{(\varepsilon)}(t_n)) \mathbb{E} g(U_\alpha^{(\varepsilon)}(t+v_1) - U_\alpha^{(\varepsilon)}(t), \ldots, U_\alpha^{(\varepsilon)}(t+v_m) - U_\alpha^{(\varepsilon)}(t)).$$

Formulas (2.94) and (2.95) together with the Lebesgue dominated convergence theorem imply that for each $t > 0$ and admissible $t_1, \ldots, t_n, v_1, \ldots, v_m$

$$\mathbb{E} f(X(t_1), \ldots, X(t_n)) g(U_\alpha(t+v_1) - U_\alpha(t), \ldots, U_\alpha(t+v_m) - U_\alpha(t))$$
$$= \mathbb{E} f(X(t_1), \ldots, X(t_n)) \mathbb{E} g(U_\alpha(t+v_1) - U_\alpha(t), \ldots, U_\alpha(t+v_m) - U_\alpha(t)). \qquad (2.96)$$

Since X and U_α are càdlàg processes, an application of the Lebesgue dominated convergence theorem shows that (2.96) holds true for each $t \geq 0$. This proves that the process $(U_\alpha(t+s) - U_\alpha(t))_{s \geq 0}$ is independent of $\mathcal{F}_t^X$ for each $t \geq 0$.

In view of (2.87), having observed $(X(s))_{s \in [0,t]}$, one can identify in a measurable way all the jumps from 0 which occur within $[0, t]$. As a consequence, one can also identify the sum of these jumps, which is equal to $\mathcal{S}^{(\theta)}(\varphi(t))$. Since $U_\alpha(t) = X(t) - \mathcal{S}^{(\theta)}(\varphi(t))$, the variable $U_\alpha(t)$ is $\mathcal{F}_t^X$-measurable. The proof of Theorem 2.3.2 is complete. □

Next, we solve a counterpart of Eq. (2.80), in which a pure-jump Lévy process has the Lévy measure $\vartheta := p\theta$ for some $p \in [0, 1]$ satisfying $\mathbb{E}^\vartheta(1 - e^{-\sigma(X)}) = p$. The counterpart boils down to (2.78) when $p = 1$. Rather than giving a resolvent characterization of a solution X like (2.77), we construct X in terms of excursion measures and entrance laws.

As before, let θ be a measure satisfying $\theta(\mathbb{R} \setminus \{0\}) = \infty$ and $\mathbb{E}^\theta(1 - e^{-\sigma(X)}) = 1$. Denote by $\hat{\mathbb{P}}_\alpha$ and $\bar{\mathbb{P}}^\theta$ the excursion measures of a symmetric α-stable Lévy process and the process X in Theorem 2.3.2, respectively.

2.3 The Skew Stable Lévy Process

Theorem 2.3.3 *Let $p \in [0, 1]$ and X be a Feller process with the excursion measure $p\bar{\mathbb{P}}^\theta + (1-p)\hat{\mathbb{P}}_\alpha$ and $X(0) = 0$. Then X is a (weak) solution to the equation*

$$X(t) = U_\alpha(t) + \mathcal{S}^{(p\theta)}(L_0^X(t)), \quad t \geq 0, \tag{2.97}$$

where U_α is a symmetric α-stable Lévy process, $\mathcal{S}^{(p\theta)}$ is a pure-jump Lévy process of locally bounded variation with the Lévy measure $p\theta$, which is independent of U_α, and L_0^X is the Blumenthal-Getoor local time of X at 0.

Remark 2.3.6 It can be checked that the measure $p\bar{\mathbb{P}}^\theta + (1-p)\hat{\mathbb{P}}_\alpha$ satisfies all compatibility assumptions of Sect. 4.6.

Proof Similarly to the proof of Theorem 2.3.2 the idea is to construct the processes X, U_α, and $\mathcal{S}^{(p\theta)}$ living on a common probability space and satisfying Eq. (2.97).

Denote by $(\rho_t)_{t>0}$, $\hat{\mathbb{P}}_\alpha$, and N_α the entrance law, the excursion measure, and the corresponding Poisson random measure of excursions of a symmetric α-stable Lévy process. Put $\theta_t := \theta P_t^0$ for $t > 0$, where $(P_t^0)_{t\geq 0}$ is the semigroup of the symmetric α-stable Lévy process killed at 0. Then $(\theta_t)_{t>0}$ is the entrance law of a Markov process (the process X in Theorem 2.3.2) with the excursion measure $\bar{\mathbb{P}}^\theta$. Denote by N the Poisson random measure of excursions of this process. Assume that N_α and N are independent. Notice that $p\bar{\mathbb{P}}^\theta + (1-p)\hat{\mathbb{P}}_\alpha$ is the excursion measure of a Markov process, which we denote by X, with the entrance law $(p\theta_t + (1-p)\rho_t)_{t>0}$ and the excursion point process $N(p\,ds, dx) + N_\alpha((1-p)ds, dx)$. Observe that X has a zero sojourn at 0 because $(p\bar{\mathbb{P}}^\theta + (1-p)\hat{\mathbb{P}}_\alpha)(1 - e^{-\sigma(X)}) = 1$.

With the Poisson random measure $N^{(\varepsilon)}$ as defined in (2.84), we use Itô's procedure, along the lines of (4.20) with $m = 0$ and (4.21) (see also Remark 4.6.1), to construct the process $X^{(\varepsilon)}$ upon the Poisson random measure $N^{(\varepsilon)}(p\,ds, dx) + N_\alpha((1-p)ds, dx)$. It follows from the construction that $X^{(\varepsilon)}$ has a zero sojourn at 0 and that

$$X^{(\varepsilon)}(t) = U_\alpha^{(\varepsilon)}(t) + \mathcal{S}^{(\theta)}(p\varphi^{(\varepsilon)}(t)), \quad t \geq 0,$$

where the process $\mathcal{S}^{(\theta)}$ is as defined in (2.88). Mimicking the proof of Theorem 2.3.2 we conclude that X satisfies the equation

$$X(t) = U_\alpha(t) + \mathcal{S}^{(\theta)}(pL_0^X(t)), \quad t \geq 0.$$

To complete the proof, observe that the process $(\mathcal{S}^{(\theta)}(pt))_{t\geq 0}$ has the same distribution as $(\mathcal{S}^{(p\theta)}(t))_{t\geq 0}$. □

Remark 2.3.7 The process U_α appearing in (2.97) is different from a symmetric α-stable Lévy process that is built upon N_α alone.

We close the section with a result providing necessary conditions for solvability of Eq. (2.76) and shedding light on uniqueness of its solutions.

Theorem 2.3.4 *Assume that X is a Feller (weak) solution to Eq. (2.76) and has a zero sojourn at 0. Then there exist $p \in [0, 1]$ and a sigma-finite measure θ on $\mathbb{R} \setminus \{0\}$ satisfying $\mathbb{E}^\theta(1 - e^{-\sigma}) = 1$, for which the triple (X, H, U_α) has the same distribution as the triple $(X, \mathcal{S}^{(p\theta)}, U_\alpha)$ appearing in Eq. (2.97) of Theorem 2.3.3.*

Remark 2.3.8 Under the assumptions of Theorem 2.3.4 there exists neither a solution to the equation

$$X(t) = X(0) + U_\alpha(t) + \gamma L_0^X(t), \quad t \geq 0,$$

for $\gamma \neq 0$, nor a solution to (2.97) for $p > 1$.

Proof Recall that according to our convention solutions to (2.76) belong to the class of Feller extensions of the minimal process associated with a symmetric α-stable Lévy process. Since X has a zero sojourn at 0 by assumption, Corollary 4.6.1 and Theorem 4.6.4 ensure the existence of a unique value of $p \in [0, 1]$ and a unique measure θ satisfying $\mathbb{E}^\theta(1 - e^{-\sigma(X)}) = 1$, for which the entrance law of X is given by $(p\,\theta\, P_t^0 + (1 - p)\rho_t)_{t>0}$. Moreover, it follows from the definition of a solution that X and U_α are adapted to some filtration $\mathcal{G}$. According to Theorem 2.3.3, there exists a process $\widetilde{X} \stackrel{d}{=} X$ satisfying

$$\widetilde{X}(t) = \widetilde{X}(0) + \widetilde{U}_\alpha(t) + \widetilde{\mathcal{S}}^{(p\theta)}(L_0^{\widetilde{X}}(t)), \quad t \geq 0.$$

Here, the processes marked with the tilde have the same distribution and satisfy the same independence assumptions as the unmarked processes.

We show next that the processes U_α and $\widetilde{U}_\alpha$ can be uniquely recovered from X and $\widetilde{X}$, respectively. Indeed, the local times L_0^X and $L_0^{\widetilde{X}}$ are a.s. continuous processes which do not increase a.s. on the sets $\{s \geq 0 : X(s) \neq 0\}$ and $\{s \geq 0 : \widetilde{X}(s) \neq 0\}$, respectively. As a consequence, given $\delta > 0$,

$$\int_{[0,t]} \mathbb{1}_{\{|X(s-)| \geq \delta\}} dH(L_0^X(s)) = 0, \quad t \geq 0 \quad \text{a.s.}$$

and

$$\int_{[0,t]} \mathbb{1}_{\{|\widetilde{X}(s-)| \geq \delta\}} d\widetilde{\mathcal{S}}^{(p\theta)}(L_0^{\widetilde{X}}(s))) = 0, \quad t \geq 0 \quad \text{a.s.}$$

2.3 The Skew Stable Lévy Process

Recall that the processes X and U_α are $\mathcal{G}$-semimartingales and $\widetilde{X}$ and $\widetilde{U}_\alpha$ are $\widetilde{\mathcal{G}}$-semimartingales, where $\widetilde{\mathcal{G}}$ is an appropriate filtration. Hence,

$$\int_{[0,t]} \mathbb{1}_{\{|X(s-)|\geq\delta\}} \mathrm{d}X(s) = \int_{[0,t]} \mathbb{1}_{\{|X(s-)|\geq\delta\}} \mathrm{d}U_\alpha(s), \quad t \geq 0 \quad \text{a.s.}$$

and

$$\int_{[0,t]} \mathbb{1}_{\{|\widetilde{X}(s-)|\geq\delta\}} \mathrm{d}\widetilde{X}(s) = \int_{[0,t]} \mathbb{1}_{\{|\widetilde{X}(s-)|\geq\delta\}} \mathrm{d}\widetilde{U}_\alpha(s), \quad t \geq 0 \quad \text{a.s.}$$

Sending $\delta \to 0+$ and using the assumption about a zero sojourn at 0, we conclude that the a.s. limits of the corresponding integrals are $U_\alpha(t)$ and $\widetilde{U}_\alpha(t)$, respectively, for each $t \geq 0$. Therefore, the deterministic map $F : D([0,\infty)) \to D([0,\infty))$ defined by

$$F(f)(t) := \lim_{\delta \to 0+} \int_{[0,t]} \mathbb{1}_{\{|f(s-)|\geq\delta\}} \mathrm{d}f(s), \quad t \geq 0$$

is well-defined at X and $\widetilde{X}$ a.s. Moreover, this map is measurable, and $\widetilde{U}_\alpha = F(\widetilde{X})$ and $U_\alpha = F(X)$ a.s. In view of $\widetilde{X} \stackrel{d}{=} X$, according to Section III.3 in [14], there exists a measurable map $G : D([0,\infty)) \to D([0,\infty))$ satisfying $L_0^{\widetilde{X}} = G(\widetilde{X})$ and $L_0^X = G(X)$. As a consequence, $(\widetilde{X}, \widetilde{U}_\alpha, L_0^{\widetilde{X}}) \stackrel{d}{=} (X, U_\alpha, L_0^X)$. Since $\widetilde{\mathcal{S}}^{(p\theta)}(L_0^{\widetilde{X}}(t)) = \widetilde{X}(t) - \widetilde{U}_\alpha(t)$ and $H(L_0^X(t)) = X(t) - U_\alpha(t)$ for $t \geq 0$, we also conclude that

$$(\widetilde{X}, \widetilde{U}_\alpha, \widetilde{\mathcal{S}}^{(p\theta)}(L_0^{\widetilde{X}}), L_0^{\widetilde{X}}) \stackrel{d}{=} (X, U_\alpha, H(L_0^X), L_0^X).$$

Hence,

$$(\widetilde{X}, \widetilde{U}_\alpha, \widetilde{\mathcal{S}}^{(p\theta)}, L_0^{\widetilde{X}}) \stackrel{d}{=} (X, U_\alpha, H, L_0^X)$$

because the processes $t \mapsto L_0^X(t)$ and $t \mapsto L_0^{\widetilde{X}}(t)$ are both a.s. continuous and $\lim_{t\to\infty} L_0^{\widetilde{X}}(t) = \lim_{t\to\infty} L_0^X(t) = \infty$ a.s. $\square$

Bibliographic Comments

The results presented in Sect. 2.3 are based on the recent paper [70]. A Markov process with density (2.52) was discussed in [135]. In particular, a martingale characterization of this process is given there.

An essential part of the construction of a diffusion in a media with semipermeable membrane is based on the result dealing with a jump of the normal derivative of a single layer potential; see [136]. It turns out that a similar result, in which the derivative has to be replaced with some nonlocal operator, holds true for a potential generated by the process

U_α. Such an approach leads to strongly continuous semigroups without nonnegativity condition [101, 118]. Thus, it cannot be used for constructing a Markov process.

Some ideas related to our investigation can be found in the papers [96, 178, 179], where various invariance principles are obtained via convergence of the excursion measures.

Functional Limit Theorems for Locally Perturbed Random Walks 3

> *An interruption, a perturbation, a small, unsettling event – such occurrences could cascade, could accumulate, could become large forces that would transform the behavior of a whole system.*
>
> James Gleick, Chaos: Making a New Science (1987)

Abstract In this chapter, we prove weak convergence of locally perturbed random walks, properly scaled, to the processes discussed in the previous chapter. Our primary focus is on the one-dimensional case, and we consider two types of local perturbations. The first type is a random walk on nonnegative integers, reflected to the right upon crossing the origin. The second type is a random walk on integers, perturbed at a finite number of points. Under various assumptions regarding the distribution tails of both the increments of the original unperturbed random walk and the perturbations, we derive limit theorems for scaled versions of these processes in the Skorokhod space. The chapter closes with a discussion of some multidimensional perturbed random walks.

3.1 Preliminaries

3.1.1 Definition of Random Walk with Membrane

Let $\xi_1, \xi_2, \ldots$ be independent copies of an integer-valued random variable ξ. Consider a standard random walk $S_\xi = (S_\xi(n))_{n \geq 0}$ defined by

$$S_\xi(0) := 0, \quad S_\xi(n) := \xi_1 + \ldots + \xi_n, \quad n \in \mathbb{N}.$$

We are interested in a derived sequence obtained by perturbing S_ξ on a subset of $\mathbb{Z}$. Here is a formal definition.

Definition 3.1.1 Let $A \subset \mathbb{Z}$ and $(\eta^{(x)})_{x \in A}$ be a collection of random variables taking values in $\mathbb{Z}$. We call a *random walk with membrane A* a Markov chain $X = (X(n))_{n \geq 0}$ with the transition probabilities given by

$$\mathbb{P}\{X(n+1) \in \cdot \mid X(n) = x\}$$
$$= \begin{cases} \mathbb{P}\{S_\xi(n+1) \in \cdot \mid S_\xi(n) = x\} = \mathbb{P}\{x + \xi \in \cdot\}, & \text{if } x \in \mathbb{Z} \setminus A, \\ \mathbb{P}\{x + \eta^{(x)} \in \cdot\}, & \text{if } x \in A \end{cases} \quad (3.1)$$

for $n \in \mathbb{N}_0$.

According to Definition 3.1.1, the jumps of X from the points of $\mathbb{Z} \setminus A$ are distributed as independent copies of ξ, and the jumps of X from the point $x \in A$ are distributed as independent copies of $\eta^{(x)}$. Definition 3.1.1 intentionally excludes the case $A = \mathbb{Z}$, which is trivial because any Markov chain on $\mathbb{Z}$ is a random walk with membrane $\mathbb{Z}$.

Our global purpose is to explore distributional convergence of the random walks with membranes. To our mind, it is natural to assume that the underlying standard random walk S_ξ has "nice" properties and then investigate the extent to which a perturbation at the points of A affects (spoils) the limit behavior of X. Keeping in mind the aforementioned purpose, we shall always assume that:

(i) The distribution of ξ belongs to the domain of attraction of an α-stable distribution with $\alpha \in (0, 2]$.
(ii) There exists a sequence $(a_n)_{n \geq 1}$ of positive numbers such that the variables $S_\xi(n)/a_n$ converge in distribution as $n \to \infty$ to a random variable with the α-stable distribution.

We refer to Sect. 4.4 in the Appendix for a background on stable distributions and stable processes. By Theorem 4.4.1, the limit distribution is $U_\alpha^{1,\beta,0}$ if $\alpha \in (0, 1)$, or $\alpha = 1$ and the distribution of ξ is symmetric, or $\alpha \in (1, 2]$ and $\mathbb{E}\xi = 0$. According to Theorem 4.4.2,

$$\left(\frac{S_\xi(\lfloor nt \rfloor)}{a_n}\right)_{t \geq 0} \implies (\mathcal{S}_\alpha(t))_{t \geq 0}, \quad n \to \infty \quad (3.2)$$

on $D([0, \infty))$, where $\mathcal{S}_\alpha = (\mathcal{S}_\alpha(t))_{t \geq 0}$ is an α-stable Lévy process with $\mathcal{S}_\alpha(1)$ having the distribution $U_\alpha^{1,\beta,0}$. We are going to investigate distributional convergence of

$$(X_n(t))_{t \geq 0} := \left(\frac{X(\lfloor nt \rfloor)}{a_n}\right)_{t \geq 0}$$

as $n \to \infty$ and to describe a distributional limit $X_\infty = (X_\infty(t))_{t \geq 0}$ whenever it exists.

3.1 Preliminaries

We mainly focus on the two cases: either A is bounded or $A = (-\infty, 0] \cap \mathbb{Z}$. In the latter case we additionally assume that, for $x \in A$, the variables $\eta^{(x)}$ are a.s. positive. As one may anticipate, this implies that the time spent in the negative halfline is negligible relative to the time spent in the positive halfline. In the case of bounded A, that is, $A \subseteq \{-m, \ldots, m\}$ for some $m > 0$, there is no loss of generality in assuming that $A = \{-m, \ldots, m\}$. Indeed, for each $x \in \{-m, \ldots, m\} \setminus A$ we can take the corresponding variable $\eta^{(x)}$ having the same distribution as ξ. Further, observe that if $A = \{-m, \ldots, m\}$, then the transition probabilities of $(X(k)/a_n)_{k \geq 0}$ only differ from those of $(S_\xi(k)/a_n)_{k \geq 0}$ on the set $\{-m/a_n, \ldots, m/a_n\}$ that contracts to 0 as $n \to \infty$. In view of this, we expect that the limit process X_∞ behaves like S_α outside 0 and may have a singular behavior or an asymmetry at 0. To avoid trivialities, we assume that the Markov chain X exits $\{-m, \ldots, m\}$ with probability one.

Now we point out two constructions of X, a random walk with membrane A. In CONSTRUCTION 1 the building blocks are independent sequences $(\xi_n)_{n \geq 1}$ and $(\eta_n^{(x)})_{n \geq 1}$ for $x \in A$ of independent identically distributed[1] integer-valued random variables, and an integer-valued random variable $X(0)$ which is independent of the sequences.

CONSTRUCTION 1 Define $(X(n))_{n \geq 1}$ recursively via

$$X(n+1) := \begin{cases} X(n) + \xi_{n+1}, & \text{if } X(n) \notin A, \\ X(n) + \eta_{n+1}^{(X(n))}, & \text{if } X(n) \in A \end{cases}$$

for $n \in \mathbb{N}_0$ or by an equivalent formula

$$X(n) := X(0) + \sum_{k=1}^{n} \left(\xi_k \mathbb{1}_{\{X(k-1) \notin A\}} + \sum_{x \in A} \eta_k^{(X(k-1))} \mathbb{1}_{\{X(k-1)=x\}} \right), \quad n \in \mathbb{N}_0. \quad (3.3)$$

Consider a possible realization of the first six elements of the sequence X with $A = \{0\}$: $X(0) = -1$, $X(1) = -1 + \xi_1 = 0$, $X(2) = \eta_2^{(0)} = 7$, $X(3) = 7 + \xi_3 = 2$, $X(4) = 2 + \xi_4 = 0$, $X(5) = \eta_5^{(0)} = X(0) + \xi_1 + \eta_2^{(0)} + \xi_3 + \xi_4 + \eta_5^{(0)}$. We observe that the variables ξ_2, ξ_5, $\eta_1^{(0)}, \eta_3^{(0)}$, and $\eta_4^{(0)}$ are missing in this realization. More generally, for a given $k \in \mathbb{N}$, any particular realization involves either ξ_k or $\eta_k^{(x)}$ for some $x \in A$ but not all of these. Thus, the existence of missing variables is an intrinsic feature of CONSTRUCTION 1, which is a drawback. CONSTRUCTION 2 does better in this respect by using all the variables (ξ_n) and $(\eta_n^{(x)})$ without gaps. Although the resulting formula is not as simple as (3.3), treatment of X constructed in this way is easier. This claim is confirmed, for instance, by the availability of relation (3.6).

---

[1] The distribution of $\eta_n^{(x)}$ may depend on x.

CONSTRUCTION 2 For $n \in \mathbb{N}$, fix a trajectory $X(0), X(1), \ldots, X(n-1)$ up to time $n-1$ and denote by $T(n)$ the number of visits of X to A up to this time:

$$T(n) := \sum_{k=0}^{n-1} \mathbb{1}_{\{X(k) \in A\}}, \quad n \in \mathbb{N}.$$

Put

$$T^{(x)}(n) := \sum_{k=0}^{n-1} \mathbb{1}_{\{X(k) = x\}}, \quad n \in \mathbb{N}, \quad x \in A,$$

that is, $T^{(x)}(n)$ is the number of visits to x (a point in the membrane) up to time $n-1$. Plainly, $T(n) = \sum_{x \in A} T^{(x)}(n)$ for $n \in \mathbb{N}$. With these at hand, define now $(X(n))_{n \geq 1}$ by

$$X(n) := X(0) + \sum_{k=1}^{n-T(n)} \xi_k + \sum_{x \in A} \sum_{k=1}^{T^{(x)}(n)} \eta_k^{(x)}$$

$$= X(0) + S_\xi(n - T(n)) + \sum_{x \in A} S_{\eta^{(x)}}(T^{(x)}(n)), \quad n \in \mathbb{N}. \quad (3.4)$$

Since $T(n)$ depends on the whole past $X(0), X(1), \ldots, X(n-1)$, it is not obvious that X is a Markov chain. The sequences obtained with the help of CONSTRUCTIONS 1 AND 2 are a measurable function of two identically distributed collections of independent random variables (the collections are comprised of ξs and ηs with and without gaps, respectively). Hence, these are identically distributed. Since the former sequence is a Markov chain, so is the latter.

Observe that in a particular case where $\eta^{(x)} = \eta$ for all $x \in A$ representation (3.4) takes a simpler form

$$X(n) := X(0) + S_\xi(n - T(n)) + S_\eta(T(n)), \quad n \in \mathbb{N}. \quad (3.5)$$

Furthermore, if A is an interval, for instance, $A = (-\infty, 0)$, then formula (3.5) holds for real-valued (not necessarily integer-valued) random variables ξ and η. This fact will be used when treating *oscillating random walks* in the proof of Theorem 3.2.1.

In many situations of interest $T(n) = o(n)$ as $n \to \infty$ in probability. As discussed in Lemma 3.1.1, the asymptotic behavior of the second summand in (3.4) brings then no surprise. This is, for instance, the case in the setting of Theorem 3.1.1.

3.1 Preliminaries

Lemma 3.1.1 *Suppose (3.2) and let $(T^*(n))_{n\geq 1}$ be any a.s. nondecreasing random sequence satisfying $T^*(n) = o(n)$ as $n \to \infty$ in probability. Then*

$$\left(\frac{S_\xi(\lfloor nt \rfloor - T^*(\lfloor nt \rfloor))}{a_n}\right)_{t\geq 0} \Longrightarrow (\mathcal{S}_\alpha(t))_{t\geq 0}, \quad n \to \infty \tag{3.6}$$

on $D([0,\infty))$.

Proof Using the a.s. monotonicity of $n \mapsto T^*(n)$ yields, for any $t_0 > 0$ and $n \in \mathbb{N}$,

$$\sup_{t\in[0,t_0]}\left|t - \frac{\lfloor nt \rfloor - T^*(\lfloor nt \rfloor)}{n}\right| = \sup_{t\in[0,t_0]}\left|\left(t - \frac{\lfloor nt \rfloor}{n}\right) + \frac{T^*(\lfloor nt \rfloor)}{n}\right|$$

$$\leq \frac{1}{n} + \frac{T^*(\lfloor nt_0 \rfloor)}{n} \quad \text{a.s.}$$

Therefore,

$$\left(\frac{\lfloor nt \rfloor - T^*(\lfloor nt \rfloor)}{n}\right)_{t\geq 0} \Longrightarrow (I(t))_{t\geq 0}, \quad n \to \infty$$

on $D([0,\infty))$, where $I(t) := t$ for $t \geq 0$. The limit function I is deterministic, continuous, and strictly increasing. Hence, by Lemma 4.1.3, there is joint convergence combining the convergence in the last limit relation and the convergence in (3.2). With this at hand, an application of Proposition 4.3.7 in combination with Remark 4.3.1 ensures (3.6). □

An asymptotic analysis of the last summand in (3.4), rescaled and normalized by a_n, is a delicate task because its outcome may depend on nontrivial properties of $\mathcal{S}_\alpha$. Nevertheless, the asymptotic behavior of the last summand can be guessed provided that there is information on the asymptotic behavior of both $S_{\eta^{(x)}}$ and $T^{(x)}(n)$. We stress that either of these variables may require normalization other than a_n.

We first consider a simple situation in which the last summand in (3.4) has no effect asymptotically. According to the statement given next, this particularly happens whenever $\alpha \in (0, 1)$.

Theorem 3.1.1 *Let $A = \{-m, \ldots, m\}$ and assume that the distribution of ξ belongs to the domain of attraction of an α-stable distribution with $\alpha \in (0, 1)$. Then*

$$\left(\frac{X(\lfloor nt \rfloor)}{a_n}\right)_{t\geq 0} \Longrightarrow (\mathcal{S}_\alpha(t))_{t\geq 0}, \quad n \to \infty$$

on $D([0,\infty))$.

Proof By Theorem 4.4.1, the assumption concerning the domain of attraction is equivalent to the fact that the function $t \mapsto \mathbb{P}\{|\xi| > t\}$ is regularly varying at ∞ of index $-\alpha$ and

$$\mathbb{P}\{\xi > t\} \sim c_+ \mathbb{P}\{|\xi| > t\} \quad \text{and} \quad \mathbb{P}\{-\xi > t\} \sim c_- \mathbb{P}\{|\xi| > t\}, \quad t \to +\infty$$

for some nonnegative c_+ and c_- summing up to 1. Furthermore, by the same theorem, the limit distribution of $S_\xi(n)/a_n$ is $U_\alpha^{1,\beta,0}$, with $\beta = c_+ - c_-$.

The claim essentially follows from transience of X that we are now going to prove. To this end, we note that the standard random walk S_ξ is transient. Although this fact is known, we provide a proof for completeness. For simplicity, we only consider the case where the distribution of ξ is 1-arithmetic, that is, it is concentrated on the set of integers $\mathbb{Z}$ and not concentrated on $d\mathbb{Z}$ for any $d \geq 2$. By Gnedenko's local limit theorem (see, for instance, Theorem 8.4.1 in [13]),

$$\lim_{n \to \infty} a_n \mathbb{P}\{S_\xi(n) = 0\} = g_\alpha(0), \tag{3.7}$$

where g_α is the density of $U_\alpha^{1,\beta,0}$. Since the characteristic function of $U_\alpha^{1,\beta,0}$ is absolutely integrable, in view of $\left|\mathbb{E}\exp(izU_\alpha^{1,\beta,0}\right| = \exp(-|z|^\alpha)$ for $z \in \mathbb{R}$, g_α is a bounded function which particularly entails that $g_\alpha(0) < \infty$. Since, by Remark 4.4.1, $a_n = c(n)$ for a function c regularly varying at infinity of index $1/\alpha > 1$, relation (3.7) implies $\sum_{n \geq 0} \mathbb{P}\{S_\xi(n) = 0\} < \infty$, thereby proving transience of S_ξ.

According to our standing assumption, the Markov chain X exits the membrane $\{-m, \ldots, m\}$ with probability one. With this at hand, transience of S_ξ entails transience of X. Hence, $T(\infty) := \lim_{n \to \infty} T(n) < \infty$ a.s. and, for any $t_0 > 0$,

$$\sup_{t \in [0, t_0]} \frac{|\sum_{|x| \leq m} S_{\eta^{(x)}}(T^{(x)}(\lfloor nt \rfloor))|}{a_n} \leq \frac{\sum_{|x| \leq m} S_{|\eta^{(x)}|}(T(\infty))}{a_n} \to 0, \quad n \to \infty \quad \text{a.s.}$$

An appeal to Slutsky's lemma together with (3.4) and (3.6) completes the proof. □

We close the introductory part with a brief remark on our definition of random walks with membranes. For the sake of transparency of exposition and simplicity of proofs we have taken a "discrete" approach by imposing the assumption that X takes values in $\mathbb{Z}$. Note that Definition 3.1.1 and CONSTRUCTION 1 carry over verbatim to a "continuous" case of real-valued variables ξ and $\eta^{(x)}$, and $A \subseteq \mathbb{R}$. CONSTRUCTION 2 lacks this advantage and obtaining its counterpart in the "continuous" setting requires some efforts.

3.1.2 Examples of Random Walks with Reflection

In this section we consider perturbed random walks on $[0, \infty)$ that bounce up, instantaneously or eventually, upon crossing 0. We start with Lindley's model as given in (2.1) and its scaling limits.

MODEL LIND (LINDLEY'S MODEL). Put

$$X(n+1) := (X(n) + \xi_{n+1})_+, \quad n \in \mathbb{N}_0,$$

where $x_+ = \max(x, 0)$ for $x \in \mathbb{R}$. Under the additional assumption that ξ is an integer-valued random variable, the sequence $X = (X(n))_{n \geq 0}$ can be used as a model of storage. For $n \in \mathbb{N}_0$, the variable $X(n)$ is then interpreted[2] as the number of goods available at time n, and the variable ξ_{n+1} is the difference between the number of goods arrived at the storage and those on demand within the period $(n, n + 1]$. The storage becomes empty if the demand exceeds the current supply plus arrivals.

As discussed in Sect. 2.1.1, Lindley's model is a precursor of the Skorokhod reflection map Γ defined in (2.3). The next result follows immediately from continuity of the Skorokhod reflection map Γ (see again Sect. 2.1.1) and the continuous mapping theorem (Corollary 4.1.1).

Theorem 3.1.2 *Assume that the distribution of a real-valued random variable ξ belongs to the domain of attraction of an α-stable distribution with $\alpha \in (0, 2]$ and that (3.2) holds (i.e., no centering is needed for S_ξ). Then, for each $x \geq 0$ and a sequence $(x_n)_{n \geq 1}$ satisfying $\lim_{n \to \infty} x_n = x$,*

$$\left(\Gamma \left(x_n + \frac{S_\xi(\lfloor nt \rfloor)}{a_n} \right) \right)_{t \geq 0} \implies (\Gamma(x + \mathcal{S}_\alpha(t)))_{t \geq 0}, \quad n \to \infty$$

on $D([0, \infty))$, where $(\mathcal{S}_\alpha(t))_{t \geq 0}$ is an α-stable Lévy process appearing in (3.2). In particular, if $\mathbb{E}\xi = 0$, $\mathrm{Var}\,\xi = \sigma^2 \in (0, \infty)$, then

$$\left(\Gamma \left(x_n + \frac{S_\xi(\lfloor nt \rfloor)}{\sqrt{n}} \right) \right)_{t \geq 0} \implies (|x + \sigma W(t)|)_{t \geq 0}, \quad n \to \infty$$

on $D([0, \infty))$, where $(W(t))_{t \geq 0}$ is a standard Brownian motion.

We proceed by introducing five models of random walks with reflection upon crossing 0. The models differ by a way of exiting the interval $(-\infty, 0]$. Apart from Lindley's model,

---

[2] The interpretation given here is different from that in Sect. 2.1.1. Of course, other interpretations are also possible.

the list includes oscillating random walk and reflected random walks, which are quite popular objects of applied probability. We provide in Bibliographic Comments a short survey of results available for both oscillating and reflected random walks.

Let $\xi_1, \xi_2, \ldots$ be independent copies of a real-valued random variable ξ, $\eta_1, \eta_2, \ldots$ independent copies of a real-valued random variable η, and $X(0)$ a nonnegative random variable, the two collections and $X(0)$ being mutually independent. Throughout this section $S_\xi(0) = 0$.

MODEL LINDREFILL (LINDLEY'S MODEL WITH REFILLING). Assume that $\eta \geq 0$ a.s. This model is a modification of MODEL LIND: if the storage is empty at time n, it is refilled with a new batch of size η_{n+1}. Namely,

$$X(n+1) := \begin{cases} X(n) + \xi_{n+1}, & \text{if } X(n) > 0 \text{ and } X(n) + \xi_{n+1} \geq 0, \\ 0, & \text{if } X(n) > 0 \text{ and } X(n) + \xi_{n+1} < 0, \\ \eta_{n+1}, & \text{if } X(n) = 0 \end{cases}$$

for $n \in \mathbb{N}_0$.

MODEL OSCRW (OSCILLATING RANDOM WALK). Assuming that $\mathbb{E}\eta > 0$, put

$$X(n+1) := \begin{cases} X(n) + \xi_{n+1}, & \text{if } X(n) \geq 0, \\ X(n) + \eta_{n+1}, & \text{if } X(n) < 0 \end{cases}$$

for $n \in \mathbb{N}_0$. The so-defined $(X_n)_{n \geq 0}$ is known in the literature as an *oscillating random walk*, hence the name of the model.

The random sequence following MODEL OSCRW can be interpreted as a model for an Internet shop with storage. All orders are accepted including those which lead to a virtually negative number of goods in the storage. If this has happened, the storage is refilled with a new batch. The variable η_{n+1} represents the size of a batch minus the sizes of all orders placed within the time interval $(n, n+1]$. The assumption $\mathbb{E}\eta > 0$ ensures that, for each $a > 0$, $\inf\{n \geq 1 : S_\eta(n) > a\} < \infty$ a.s., that is, the number of goods in the storage becomes positive eventually with probability one.

MODEL REPULS (MODEL WITH REPULSIVE HALF-PLANE). Assuming that $\mathbb{E}\eta > 0$, put

$$X(n+1) := \begin{cases} X(n) + \xi_{n+1}, & \text{if } X(n) > 0, \\ 0, & \text{if } X(n) < 0, \\ \eta_{n+1}, & \text{if } X(n) = 0. \end{cases}$$

3.1 Preliminaries

The name of the model is more or less self-explained. Whenever the Markov chain hits an open lower half-plane, it is set to be 0 at the next step.

Assume that $X(0)$, ξ, and η are integer-valued random variables. Then the random walks with reflection defined by MODELS OSCRW and REPULS are random walks with membranes $A = -\mathbb{N}$ (the set of negative integers) and $A = -\mathbb{N}_0$ (the set of nonpositive integers), respectively, in the sense of Definition 3.1.1. For the former model, $\eta^{(x)} = \eta$ for $x \in -\mathbb{N}$, whereas for the latter model, $\eta^{(x)} = -x$ for $x \in -\mathbb{N}$ and $\eta^{(0)} = \eta$. Allowing X_0, ξ and η to be real-valued, rather than integer-valued, the random walk with reflection given by MODEL OSCRW can be thought of as a random walk with membrane $A = (-\infty, 0)$.

MODEL REFLRW (REFLECTED RANDOM WALK). Put

$$X(n+1) := |X(n) + \xi_{n+1}|, \quad n \in \mathbb{N}_0.$$

Any jump into negative halfline is forbidden, and the next position of the Markov chain is set to be the size of the intended overshoot at 0 into the negative halfline. The so-defined $(X_n)_{n \geq 0}$ is known in the literature as a *reflected random walk*, hence the name of the model. If the distribution of ξ is symmetric, then $(X(n))_{n \geq 0}$ has the same distribution as $(|X(0) + S_\xi(n)|)_{n \geq 0}$. If the distribution of ξ is not symmetric, then the distributions of $(X(n))_{n \geq 0}$ and $(|X(0) + S_\xi(n)|)_{n \geq 0}$ are different.

MODEL FREEZE (MODEL WITH FREEZING). Put

$$X(n+1) := \begin{cases} X(n) + \xi_{n+1}, & \text{if } X(n) + \xi_{n+1} \geq 0, \\ X(n), & \text{if } X(n) + \xi_{n+1} < 0 \end{cases}$$

for $n \in \mathbb{N}_0$. The Markov chain stays still when attempting to jump into the negative halfline. Formally, the model exhibits a reflection, with the fictitious jump from 0 being equal to the previous position.

In all the models, $X(n+1) := X(n) + \xi_{n+1}$ if $X(n) + \xi_{n+1} \geq 0$ or $X(n) > 0$. The reflection in MODEL OSCRW is eventual, whereas the reflection in MODEL REPULS is instantaneous if $\eta > 0$ a.s., and it is instantaneous in all the other models without further conditions.

Now we discuss the simplest case where $\mathbb{E}\xi = 0$ and $\text{Var}\,\xi = 1$, so that the stochastic processes $(S_\xi(\lfloor nt \rfloor)/\sqrt{n})_{t \geq 0}$ converge in distribution to a standard Brownian motion. Thus, we expect that the limit process for $(X(\lfloor nt \rfloor)/\sqrt{n})_{t \geq 0}$ behaves like a Brownian motion on $(0, \infty)$ and admits some kind of reflection upon visits to 0. Here is a brief summary of the results:

(a) If η has a finite mean, then Donsker's scaling limit is a reflected Brownian motion.
(b) If the distribution of η belongs to the domain of attraction of a β-stable distribution with $\beta \in (0, 1)$, then Donsker's scaling limit is a Brownian motion with jump-type exit from 0 and reflection governed by a β-stable subordinator; see Sect. 2.1.3.
(c) If the distribution tail of η is slowly varying at infinity, then there is no any scaling limit.

3.2 Functional Limit Theorems for Random Walks with Reflection

3.2.1 Perturbations with Finite Mean

For each $n \in \mathbb{N}$, let $(X^{(n)}(k))_{k \geq 0}$ be a Markov chain having the same transition probabilities as $(X(k))_{k \geq 0}$ but possibly satisfying a different initial condition. For each $n \in \mathbb{N}$, put

$$X_n(t) := \frac{X^{(n)}(\lfloor nt \rfloor)}{\sqrt{n}}, \quad t \geq 0.$$

Theorem 3.2.1 *Let $x \geq 0$ and assume that $X_n(0)$ converges in probability to x as $n \to \infty$. If $\mathbb{E}\xi = 0$, $\operatorname{Var} \xi = 1$, then for* MODELS REFLRW *and* FREEZE,

$$(X_n(t))_{t \geq 0} \implies (|x + W(t)|)_{t \geq 0}, \quad n \to \infty, \tag{3.8}$$

on $D([0, \infty))$, where $(W(t))_{t \geq 0}$ is a standard Brownian motion. Under the additional assumptions that $\eta \geq 0$ a.s. and $\mathbb{E}\eta \in (0, \infty)$, (3.8) holds for MODELS LINDREFILL, OSCRW, *and* REPULS.[3]

Remark 3.2.1 An inspection of the proof that follows in combination with Corollary 2.1.1 and Remark 2.1.3 reveals that there is even the joint convergence

$$(X_n(t), L_n(t))_{t \geq 0} \implies (|x + W(t)|, L_0^{x+W}(t))_{t \geq 0}, \quad n \to \infty$$

in the J_1-topology on $D([0, \infty), \mathbb{R}^2)$. Here, L_n is an appropriate additive functional of X_n, formula (2.19) is used to identify the limit process, and L_0^{x+W} is the symmetric semimartingale local time at 0 of the process $(x + W(t))_{t \geq 0}$. For instance, an appropriate choice of L_n for MODEL OSCRW is

$$L_n(t) := \frac{\sum_{k=0}^{\lfloor nt \rfloor - 1} \eta_k \mathbb{1}_{\{X^{(n)}(k) < 0\}}}{\sqrt{n}}, \quad t \geq 0.$$

---

[3] The assumption $\eta \geq 0$ a.s. is included in the definition of MODEL LINDREFILL.

3.2 Functional Limit Theorems for Random Walks with Reflection

Also, a perusal of our proof makes it clear that Theorem 3.2.1 continues to hold upon replacing $(-\infty, 0)$ with $(-\infty, 0]$ in MODELS OSCRW and FREEZE.

Proof Throughout the proof we use standard random walks S_ξ and S_η starting at 0, that is, $S_\xi(0) = S_\eta(0) = 0$. Without loss of generality we can and do assume that, for each $n \in \mathbb{N}$, the random variables $(X^{(n)}(k))_{k \geq 1}$ are functions of S_ξ, S_η, and $X^{(n)}(0)$.

We first treat MODELS REFLRW and FREEZE. To this end, we use a decomposition: for each $n \in \mathbb{N}$,

$$X^{(n)}(k) = X^{(n)}(0) + S_\xi(k) + L^{(n)}(k), \quad k \in \mathbb{N}_0,$$

where $L^{(n)}(0) = 0$ and, for $k \in \mathbb{N}$,

$$L^{(n)}(k) := \begin{cases} \sum_{i=0}^{k-1} 2|X^{(n)}(i) + \xi_{i+1}|\mathbb{1}_{\{X^{(n)}(i)+\xi_{i+1}<0\}} & \text{for MODEL REFLRW,} \\ \sum_{i=0}^{k-1} |\xi_{i+1}|\mathbb{1}_{\{X^{(n)}(i)+\xi_{i+1}<0\}} & \text{for MODEL FREEZE.} \end{cases}$$

For $n \in \mathbb{N}$, put

$$W_n(t) := \frac{X^{(n)}(0) + S_\xi(\lfloor nt \rfloor)}{\sqrt{n}}, \quad L_n(t) := \frac{L^{(n)}(\lfloor nt \rfloor)}{\sqrt{n}}, \quad t \geq 0.$$

We intend to apply Corollary 2.1.1, with the modification stated in Remark 2.1.3. The triple (X_n, Y_n, L_n) in the notation of Corollary 2.1.1 is (X_n, W_n, L_n). Condition (I) of the corollary holds according to our assumption concerning $X^{(n)}(0)$ and Donsker's theorem (Theorem 4.3.1). Condition (IIIa) follows from the facts that, for each $n \in \mathbb{N}$, the sequence $(L^{(n)}(k))_{k \geq 0}$ is a.s. nondecreasing with $L^{(n)}(0) = 0$. Condition (IIIb) is secured by $X^{(n)}(k) \geq 0$ a.s. for $n \in \mathbb{N}$ and $k \in \mathbb{N}_0$.

CHECKING THE CONDITION IN REMARK 2.1.3. We are going to show that the sequence of distributions of $(X_n)_{n \geq 1}$ is tight. By the assumption, $X_n(0)$ converges in probability to x as $n \to \infty$. Hence, according to Corollary on p. 142 in [12], it is enough to prove that, for all $t_0 > 0$ and all $\varepsilon > 0$,

$$\lim_{\delta \to 0+} \limsup_{n \to \infty} \mathbb{P}\left\{\sup_{|t-s| \leq \delta, \, t, s \in [0, t_0]} |X_n(t) - X_n(s)| > \varepsilon\right\} = 0$$

or equivalently

$$\lim_{\delta \to 0+} \limsup_{n \to \infty} \mathbb{P}\left\{\sup_{|l-k| \leq \lfloor \delta n \rfloor, \, l, k \in [0, \lfloor nt_0 \rfloor]} |X^{(n)}(l) - X^{(n)}(k)| > \varepsilon \sqrt{n}\right\} = 0. \quad (3.9)$$

Let $k, l \geq 0$, $k < l$ be fixed integers. To prove (3.9), observe that the original definitions of the models ensure that

$$X^{(n)}(l) - X^{(n)}(k) = S_\xi(l) - S_\xi(k)$$

on the event $\{\min_{k \leq j \leq l-1}(X^{(n)}(j) + \xi_{j+1}) \geq 0\}$. On the complementary event, we use a trivial estimate

$$|X^{(n)}(l) - X^{(n)}(k)| \leq |X^{(n)}(l)| \vee |X^{(n)}(k)| = X^{(n)}(l) \vee X^{(n)}(k)$$

and put $k_+^* := \min\{k \leq i \leq l : X^{(n)}(i) + \xi_{i+1} < 0\}$ and

$$l_-^* := \max\{k \leq i \leq l : X^{(n)}(i) + \xi_{i+1} < 0\}.$$

Then, on the complementary event,

$$0 > X^{(n)}(k_+^*) + \xi_{k_+^*+1} = X^{(n)}(k) + S_\xi(k_+^* + 1) - S_\xi(k),$$

whence

$$X^{(n)}(k) < S_\xi(k) - S_\xi(k_+^* + 1) \leq \max_{k \leq i,j \leq l} |S_\xi(i) - S_\xi(j)|.$$

Further, on the complementary event, $X^{(n)}(l) < |\xi_{l+1}| \leq \max_{1 \leq i \leq l+1} |\xi_i|$ if $l = l_-^*$ or $l = l_-^* + 1$. The (intermediate) estimates of $X^{(n)}(l)$ are model-dependent for $l > l_-^* + 1$. For MODEL REFLRW,

$$X^{(n)}(l) = -X^{(n)}(l_-^*) - \xi_{l_-^*+1} + S_\xi(l) - S_\xi(l_-^* + 1) \leq |\xi_{l_-^*+1}| + |S_\xi(l) - S_\xi(l_-^* + 1)|$$

$$\leq \max_{1 \leq i \leq l} |\xi_i| + \max_{k \leq i,j \leq l} |S_\xi(i) - S_\xi(j)|.$$

For MODEL FREEZE, $X^{(n)}(l_-) = X^{(n)}(l_- + 1)$ and thereupon

$$X^{(n)}(l) = X^{(n)}(l_-^*) - \xi_{l_-^*+1} + S_\xi(l) - S_\xi(l_-^* + 2) \leq |\xi_{l_-^*+1}| + |S_\xi(l) - S_\xi(l_-^* + 2)|$$

$$\leq \max_{1 \leq i \leq l} |\xi_i| + \max_{k \leq i,j \leq l} |S_\xi(i) - S_\xi(j)|.$$

The argument in the case $k \geq l$ is analogous. Summarizing, we have proved that

$$\sup_{|l-k| \leq \lfloor \delta n \rfloor, \, l,k \in [0, \lfloor nt_0 \rfloor]} |X^{(n)}(l) - X^{(n)}(k)| \leq \sup_{|l-k| \leq \lfloor \delta n \rfloor, \, l,k \in [0, \lfloor nt_0 \rfloor]} |S_\xi(l) - S_\xi(k)|$$

$$+ \max_{1 \leq i \leq \lfloor nt_0 \rfloor + 1} |\xi_i|.$$

3.2 Functional Limit Theorems for Random Walks with Reflection

As a consequence of Donsker's theorem and particularly a.s. continuity of the limit Brownian motion,

$$\lim_{\delta \to 0+} \limsup_{n \to \infty} \mathbb{P}\left\{ \sup_{|l-k| \leq \lfloor \delta n \rfloor, l,k \in [0, \lfloor n t_0 \rfloor]} |S_\xi(l) - S_\xi(k)| > \varepsilon \sqrt{n} \right\} = 0.$$

In view of $\mathbb{E}\xi^2 < \infty$, $\lim_{n \to \infty} n^{-1/2} |\xi_n| = 0$ a.s. by the Borel-Cantelli lemma, whence, for all $t_0 > 0$,

$$\lim_{n \to \infty} n^{-1/2} \max_{1 \leq i \leq \lfloor n t_0 \rfloor + 1} |\xi_i| = 0 \quad \text{a.s.}$$

This completes the proof of (3.9). Thus, the sequence of distributions of $(X_n)_{n \geq 1}$ is tight, and, by Corollary on p. 142 in [12], every distributional limit point of $(X_n)_{n \geq 1}$ is a.s. continuous. Since the distributional limit of (W_n) is a.s. continuous (Brownian motion) and $L_n = X_n - W_n$, we conclude that every distributional limit point of $(L_n)_{n \geq 1}$ is a.s. continuous.

CHECKING CONDITION (IIIC) Our purpose is to prove that, for all $t_0 > 0$ and all $\delta > 0$,

$$\int_{[0, t_0]} \mathbb{1}_{\{X_\infty(s) > \delta\}} dL_\infty(s) = 0 \quad \text{a.s.} \tag{3.10}$$

for every distributional limit point (X_∞, L_∞) of $((X_n, L_n))_{n \geq 1}$. Recall that this limit point is a.s. continuous.

Let $(n_k)_{k \geq 1}$ be a diverging sequence of positive numbers such that

$$(X_{n_k}, L_{n_k}) \implies (X_\infty, L_\infty), \quad k \to \infty$$

in the product topology on $(D([0, \infty), \mathbb{R}))^2$. By the Skorokhod representation theorem (Theorem 4.1.2) there exist $((\widetilde{X}_{n_k}, \widetilde{L}_{n_k}))_{k \geq 1}$ versions of $(X_{n_k}, L_{n_k})_{k \geq 1}$ and $(\widetilde{X}_\infty, \widetilde{L}_\infty)$ a version of (X_∞, L_∞) (all the versions being defined on a common probability space) such that

$$\lim_{k \to \infty} (\widetilde{X}_{n_k}, \widetilde{L}_{n_k}) = (\widetilde{X}_\infty, \widetilde{L}_\infty) \quad \text{locally uniformly a.s.} \tag{3.11}$$

To prove (3.10), with t_0 and δ fixed, it is sufficient to show that, for all intervals $[a, b] \subseteq [0, t_0]$,

$$\mathbb{P}\left\{ \min_{t \in [a,b]} \widetilde{X}_\infty(t) > \delta \text{ and } \widetilde{L}_\infty(a) < \widetilde{L}_\infty(b) \right\} = 0.$$

Fix $\omega \in \Omega$ such that (3.11) holds and fix the interval $[a, b] \in [0, t_0]$ such that $\min_{t \in [a,b]} \widetilde{X}_\infty(t, \omega) > \delta$. Then for k large enough $\widetilde{X}_{n_k}(l/n_k, \omega) > \delta$ for all $l \in \mathbb{N}$ satisfying $a \leq (l-1)/n_k < (l+1)/n_k \leq b$. Further, observe that $\widetilde{L}_\infty(a, \omega) < \widetilde{L}_\infty(b, \omega)$ entails $\widetilde{L}_{n_k}(a, \omega) < \widetilde{L}_{n_k}(b, \omega)$ for k large enough. Then there exists l^* satisfying $a \leq (l^*-1)/n_k < (l^*+1)/n_k \leq b$ and $\widetilde{L}_{n_k}((l^*+1)/n_k, \omega) - \widetilde{L}_{n_k}(l^*/n_k, \omega) > 0$. It follows from the definitions of MODELS REFLRW and FREEZE that the latter inequality entails

$$\widetilde{L}_{n_k}((l^*+1)/n_k, \omega) - \widetilde{L}_{n_k}(l^*/n_k, \omega) > \widetilde{X}_{n_k}((l^*+1)/n_k, \omega) > \delta.$$

The left-hand side converges to 0 because the convergence $\lim_{k \to \infty} \widetilde{L}_{n_k}(\cdot, \omega) = \widetilde{L}_\infty(\cdot, \omega)$ is locally uniform, and $\widetilde{L}_\infty(\cdot, \omega)$ is a continuous function. This contradiction shows that (3.10) does indeed hold.

An appeal to Corollary 2.1.1, with the modification stated in Remark 2.1.3, in combination with formula (2.19) (used to identify the limit process) completes the proof for MODELS REFLRW and FREEZE.

Since our arguments for MODELS LINDREFILL and REPULS follow almost the same path, we only give a sketch of a proof for MODEL LINDREFILL. First, we shall construct a random sequence $(\widetilde{X}^{(n)}(k))_{k \geq 0}$ which has the same distribution as $(X^{(n)}(k))_{k \geq 0}$, for each $n \in \mathbb{N}$, and satisfies a formula similar to (3.4). Fix $n \in \mathbb{N}$. The sequence is constructed inductively by putting first $\widetilde{X}^{(n)}(0) = X^{(n)}(0)$. Assuming that $\widetilde{X}^{(n)}(0), \ldots, \widetilde{X}^{(n)}(k-1)$ have already been constructed, for some $k \geq 1$, put

$$T^{(0,n)}(k) := \sum_{i=0}^{k-1} \mathbb{1}_{\{\widetilde{X}^{(n)}(i) = 0\}},$$

and then

$$\widetilde{X}^{(n)}(k) := \widetilde{X}^{(n)}(0) + S_\xi(k - T^{(0,n)}(k))$$
$$+ \sum_{i=1}^{k} |\widetilde{X}^{(n)}(i-1) + \xi_i| \mathbb{1}_{\{\widetilde{X}^{(n)}(i-1) > 0, \widetilde{X}^{(n)}(i-1) + \xi_i < 0\}} + S_\eta(T^{(0,n)}(k))$$
$$=: \widetilde{X}^{(n)}(0) + S_\xi(k - T^{(0,n)}(k)) + L^{(n)}(k). \tag{3.12}$$

For each $n \in \mathbb{N}$, put

$$\widetilde{X}_n(t) := \frac{\widetilde{X}^{(n)}(\lfloor nt \rfloor)}{\sqrt{n}}, \quad t \geq 0.$$

A counterpart of equality (3.10) can be obtained along the same lines as for MODELS REFLRW and FREEZE. We shall only find Donsker's scaling limit of $(S_\xi(k - T^{(0,n)}(k)))$

3.2 Functional Limit Theorems for Random Walks with Reflection

and prove that the sequence of distributions of $(\tilde{X}_n)_{n\geq 1}$ is tight. A proof of tightness requires an additional argument in comparison to the corresponding proofs for MODELS REFLRW and FREEZE.

By the same reasoning as in the previous part of the proof, for all $\delta > 0$ and all $t_0 > 0$,

$$\sup_{|l-k|\leq \lfloor \delta n\rfloor,\, l,k\in [0,\lfloor nt_0\rfloor]} |\tilde{X}^{(n)}(l) - \tilde{X}^{(n)}(k)| \leq 2 \sup_{|l-k|\leq \lfloor \delta n\rfloor,\, l,k\in [0,\lfloor nt_0\rfloor]} |S_\xi(l) - S_\xi(k)|$$
$$+ 2\max_{1\leq i\leq \lfloor nt_0\rfloor} |\xi_i| + \max_{1\leq j\leq T^{(0,n)}(\lfloor nt_0\rfloor)} \eta_j.$$

Treatment of the first two terms on the right-hand side has already been undertaken in the proof for MODELS REFLRW and FREEZE. Thus, we are left with showing that

$$\frac{\max_{1\leq j\leq T^{(0,n)}(\lfloor nt_0\rfloor)} \eta_j}{\sqrt{n}} \xrightarrow{\mathbb{P}} 0, \quad n\to\infty. \tag{3.13}$$

While doing so, we write $T^{(0)}(k)$ and $\tilde{X}(k)$ for $T^{(0,n)}(k)$ and $\tilde{X}^{(n)}(k)$.

As a preparation, put

$$\nu(k) := \inf\{i\geq 1 : S_\eta(i) > k\} \quad \text{and} \quad M(k) := -\min_{0\leq j\leq k} S_\xi(j), \quad k\in\mathbb{N}_0.$$

We first prove that, for each $n\in\mathbb{N}$,

$$T^{(0)}(k) \leq \nu(M(k)), \quad k\in\mathbb{N} \quad \text{a.s.} \tag{3.14}$$

The argument that follows is valid for deterministic sequences. In other words, we work with a fixed ω. Let k and $L = L(k)$ be positive integers satisfying $S_\eta(L) > M(k)$. We claim that $T^{(0)}(k) \leq L$. To justify the claim, put $j_0 := \min\{i\geq 1 : T^{(0)}(i) = L\}$. By monotonicity of $T^{(0)}$, $T^{(0)}(k) \leq L$ if $k \leq j_0$. We intend to show that $T^{(0)}(k) = L$ if $k > j_0$. Since $T^{(0)}(j_0) = T^{(0)}(k)$ provided that $\min_{j_0\leq i\leq k-1} \tilde{X}(i) > 0$, it suffices to show that the latter inequality is implied by $S_\eta(L) > M(k)$. Let i be an integer number satisfying $j_0 \leq i \leq k-1$. Using (3.12), with i replacing k, and the fact that $\tilde{X}(0) \geq 0$ we infer

$$\tilde{X}(i) \geq S_\xi(i - T^{(0)}(i)) + S_\eta(T^{(0)}(i)) \geq \min_{0\leq j\leq k} S_\xi(j) + S_\eta(L) > 0$$

having utilized $\eta \geq 0$ and $T^{(0)}(i) \geq T^{(0)}(j_0) = L$ which follows by monotonicity of $T^{(0)}$. Thus, the claim has been justified. Getting back to random world, put $L = \nu(M(k))$. Since $S_\eta(\nu(M(k))) > M(k)$ for each $k\in\mathbb{N}$, inequality (3.14) follows.

Donsker's theorem implies that $M(k)/\sqrt{k} \xrightarrow{d} -\min_{s\in[0,1]} W(s)$ as $k\to\infty$. It is known that $-\min_{s\in[0,1]} W(s)$ has the same distribution as $|\mathcal{N}(0,1)|$, where $\mathcal{N}(0,1)$ is a random variable with the standard normal distribution. According to the strong law of large numbers for standard random walks and renewal processes

$$\lim_{k\to\infty} \frac{S_\eta(k)}{k} = \mathbb{E}\eta, \quad \lim_{k\to\infty} \frac{\max_{1\le i\le k} \eta_i}{k} = 0, \quad \text{and} \quad \lim_{k\to\infty} \frac{\nu(k)}{k} = \frac{1}{\mathbb{E}\eta} \quad \text{a.s.}$$

Thus, for all $\delta > 0$ and $C > 0$,

$$\limsup_{k\to\infty} \mathbb{P}\left\{\frac{\max_{1\le j\le T^{(0)}(k)} \eta_j}{\sqrt{k}} > \delta\right\}$$

$$\le \limsup_{k\to\infty} \mathbb{P}\left\{\frac{\max_{1\le j\le C\sqrt{k}} \eta_j}{\sqrt{k}} > \delta\right\} + \limsup_{k\to\infty} \mathbb{P}\{T^{(0)}(k) > C\sqrt{k}\}$$

$$= \limsup_{k\to\infty} \mathbb{P}\{T^{(0)}(k) > C\sqrt{k}\}$$

$$\le \limsup_{k\to\infty} \mathbb{P}\left\{\nu(M(k)) > C\sqrt{k}\right\}$$

$$= \limsup_{k\to\infty} \mathbb{P}\left\{\frac{\nu(M(k))}{M(k)} \frac{M(k)}{\sqrt{k}} > C\right\} = \mathbb{P}\{|\mathcal{N}(0,1)| > C\mathbb{E}\eta\}.$$

We have used (3.14) for the last inequality. The last expression converges to 0 as $C \to \infty$. This completes the proof of (3.13) and the tightness of distributions of $(\tilde{X}_n)_{n\ge 1}$.

It follows from the last chain of inequalities that $T^{(0)}(k) = o(k)$ in probability as $k \to \infty$. Hence, by Lemma 3.1.1,

$$\left(\frac{S_\xi(\lfloor nt\rfloor) - T^{(0)}(\lfloor nt\rfloor)}{\sqrt{n}}\right)_{t\ge 0} \Longrightarrow (W(t))_{t\ge 0}, \quad n \to \infty$$

on $D([0,\infty))$.

Now we give a partial proof for MODEL OSCRW. As has already been mentioned, the Markov chain following this model is a random walk with membrane $A = (-\infty, 0)$ and $\eta^{(x)} = \eta$ for all $x \in A$. Thus, for each $n \in \mathbb{N}$, $(X^{(n)}(k))_{k\ge 0}$ has the same distribution as the random sequence given by formula (3.5). This representation can be used to check that the conditions (I) and (III) of Corollary 2.1.1 hold true in the present context.

We only show that relation (3.9) holds, thereby proving tightness of the sequence of distributions of $(X_n)_{n\ge 1}$. While doing so we only exploit the original definition of MODEL OSCRW, whereas representation (3.5) is not used. Let $k, l \ge 0$, $k < l$ be fixed integers. The equality $X^{(n)}(l) - X^{(n)}(k) = S_\xi(l) - S_\xi(k)$ holds on the event $\{\min_{k\le j\le l} X(j) \ge 0\}$. On the complementary event, $X^{(n)}(j) < 0$ for some integer $j \in [k, l]$, and the random variables $k_+ := \min\{i \ge k : X^{(n)}(i) < 0\}$ and $l_- := \max\{i \le l : X^{(n)}(i) < 0\}$

3.2 Functional Limit Theorems for Random Walks with Reflection

are well-defined. We suppress in the notation the dependence of k_+ and l_- on n. Thus, assuming that $k_+ > k$ and $l_+ < l$ we obtain on the complementary event

$$|X^{(n)}(l) - X^{(n)}(k)| \leq |X^{(n)}(l) - X^{(n)}(l_- + 1)| + |X^{(n)}(l_- + 1) - X^{(n)}(l_-)|$$
$$+ |X^{(n)}(l_-) - X^{(n)}(k_+)| + |X^{(n)}(k_+) - X^{(n)}(k)|$$
$$\leq |X^{(n)}(l) - X^{(n)}(l_- + 1)| + |X^{(n)}(l_- + 1) - X^{(n)}(l_-)|$$
$$+ |X^{(n)}(l_-)| + |X^{(n)}(k_+)| + |X^{(n)}(k_+) - X^{(n)}(k)|$$
$$\leq |S_\xi(l) - S_\xi(l_- + 1)| + |\eta_{l_-+1}| + |\xi_{l_-+1}| + |\xi_{k_+}| + |S_\xi(k_+) - S_\xi(k)|$$

having utilized $X^{(n)}(l) - X^{(n)}(l_- + 1) = S_\xi(l) - S_\xi(l_- + 1)$, $X^{(n)}(l_- + 1) - X^{(n)}(l_-) = \eta_{l_-+1}$, $|X^{(n)}(l_-)| = |X^{(n)}(l_- + 1) - \xi_{l_-+1}| \leq |\xi_{l_-+1}|$, $|X^{(n)}(k_+)| = |X^{(n)}(k_+ - 1) + \xi_{k_+}| \leq |\xi_{k_+}|$ and $X^{(n)}(k_+) - X^{(n)}(k) = S_\xi(k_+) - S_\xi(k)$. We stress that the assumption $\eta \geq 0$ a.s. has not come into play yet. This assumption will now be used when treating the boundary cases $k_+ = k$ and $l_- = l$. Let $k_+ = k \geq 1$ (k_+ cannot be equal to 0 because $X^{(n)}(0) \geq 0$ by assumption). Then one possibility is that $X^{(n)}(k_+ - 1) \geq 0$ and $0 > X^{(n)}(k_+) = X^{(n)}(k_+ - 1) + \xi_{k_+}$, whence $|X^{(n)}(k_+)| \leq |\xi_{k_+}|$. The other possibility is $X^{(n)}(k_+ - 1) < 0$, so that there exists $i^* := \min\{1 \leq i \leq k_+ - 1 : \max_{i \leq j \leq k_+ - 1} X^{(n)}(j) < 0\}$. Then $X^{(n)}(i^* - 1) \geq 0$, and, since $\eta \geq 0$ and $X^{(n)}(k_+) < 0$, we infer

$$|X^{(n)}(k_+)| = |X^{(n)}_{i^*-1} + \xi_{i^*} + \eta_{i^*+1} + \ldots + \eta_{k^*}| \leq |\xi_{i^*}|.$$

Thus, in the case $k_+ = k \geq 1$ and $l_+ < l$,

$$|X^{(n)}(l) - X^{(n)}(k)| \leq |S_\xi(l) - S_\xi(l_- + 1)| + \eta_{l_-+1} + |\xi_{l_-+1}|$$
$$+ |\xi_{k_+}| \mathbb{1}_{\{X^{(n)}(k_+-1) \geq 0\}} + |\xi_{i^*}| \mathbb{1}_{\{X^{(n)}(k_+-1) < 0\}}.$$

Arguing similarly we obtain in the case $l_- = l$ that

$$|X^{(n)}(l_-)| \leq |\xi_{l_-}| \mathbb{1}_{\{X^{(n)}(l_--1) \geq 0\}} + |\xi_{r^*}| \mathbb{1}_{\{X^{(n)}(l_--1) < 0\}},$$

where $r^* := \min\{1 \leq i \leq l_- - 1 : \max_{i \leq j \leq l_- - 1} X^{(n)}(j) < 0\}$. We omit further details. In the case $k \geq l$ the estimates are analogous. Hence, we conclude that, for all $t_0 > 0$ and all $\delta > 0$,

$$\sup_{|l-k| \leq \lfloor \delta n \rfloor, \, l,k \in [0, \lfloor nt_0 \rfloor]} |X^{(n)}(l) - X^{(n)}(k)| \leq 2 \sup_{|l-k| \leq \lfloor \delta n \rfloor, \, l,k \in [0, \lfloor nt_0 \rfloor]} |S_\xi(l) - S_\xi(k)|$$
$$+ 2 \max_{1 \leq i \leq \lfloor nt_0 \rfloor} |\xi_i| + \max_{1 \leq r \leq T(\lfloor nt_0 \rfloor)} |\eta_r|,$$

where $T(n) = \sum_{k=0}^{n-1} \mathbb{1}_{\{X^{(n)}(k)<0\}}$ for $n \in \mathbb{N}$. To treat the first two terms on the right-hand side, we invoke the argument used for MODELS REFLRW and FREEZE. To treat the last term on the right-hand side, we invoke the argument used for MODEL LINDREFILL. In this way we arrive at (3.9). □

The arguments used in this section are based on elementary methods and continuity of the Skorokhod problem. In the next section we work out a more advanced approach. Although the model formulation and the result may seem boring, applications to concrete models are much simpler.

3.2.2 On Two-Stage Models and the Generalized Reflection

The models of random walks with reflection considered in Sect. 3.2.1 can be informally described as follows. There is a basic mode, mode A say, acting at the nonnegative halfline. Assume that Donsker's scaling limit of some Markov chain Y exists in mode A. The other mode, mode B say, initiates once the Markov chain Y has hit the negative halfline. An assumption is made which ensures that the time spent by Y during the period $[0, n]$ in mode B is asymptotically negligible as $n \to \infty$ in comparison to the time spent during the same period in mode A. In particular, the assumption guarantees that the increments eventually push Y out of the negative halfline. For a simple example, consider MODEL OSCRW, in which the condition $\mathbb{E}\eta > 0$ does the job, as follows from Theorem 3.2.1.

If a scaling limit in mode A is a.s. continuous, for instance, a standard Brownian motion, then it is natural to expect that the scaling limit on the whole line is a nonnegative process which behaves like a Brownian motion while it is positive and spends zero time at 0 a.s. It only remains to describe the way of exiting 0.

Below we introduce a two-stage *deterministic* switching model, with a main mode acting at the nonnegative halfline and an auxiliary mode acting at the negative halfline and eventually pushing a sequence in focus out of the negative halfline. The main result of this section, Theorem 3.2.2, describes limits of solutions to two-stage switching problems. It can be applied to a majority of random walks with reflection and, more generally, of locally perturbed random walks with nonnegative perturbations at the negative halfline. Also, the result explains how different scalings in the main and the auxiliary modes interact in order to produce limits of solutions to some reflection problems. A continuous Skorokhod's reflection, a jump-type reflection, and a stickiness at 0 turn out to be special cases of the general result.

Let y and F be càdlàg (nonrandom) functions, $y(0) \geq 0$, $F(0) = 0$, and $\delta > 0$. To avoid trivialities, assume that $y(t) < 0$ for some $t \in \mathbb{R}$. We construct the graph of a function x by merging consequently fragments of the graphs of y and F with the help of translations. To facilitate presentation we interpret the argument of the functions involved as the time. The first fragment of the graph of x, on the interval $[0, \rho_1]$, coincides with that

3.2 Functional Limit Theorems for Random Walks with Reflection

of y (mode A), where ρ_1 is the first exit time of y from $(-\delta, \infty)$. The second fragment of the graph of x, on the interval $(\rho_1, \tau_1]$, is obtained from that of F on the interval $(0, \tau_1 - \rho_1]$ (mode B) by translation (the origin of the coordinate system in which the graph of F is depicted is shifted to the point $(\rho_1, x(\rho_1))$), where τ_1 is the first exit time of a time-changed version of F from $(-\infty, -x(\rho_1))$. The third fragment of the graph of x, on the interval $(\tau_1, \rho_2]$, is obtained from that of y on the interval $(\rho_1, \rho_1 + \rho_2 - \tau_1]$ (mode A) by translation, where ρ_2 is the first exit time of a time-changed and translated version of x from $(-\delta - x(\tau_1), \infty)$, and so on. Now we give a rigorous formulation of the *switch problem with the noise y, the regulator F, and the gap δ*. To this end, we define recursively the time epochs $0 = \tau_0 \leq \rho_1 \leq \tau_1 \leq \rho_2 \leq \cdots \leq \infty$, the accompanying sequences

$$T_0^A := T_0^B := 0, \quad T_k^A := \sum_{i=0}^{k-1}(\rho_{i+1} - \tau_i), \quad T_k^B := \sum_{i=1}^{k}(\tau_i - \rho_i), \quad k \in \mathbb{N},$$

and the function x as follows. Put $\rho_1 := \inf\{t \geq 0 : y(t) \leq -\delta\}$ and then $x(t) = y(t)$ for $t \in [0, \rho_1]$; $\tau_1 := \inf\{t \geq \rho_1 : x(\rho_1) + F(t - \rho_1) \geq 0\}$ and then $x(t) = x(\rho_1) + F(t - \rho_1)$ for $t \in (\rho_1, \tau_1]$. Further, put, for $k \in \mathbb{N}$,

$$\rho_{k+1} := \inf\{t \geq \tau_k : x(\tau_k) + (y(T_k^A + t - \tau_k) - y(T_k^A)) \leq -\delta\},$$

$$x(t) := x(\tau_k) + (y(T_k^A + t - \tau_k) - y(T_k^A)), \quad t \in (\tau_k, \rho_{k+1}],$$

$$\tau_{k+1} := \inf\{t \geq \rho_{k+1} : x(\rho_{k+1}) + (F(T_k^B + t - \rho_{k+1}) - F(T_k^B)) \geq 0\},$$

$$x(t) := x(\rho_{k+1}) + (F(T_k^B + t - \rho_{k+1}) - F(T_k^B)), \quad t \in (\rho_{k+1}, \tau_{k+1}].$$

It is important that the sequence $\{\tau_0, \rho_1, \tau_1, \rho_2, \ldots\}$ does not have finite accumulation points because y and F are càdlàg and $\delta > 0$. Thus, the process of constructing x is well-defined. This construction naturally extends to the case where some τ_k or ρ_k are infinite. In what follows when writing ρ_k or τ_k we tacitly assume that these are finite.

We note that, for $k \in \mathbb{N}$,

$$x(\rho_k) - x(\rho_k-) = y(T_k^A) - y(T_k^A-) \quad \text{and} \quad x(\tau_k) - x(\tau_k-) = F(T_k^B) - F(T_k^B-).$$

This means that the jump at the time of transition to a new mode can only be caused by the previous mode.

Definition 3.2.1 We call x a *solution to the switch problem with the noise y, the regulator F, and the gap δ* and use the notation $x = \mathcal{G}_\delta(y, F)$.

Since "a picture is worth a thousand words," an example of $x = \mathcal{G}_\delta(y, F)$ is given in Fig. 3.1.

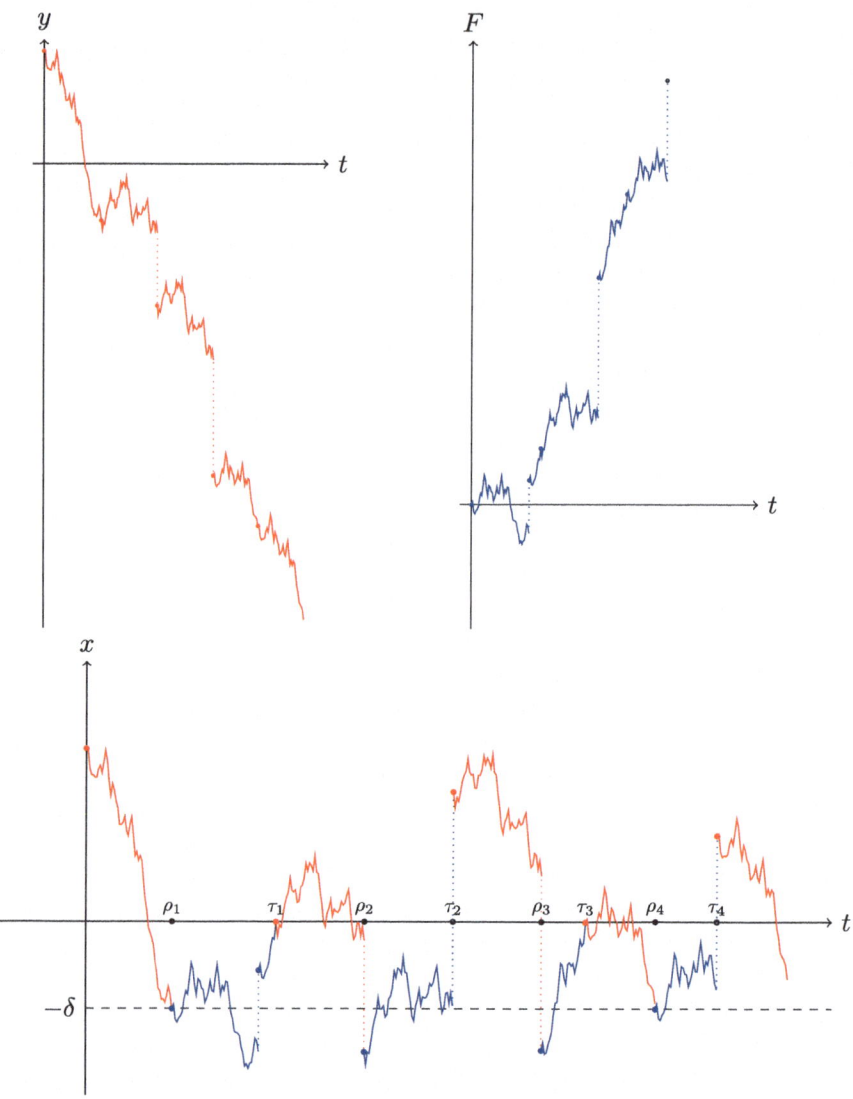

Fig. 3.1 A noise y (red, top left), a regulator F (blue, top right), and the solution $x = \mathcal{G}_\delta(y, F)$ (bottom) to the switch problem with the noise y and the regulator F. In order to recover the noise y (respectively, regulator F) from x one should merge the red (respectively, blue) parts keeping the jumps (dotted vertical lines) of the respective color. Observe that, by construction, the pieces of the trajectory of the solution coming from the noise (red parts) always lie above the horizontal line $x = -\delta$, whereas the pieces coming from the regulator (blue parts) always lie below the abscissa axis

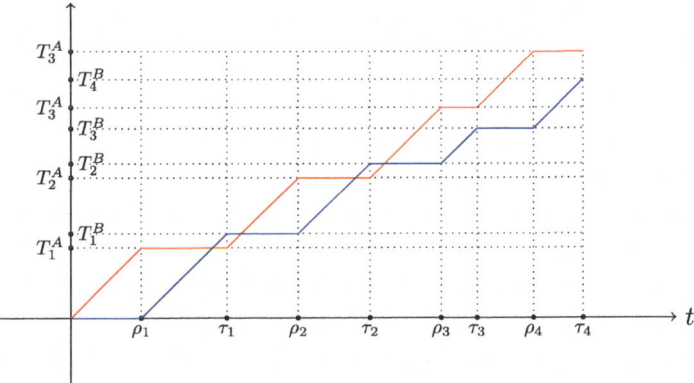

Fig. 3.2 The functions T_A (red) and T_B (blue) constructed from the functions y and F on Fig. 3.1

We say that the set $\cup_{k\geq 0}[\tau_k, \rho_{k+1})$ forms mode A and the set $\cup_{k\geq 1}[\rho_k, \tau_k)$ forms mode B. Define the functions T_A and T_B by

$$T_A(t) := \text{LEB}([0,t] \cap A) \quad \text{and} \quad T_B(t) := \text{LEB}([0,t] \cap B), \quad t \geq 0.$$

Alternatively, T_A (respectively, T_B) is the linear interpolation of the ordered set of points (τ_k, T_k^A), (ρ_{k+1}, T_{k+1}^A), $k \geq 0$ (respectively, (τ_k, T_k^B), (ρ_{k+1}, T_k^B), $k \geq 0$); see Fig. 3.2. The values $T_A(t)$ and $T_B(t)$ can be interpreted as the amount of time spent by y during the period $[0, t]$ in modes A and B, respectively. In particular, $T_k^A = T_A(\rho_k) = T_A(\tau_k)$ and $T_k^B = T_B(\tau_k) = T_B(\rho_{k+1})$ for $k \in \mathbb{N}$. The functions T_A and T_B are nondecreasing, Lipschitz continuous with the Lipschitz constant 1, and $T_A(t) + T_B(t) = t$ for $t \geq 0$. Using T_A and T_B yields a compact formula for x:

$$x(t) = y(T_A(t)) + F(T_B(t)), \quad t \geq 0. \tag{3.15}$$

The functions T_A and $y \circ T_A$, where $\circ$ denotes composition, take constant values on each interval $[\rho_k, \tau_k]$, and the functions T_B and $F \circ T_B$ do so on each interval $[\tau_k, \rho_{k+1}]$. Also, $x(\tau_k) \geq 0$, $x(t) \leq 0$ for all $t \in [\rho_k, \tau_k)$ and $x(\rho_k) \leq -\delta$, $x(t) \geq -\delta$ for all $t \in [\tau_{k-1}, \rho_k)$. Using induction in k in combination with these facts we infer

$$F(T_B(\tau_k)) = F(T_k^B) = \max_{s \in [0, T_k^B]} F(s) = \max_{t \in [0, \tau_k]} F(T_B(t)), \quad k \geq 0 \tag{3.16}$$

and

$$y(T_A(\rho_k)) = y(T_k^A) = \min_{s \in [0, T_k^A]} y(s) = \min_{t \in [0, \rho_k]} y(T_A(t)), \quad k \geq 1. \tag{3.17}$$

Recalling a familiar notation $m(t) = \sup_{s \in [0,t]} ((y(s))_-)$ for $t \geq 0$ and using that the terms in (3.17) are negative, we conclude also that

$$y(T_A(\rho_k)) = -m(T_A(\rho_k)), \quad k \geq 1. \tag{3.18}$$

We proceed by giving a key inequality linking $m(T_A)$ and $F(T_B)$. The inequality will be essentially used in the proof of the main result of this section.

Lemma 3.2.1 *For each $t \geq 0$,*

$$F((T_B(t))-) \leq m(T_A(t)) \leq F(T_B(t)) + \delta + \sup_{s \in [0, T_A(t)]} (y(s-) - y(s))$$

$$+ \sup_{0 \leq s_1 \leq s_2 \leq T_B(t)} (F(s_1) - F(s_2)). \tag{3.19}$$

Proof Recall that

$$A = \bigcup_{k \geq 0} [\tau_k, \rho_{k+1}), \quad B = \bigcup_{k \geq 1} [\rho_k, \tau_k).$$

Observe that $x(t) \leq 0$ for $t \in B$, whence $x(t-) \leq 0$ for $t \in (\rho_k, \tau_k]$.

We first prove the left-hand inequality in (3.19). If $t \in (\rho_k, \tau_k]$ for some $k \in \mathbb{N}$, then

$$F((T_B(t))-) = F(T_B(t-)) = x(t-) - y(T_A(t-)) \leq -y(T_A(t-))$$

$$\leq m(T_A(t-)) \leq m(T_A(t)),$$

where the second equality follows from (3.15), and the first inequality is a consequence of $x(t-) \leq 0$. If $t \in (\tau_k, \rho_{k+1}]$ for some $k \in \mathbb{N}_0$, then

$$F((T_B(t))-) = F((T_B(\tau_k))-) = F(T_B(\tau_k-)) = x(\tau_k-) - y(T_A(\tau_k-))$$

$$\leq -y(T_A(\tau_k-)) = -y(T_A(\rho_k)) \stackrel{(3.18)}{=} m(T_A(\rho_k)) \leq m(T_A(t)).$$

In order to prove the right-hand inequality in (3.19) we argue as follows. If $t \in [\rho_k, \tau_k)$ for some $k \in \mathbb{N}$, then

$$F(T_B(t)) = \big(F(T_B(\rho_k-)) + y(T_A(\rho_k-))\big) + \big(F(T_B(t)) - F(T_B(\rho_k-))\big)$$

$$+ \big(y(T_A(\rho_k)) - y(T_A(\rho_k-))\big) - y(T_A(\rho_k))$$

$$= x(\rho_k-) + \big(F(T_B(t)) - F(T_B(\rho_k-))\big)$$

$$+ \big(y(T_A(\rho_k)) - y(T_A(\rho_k-))\big) - y(T_A(\rho_k))$$

3.2 Functional Limit Theorems for Random Walks with Reflection

$$\geq -\delta - \sup_{0 \leq s_1 \leq s_2 \leq T_B(t)} (F(s_1) - F(s_2))$$

$$+ \big(y(T_A(\rho_k)) - y(T_A(\rho_k-))\big) - y(T_A(\rho_k))$$

$$\stackrel{(3.18)}{=} -\delta - \sup_{0 \leq s_1 \leq s_2 \leq T_B(t)} (F(s_1) - F(s_2))$$

$$+ \big(y(T_A(\rho_k)) - y(T_A(\rho_k-))\big) + m(T_A(\rho_k))$$

$$= -\delta - \sup_{0 \leq s_1 \leq s_2 \leq T_B(t)} (F(s_1) - F(s_2))$$

$$+ \big(y(T_A(\rho_k)) - y(T_A(\rho_k-))\big) + m(T_A(t))$$

$$\geq -\delta - \sup_{0 \leq s_1 \leq s_2 \leq T_B(t)} (F(s_1) - F(s_2))$$

$$- \sup_{s \in [0, T_A(t)]} (y(s-) - y(s)) + m(T_A(t)).$$

Assume now that $t \in [\tau_k, \rho_{k+1})$ for some $k \in \mathbb{N}_0$. Hence, for all $s \in [\tau_k, t]$,

$$-\delta \leq x(s) = F(T_B(s)) + y(T_A(s)) = F(T_B(\tau_k)) + y(T_A(s)).$$

Passing to minimum over $s \in [\tau_k, t]$ and using that

$$\min_{s \in [\tau_k, t]} y(T_A(s)) = \min_{s \in [\rho_k, t]} y(T_A(s)) \leq y(T_A(\rho_k)) \stackrel{(3.17)}{=} \min_{s \in [0, \rho_k]} y(T_A(s)),$$

we obtain

$$-\delta \leq F(T_B(\tau_k)) + \min_{s \in [\tau_k, t]} y(T_A(s)) \leq F(T_B(\tau_k)) + \min_{s \in [0, t]} y(T_A(s))$$

$$= F(T_B(\tau_k)) - m(T_A(t)). \quad (3.20)$$

Thus, (3.20) implies $m(T_A(t)) \leq F(T_B(\tau_k)) + \delta = F(T_B(t)) + \delta$ for $t \in [\tau_k, \rho_{k+1})$. This inequality entails the right-hand inequality in (3.19) because both suprema in (3.19) are nonnegative. The proof of Lemma 3.2.1 is complete. □

Remark 3.2.2 In general, it is not possible to define the switch problem with $\delta = 0$. Indeed, doing this formally may run into trouble with points t satisfying $x(t) = 0$. Even if x never hits 0, the sequence $\{\tau_0, \rho_1, \tau_1, \rho_2, \dots\}$ may have a finite accumulation point, which makes unclear how to define $x(t)$ for $t > \lim_{n \to \infty} \rho_n$. On the other hand, the switch problem with $\delta = 0$ can be defined in some particular cases. This is the case, for instance, if y and F are piecewise constant functions with finitely many jumps on each compact set. The modes are switched either upon entering $(-\infty, 0)$ and $[0, \infty)$ or upon entering

$(-\infty, 0]$ and $(0, \infty)$. In what follows, when writing $x = \mathcal{G}_0(y, F)$ we tacitly assume that the corresponding problem is well-defined in some sense. Of course, each such a model incorporates the aforementioned switching rules of alternating between open and semi-open halflines or vice versa. The switch models with the gap 0 will be essentially used in our proof of Theorem 3.2.3 in Sect. 3.2.3 and Theorem 3.2.4 in Sect. 3.2.4. We stress that all the results related to the switch models that have been presented so far, particularly Lemma 3.2.1, remain valid in the case $\delta = 0$ provided that the corresponding model is well-defined.

The following theorem is the main result of this section. To formulate it we need a definition. A point $t_0 > 0$ is called[4] *point of growth* of a nondecreasing function $f : [0, \infty) \to \mathbb{R}$ if $f(t) > f(t_0)$ for $t > t_0$ and $f(t) < f(t_0)$ for $t \in [0, t_0)$. Also, we recall that $\widetilde{\Gamma}$ denotes the generalized Skorokhod map introduced in Definition 2.1.5.

Theorem 3.2.2 *For $n \in \mathbb{N}_0$, let $y_n, F_n \in D([0, \infty))$ and, for $n \in \mathbb{N}$, $\delta_n \geq 0$ and $\varrho_n > 0$ satisfy*

(I) $\lim_{n \to \infty} y_n = y_0$ and $\lim_{n \to \infty} F_n = F_0$ on $D([0, \infty))$.
(II) $y_0(0) \geq 0$ and $F_n(0) = 0$ for $n \in \mathbb{N}_0$.
(III) $\lim_{n \to \infty} \delta_n = 0$.
(IV) $\lim_{n \to \infty} \varrho_n = \varrho_0$ for some $\varrho_0 \in [0, \infty]$.
(V) F_0 is strictly increasing and $\lim_{t \to \infty} F_0(t) = \infty$.
(VI) y_0 does not have negative jumps.
(VII) *If $u_0, t_0 > 0$ are such that $F_0(u_0) \neq F_0(u_0-) = m_0(t_0)$, then t_0 is a point of growth of m_0. Here, $m_0(t) = \sup_{s \in [0,t]} ((y_0(s))_-)$ for $t \geq 0$.*

For each $n \in \mathbb{N}$, define the function x_n by

$$x_n(t) := \mathcal{G}_{\delta_n}(y_n(t), F_n(t/\varrho_n)), \quad t \geq 0.$$

Then $\lim_{n \to \infty} x_n = x_0$ on $D([0, \infty))$. Depending on the value of ϱ_0 the limit x_0 is given as follows:

- *If $\varrho_0 = 0$ (generalized reflection), then $x_0 = \widetilde{\Gamma}(y_0, F_0)$.*
- *If $\varrho_0 \in (0, \infty)$ (delayed generalized reflection), then*

$$x_0(t) = \big(\widetilde{\Gamma}(y_0, F_0)\big)(C_{\varrho_0}^{-1}(t)), \quad C_{\varrho_0}(t) := t + \varrho_0 F_0^{\leftarrow}(m_0(t)), \quad t \geq 0,$$

where $C_{\varrho_0}^{-1}$ is the usual inverse function.

---

[4] We stress that the point of growth may be different from the point of increase s_0, which is defined by the requirement that $f(b) - f(a) > 0$ for any interval (a, b) containing s_0.

3.2 Functional Limit Theorems for Random Walks with Reflection

- If $\varrho_0 = \infty$ *(absorption at 0) and the equality*

$$\inf\{s \geq 0 : y_0(s) = 0\} = \inf\{s \geq 0 : y_0(s) < 0\} \tag{3.21}$$

holds, then $x_0(t) = \big(\widetilde{\Gamma}(y_0, F_0)\big)(t \wedge \sigma)$ *for* $t \geq 0$, *where* $\sigma := \inf\{s \geq 0 : y_0(s) = 0\}$.

Remark 3.2.3 A perusal of the proof reveals that assumption (VII) of the theorem is not used in the case $\varrho_0 = \infty$.

Remark 3.2.4 The function C_{ϱ_0} is continuous, strictly increasing with $C_{\varrho_0}(0) = 0$ and $\lim_{t \to \infty} C_{\varrho_0}(t) = \infty$. Hence, the (usual) inverse function $C_{\varrho_0}^{-1}$ is well-defined. The only nontrivial statement here is continuity, which we now justify. The function $F_0^{\leftarrow}$ is continuous because F_0 is strictly increasing; see Proposition 4.3.13(VII). The function m_0 is continuous because, according to assumption (VI) of Theorem 3.2.2, y_0 does not have negative jumps; see the proof of Theorem 2.1.2.

Remark 3.2.5 As has been discussed in Sect. 2.1.2 the function $F_0^{\leftarrow}(m_0)$ may only increase on the set $\{t \geq 0 : x_0(t) = 0\}$. Hence, the transformation $C_{\varrho_0}^{-1}$ "stretches" time, while x_0 is equal to 0 and acts as identity maps with shifts while x_0 is positive.

Remark 3.2.6 The limit function x_0 can be formally written as a composition $\big(\widetilde{\Gamma}(y_0, F_0)\big)(C_{\varrho_0}^{-1}(\cdot))$ in all the three cases because

$$\lim_{\varrho_0 \to 0+} C_{\varrho_0}^{-1}(t) = t \quad \text{and} \quad \lim_{\varrho_0 \to \infty} C_{\varrho_0}^{-1}(t) = t \wedge \sigma.$$

Corollary 3.2.1 *Let* $(\delta_n)_{n \geq 1}$ *and* $(\varrho_n)_{n \geq 1}$ *be deterministic sequences as in Theorem 3.2.2, and* $(y_n)_{n \geq 0}$ *and* $(F_n)_{n \geq 0}$ *sequences of càdlàg processes, whose paths a.s. satisfy assumptions (II)–(VII) of Theorem 3.2.2. Assume that* $(y_n, F_n) \Longrightarrow (y_0, F_0)$ *as* $n \to \infty$ *in the product topology on* $(D([0, \infty)))^2$. *Then* $x_n \Longrightarrow x_0$ *as* $n \to \infty$ *on* $D([0, \infty))$, *where* x_n *and* x_0 *are random analogues of those appearing in Theorem 3.2.2. Furthermore, if* y_0 *and* F_0 *are a.s. continuous, then*

$$\Big(y_n, y_n(T_{A_n}), x_n, F_n(\varrho_n^{-1} T_{B_n})\Big) \Longrightarrow \Big(y_0, y_0(C_{\varrho_0}^{-1}), x_0, m_0(C_{\varrho_0}^{-1})\Big), \quad n \to \infty \tag{3.22}$$

in the product topology on $(D([0, \infty)))^4$.

A sketch of a proof of Corollary 3.2.1 will be given after the proof of Theorem 3.2.2.

Proof of Theorem 3.2.2 For each $n \in \mathbb{N}$, associated with $(y_n, F_n(\cdot/\varrho_n), \delta_n)$ is the switch problem with the noise y_n, the regulator $F_n(\cdot/\varrho_n)$, and the gap δ_n. Denote by A_n and B_n the counterparts of sets A and B. For each $n \in \mathbb{N}$, the functions T_{A_n} and T_{B_n} are globally

Lipschitz continuous with the Lipschitz constant 1, and $T_{A_n}(0) = T_{B_n}(0) = 0$. As a consequence, the sequences $(T_{A_n})_{n\geq 1}$ and $(T_{B_n})_{n\geq 1}$ are relatively compact in $C([0,\infty))$ by the Arzela-Ascoli theorem. Let $(n_k)_{k\geq 1}$ be a sequence such that the locally uniform limits $\lim_{k\to\infty} T_{A_{n_k}}$ and $\lim_{k\to\infty} T_{B_{n_k}}$ exist. Denote these limits by T_{A_0} and T_{B_0}, respectively. Plainly, $T_{A_0}(t) + T_{B_0}(t) = t$ and $T_{A_0}(t), T_{B_0}(t) \in [0, t]$ for each $t \geq 0$. For $n \in \mathbb{N}_0$, put

$$m_n(t) := \sup_{s\in[0,t]} ((y_n(s))_-), \quad t \geq 0. \tag{3.23}$$

By Lemma 3.2.1,

$$F_{n_k}((\varrho_{n_k}^{-1}T_{B_{n_k}}(t))-) \leq m_{n_k}(T_{A_{n_k}}(t)) \leq F_{n_k}(\varrho_{n_k}^{-1}T_{B_{n_k}}(t)) + \delta_{n_k}$$
$$+ \sup_{s\in[0,T_{A_{n_k}}(t)]} (y_{n_k}(s-) - y_{n_k}(s)) + \sup_{0\leq s_1\leq s_2\leq T_{B_{n_k}}(t)} (F_{n_k}(\varrho_{n_k}^{-1}s_1) - F_{n_k}(\varrho_{n_k}^{-1}s_2)).$$
$$\tag{3.24}$$

In view of $\lim_{n\to\infty} y_n = y_0$ on $D([0,\infty))$ (assumption (I)), we conclude with the help of Proposition 4.3.11 that $\lim_{n\to\infty} m_n = m_0$ on $D([0,\infty))$. The function m_0 is continuous because y_0 does not have negative jumps (assumption (VI)); see the proof of Theorem 2.1.2. Since the function T_{A_0} is also continuous, so is the composition $m_0(T_{A_0})$. It follows from Propositions 4.3.7 and 4.3.4 that

$$\lim_{k\to\infty} m_{n_k}(T_{A_{n_k}}) = m_0(T_{A_0}) \tag{3.25}$$

locally uniformly.

Using $T_{A_{n_k}}(t) \in [0,t]$ we infer, for each $t > 0$,

$$0 \leq \sup_{s\in[0,T_{A_{n_k}}(t)]} (y_{n_k}(s-) - y_{n_k}(s)) \leq \sup_{s\in[0,t]} (y_{n_k}(s-) - y_{n_k}(s)). \tag{3.26}$$

The right-hand side converges to zero, as $k \to \infty$, for each $t > 0$ which is a point of continuity of y_0 by Proposition 4.3.12, since $\lim_{k\to\infty} y_{n_k} = y_0$ on $D([0,\infty))$ and y_0 does not have negative jumps. By monotonicity the limit relation holds for all $t > 0$.

Let $t > 0$ be a point of continuity of F_0. In view of $\lim_{k\to\infty} F_{n_k} = F_0$ on $D([0,\infty))$ (assumption (I)), for each $k \in \mathbb{N}$, there exists a strictly increasing continuous function λ_{n_k} on $[0,t]$ satisfying $\lambda_{n_k}(0) = 0$, $\lambda_{n_k}(t) = t$ and

$$\lim_{k\to\infty} \sup_{s\in[0,t]} |F_{n_k}(\lambda_{n_k}(s)) - F_0(s)| = 0.$$

3.2 Functional Limit Theorems for Random Walks with Reflection

Hence,

$$\sup_{0 \le s_1 \le s_2 \le t} (F_{n_k}(s_1) - F_{n_k}(s_2)) = \sup_{0 \le s_1 \le s_2 \le t} (F_{n_k}(\lambda_{n_k}(s_1)) - F_{n_k}(\lambda_{n_k}(s_2)))$$

$$\to \sup_{0 \le s_1 \le s_2 \le t} (F_0(s_1) - F_0(s_2)) = 0$$

as $k \to \infty$, where the limit relation is justified by the aforementioned uniform convergence and the equality to zero is guaranteed by monotonicity of F_0 (assumption (V)). By monotonicity of the left-hand side, this holds true for all $t > 0$ (rather than just for points t of continuity of F_0). Since $T_{B_{n_k}}(t) \in [0, t]$, we conclude that, for each $t > 0$,

$$0 \le \sup_{0 \le s_1 \le s_2 \le T_{B_{n_k}}(t)} (F_{n_k}(s_1) - F_{n_k}(s_2)) \le \sup_{0 \le s_1 \le s_2 \le t} (F_{n_k}(s_1) - F_{n_k}(s_2)) \to 0 \quad (3.27)$$

as $k \to \infty$.

Thus, by virtue of $\lim_{k \to \infty} \delta_{n_k} = 0$ (assumption (III)), (3.26) and (3.27), the limits of the last three terms on the right-hand side of (3.24) are equal to 0. Taking this into account we obtain with the help of (3.25) a limit form of (3.24)

$$\limsup_{k \to \infty} F_{n_k}((\varrho_{n_k}^{-1} T_{B_{n_k}}(t))-) \le m_0(T_{A_0}(t)) \le \liminf_{k \to \infty} F_{n_k}(\varrho_{n_k}^{-1} T_{B_{n_k}}(t)), \quad t \ge 0.$$
(3.28)

Formula (3.28) is a basis for what follows.

Passing to the main part of our proof we treat three cases separately.

CASE $\varrho_0 \in (0, \infty)$. Fix any $t > 0$. In view of $\lim_{k \to \infty} y_{n_k} = y_0$ on $D([0, \infty))$ and $\lim_{k \to \infty} T_{B_{n_k}}(t) = T_{B_0}(t)$ invoking Proposition 4.3.3 we conclude that each limit point of the sequence $(F_{n_k}((\varrho_{n_k}^{-1} T_{B_{n_k}}(t))-))_{k \ge 1}$ and each limit point of $(F_{n_k}(\varrho_{n_k}^{-1} T_{B_{n_k}}(t)))_{k \ge 1}$ belong to the set $\{F_0((\varrho_0^{-1} T_{B_0}(t))-), F_0(\varrho_0^{-1} T_{B_0}(t))\}$. Using this fact in combination with monotonicity of F_0 and passing to the limit (along further subsequences) in (3.28) we obtain

$$F_0((\varrho_0^{-1}(t - T_{A_0}(t)))-) = F_0((\varrho_0^{-1} T_{B_0}(t))-) \le m_0(T_{A_0}(t))$$

$$\le F_0(\varrho_0^{-1} T_{B_0}(t)) = F_0(\varrho_0^{-1}(t - T_{A_0}(t))).$$

Hence, $\varrho_0^{-1}(t - T_{A_0}(t)) = F_0^{\leftarrow}(m_0(T_{A_0}(t)))$ for $t \ge 0$ (we have only proved the equality for $t > 0$; if $t = 0$, then both sides are equal to 0). Equivalently,

$$T_{A_0}(t) + \varrho_0 F_0^{\leftarrow}(m_0(T_{A_0}(t))) = t, \quad t \ge 0, \quad (3.29)$$

whence $T_{A_0}(t) = C_{\varrho_0}^{-1}(t)$ for $t \ge 0$ because $C_{\varrho_0}(t) = t + \varrho_0 F_0^{\leftarrow}(m_0(t))$.

Since the right-hand side does not depend on the sequence $(n_k)_{k\geq 1}$, we conclude that, locally uniformly,

$$T_{A_n}(t) \to T_{A_0}(t), \quad T_{B_n}(t) = t - T_{A_n}(t) \to t - T_{A_0}(t) = T_{B_0}(t), \quad n \to \infty$$

and

$$\lim_{n\to\infty} m_n(T_{A_n}) = m_0(T_{A_0}). \tag{3.30}$$

It follows from (3.29) that

$$T_{B_0}(t) = t - T_{A_0}(t) = \varrho_0 F_0^{\leftarrow}(m_0(T_{A_0}(t))) = \varrho_0 F_0^{\leftarrow}(m_0(C_{\varrho_0}^{-1}(t))), \quad t \geq 0.$$

Recalling that, for each $n \in \mathbb{N}$,

$$x_n(t) = y_n(T_{A_n}(t)) + F_n(\varrho_n^{-1} T_{B_n}(t)), \quad t \geq 0, \tag{3.31}$$

we are going to prove the following limit relations on $D([0, \infty))$:

$$\lim_{n\to\infty} y_n(T_{A_n}) = y_0(T_{A_0}) = y_0(C_{\varrho_0}^{-1}), \quad n \to \infty, \tag{3.32}$$

$$\lim_{n\to\infty} F_n(\varrho_n^{-1} T_{B_n}) = F_0(\varrho_0^{-1} T_{B_0}) = F_0(F_0^{\leftarrow}(m_0(C_{\varrho_0}^{-1}))), \quad n \to \infty \tag{3.33}$$

and

$$\lim_{n\to\infty} \left(y_n(T_{A_n}(t)) + F_n(\varrho_n^{-1} T_{B_n}(t)) \right) = y_0(C_{\varrho_0}^{-1}) + F_0(F_0^{\leftarrow}(m_0(C_{\varrho_0}^{-1})))$$
$$= \left(y_0 + F_0(F_0^{\leftarrow}(m_0)) \right) \circ C_{\varrho_0}^{-1} = \widetilde{\Gamma}(y_0, F_0) \circ C_{\varrho_0}^{-1}, \quad n \to \infty. \tag{3.34}$$

Since $T_{A_0} = C_{\rho_0}^{-1}$ is continuous and strictly increasing, formula (3.32) is secured by Proposition 4.3.7. Now we prove (3.33). To this end, we intend to apply Proposition 4.3.3. Fix any $t_0 \geq 0$ and let $(t_n)_{n\geq 1}$ be a sequence of positive numbers such that $\lim_{n\to\infty} t_n = t_0$ and the limit $\lim_{n\to\infty} F_n(\varrho_n^{-1} T_{B_n}(t_n))$ exists (we can assume the latter without loss of generality passing if needed to subsequences). The convergence $v_n := \varrho_n^{-1} T_{B_n}(t_n) \to \varrho^{-1} T_{B_0}(t_0) =: v_0$ holds true because $\lim_{n\to\infty} T_{B_n} = T_{B_0}$ locally uniformly. Hence, by Proposition 4.3.3 with $f_n = F_n$ and $u_n = v_n$ for $n \in \mathbb{N}_0$, the assumed convergence $\lim_{n\to\infty} F_n = F_0$ on $D([0, \infty))$ ensures that

$$\lim_{n\to\infty} F_n(\varrho_n^{-1} T_{B_n}(t_n))) \in \{F_0(u_0-), F_0(u_0)\} = \{F_0((\varrho_0^{-1} T_{B_0}(t_0))-), F_0(\varrho_0^{-1} T_{B_0}(t_0))\}$$
$$= \{F_0((F_0^{\leftarrow}(m_0(C_{\varrho_0}^{-1}(t_0))))-), F_0(F_0^{\leftarrow}(m_0(C_{\varrho_0}^{-1}(t_0))))\}. \tag{3.35}$$

3.2 Functional Limit Theorems for Random Walks with Reflection

From now on we apply Proposition 4.3.3 with $f_n = F_n(\varrho_n^{-1} T_{B_n})$ and $u_n = t_n$ for $n \in \mathbb{N}_0$. In particular, condition (I) of that proposition reads

$$\lim_{n \to \infty} F_n(\varrho_n^{-1} T_{B_n}(t_n))) \in \{F_0(\varrho_0^{-1} T_{B_0}(t_0-)), F_0(\varrho_0^{-1} T_{B_0}(t_0))\}. \tag{3.36}$$

We stress that this inclusion is slightly different from (3.35) because T_{B_0} is nondecreasing but not strictly increasing. As a consequence, the equality

$$F_0((\varrho_0^{-1} T_{B_0}(t_0))-) = F_0(\varrho_0^{-1} T_{B_0}(t_0-)) \tag{3.37}$$

is not granted and does require a proof.

Proof of (3.36) If $v_0 = \varrho_0^{-1} T_{B_0}(t_0)$ is a point of continuity of F_0, then (3.36) trivially holds. Assume now that v_0 is a jump point of F_0. Put $\alpha := F_0(v_0-) < F_0(v_0) =: \beta$. By (3.35), the limit $\lim_{n \to \infty} F_n(\varrho_n^{-1} T_{B_n}(t_n)))$ is equal to either α or β. Observe that

$$F_0^{\leftarrow}(z) \begin{cases} < v_0, & \text{if } z < \alpha, \\ = v_0, & \text{if } z \in [\alpha, \beta], \\ > v_0, & \text{if } z > \beta. \end{cases} \tag{3.38}$$

This entails $m_0(C_{\varrho_0}^{-1}(t_0)) \in [\alpha, \beta]$ because $v_0 = \varrho_0^{-1} T_{B_0}(t_0) = F_0^{\leftarrow}(m_0(C_{\varrho_0}^{-1}(t_0)))$. It follows from (3.28) with $n_k = n$ that

$$\lim_{n \to \infty} F_n(\varrho_n^{-1} T_{B_n}(t_n))) \geq m_0(T_{A_0}(t_0)) = m_0(C_{\varrho_0}^{-1}(t_0))$$

having utilized (3.30) and $T_{A_0} = C_{\varrho_0}^{-1}$. If $m_0(C_{\varrho_0}^{-1}(t_0)) \in (\alpha, \beta]$, then the limit $\lim_{n \to \infty} F_n(\varrho_n^{-1} T_{B_n}(t_n)))$, being larger than α, has to be equal to $\beta = F_0(v_0)$. Thus, (3.36) holds true in this case. Left with the situation where $m_0(C_{\varrho_0}^{-1}(t_0)) = \alpha = F_0(v_0-)$ we recall that the function $C_{\varrho_0}^{-1}$ is strictly increasing and note that $C_{\varrho_0}^{-1}(t_0)$ is a point of growth of m_0 (assumption (VII)). Thus, $m_0(C_{\varrho_0}^{-1}(t)) < \alpha$ and $F_0^{\leftarrow}(m_0(C_{\varrho_0}^{-1}(t))) < v_0$ for $t < t_0$, which proves (3.37).

The conditions (II) and (III) of Proposition 4.3.3 can be checked along similar lines, and we omit details. The proof of relation (3.33) is complete.

According to Proposition 4.3.6, (3.34) holds true if we can show that $y_0(C_{\rho_0}^{-1})$ and $F_0(F_0^{\leftarrow}(m_0(C_{\rho_0}^{-1})))$ do not have common points of discontinuity, or equivalently that neither do y_0 and $F_0(F_0^{\leftarrow}(m_0))$. Assume that $F_0(F_0^{\leftarrow}(m_0))$ is discontinuous at some $t_0 > 0$. Then F_0 has a positive jump at point $w_0 := F_0^{\leftarrow}(m_0(t_0))$. Since m_0 is a nondecreasing continuous function, it can be checked with the help of (3.38), in which we replace v_0 by w_0, that $m_0(t_0) = F_0(u_0-)$. Hence, according to assumption (VII), t_0 is a point of growth of m_0. Then $(y_0)_-$ does not have a negative jump at t_0 and, as such, is continuous at t_0, for $(y_0)_-$ does not have positive jumps (assumption (VI)). Thus, we have proved that y_0 is continuous at t_0.

The proof in the case $\varrho_0 \in (0, \infty)$ is complete.

CASE $\varrho_0 = 0$. Recall that $\lim_{k \to \infty} T_{B_{n_k}} = T_{B_0}$ locally uniformly. Assumption (V) yields $\lim_{s \to +\infty} F_0(s) = \infty$. Therefore, the left-hand inequality in (3.28) entails $T_{B_0}(t) = 0$ for all $t \geq 0$ and also boundedness of the sequence $(\varrho_{n_k}^{-1} T_{B_{n_k}}(t))_{k \geq 1}$ for each fixed $t \geq 0$. Since T_{B_0} does not depend on $(n_k)_{k \geq 1}$ we conclude that

$$\lim_{n \to \infty} T_{B_n} = T_{B_0} \quad \text{and} \quad \lim_{n \to \infty} T_{A_n} = T_{A_0}$$

locally uniformly, and $T_{A_0}(t) = t$ for $t \geq 0$. Thus, (3.28) reads

$$\limsup_{n \to \infty} F_n((\widetilde{T}_{B_n}(t))-) \leq m_0(t) \leq \liminf_{n \to \infty} F_n(\widetilde{T}_{B_n}(t)), \quad t \geq 0,$$

where $\widetilde{T}_{B_n}(t) := \varrho_n^{-1} T_{B_n}(t)$ for $n \in \mathbb{N}$ and $t \geq 0$. Fix $t \geq 0$. The convergence $\lim_{n \to \infty} F_n = F_0$, strict monotonicity of F_0, and boundedness of the sequence $((\widetilde{T}_{B_n}(t))-)_{n \geq 1}$ in combination with Proposition 4.3.3 imply that each limit point $\widetilde{t}$ of $(\widetilde{T}_{B_n}(t))_{n \geq 1}$ satisfies

$$F_0(\widetilde{t}-) \leq m_0(t) \leq F_0(\widetilde{t}).$$

Therefore, $\widetilde{t} = F_0^{\leftarrow}(m_0(t))$. Since the right-hand side does not depend on a subsequence along which the limit point $\widetilde{t}$ has been obtained, we infer

$$\lim_{n \to \infty} \widetilde{T}_{B_n}(t) = F_0^{\leftarrow}(m_0(t))$$

for all $t \geq 0$, Furthermore, the convergence is locally uniform as a pointwise convergence of a sequence of nondecreasing functions to a continuous limit.[5]

Since, for each $n \in \mathbb{N}$,

$$x_n(t) = y_n(T_{A_n}(t)) + F_n(\widetilde{T}_{B_n}(t)), \quad t \geq 0,$$

with T_{A_n} and $\widetilde{T}_{B_n}$ converging locally uniformly to the limits specified above, the remainder of the proof is similar to the corresponding part of the proof in the case $\varrho_0 \in (0, \infty)$. Namely, counterparts of formulae (3.32), (3.33), and (3.34) have to be obtained. We omit further details.

---

[5] This statement can be found, for instance, on p. 113 in [16]. It is known in the literature as Polya's theorem or the second Dini's theorem or Polya's companion to Dini's theorem.

3.2 Functional Limit Theorems for Random Walks with Reflection

CASE $\varrho_0 = \infty$. Since $m_0(t \wedge \sigma) = 0$, we conclude that $\big(\widetilde{\Gamma}(y_0, F_0)\big)(t \wedge \sigma) = y_0(t \wedge \sigma)$ for $t \geq 0$; see Definition 2.1.5.

In view of $T_{B_n}(t) \in [0, t]$, we infer $\lim_{n \to \infty} \varrho_n^{-1} T_{B_n} = \Theta$ locally uniformly, where $\Theta(t) := 0$ for $t \geq 0$. By Proposition 4.3.9,

$$\lim_{n \to \infty} F_n(\varrho_n^{-1} T_{B_n}) = F_0(\Theta) = \Theta \tag{3.39}$$

on $D([0, \infty))$. By Proposition 4.3.4, the latter convergence is locally uniform because the limit function is continuous. In view of representation (3.31) and Proposition 4.3.6 we are left with showing that

$$\lim_{n \to \infty} y_n(T_{A_n}(t)) = y_0(t \wedge \sigma) \tag{3.40}$$

on $D([0, \infty))$.

As a preparation for the proof of (3.40), we now check that

$$\lim_{n \to \infty} T_{A_n} = T_{A_0} \tag{3.41}$$

locally uniformly, where $T_{A_0}(t) = t \wedge \sigma$ for $t \geq 0$. Since the pointwise convergence of nondecreasing functions to a continuous limit is locally uniform, it is enough to prove the pointwise convergence in (3.41). Plainly, $T_{A_n}(t) = t$ for each $t \in [0, \sigma)$ and sufficiently large n. On the other hand, by (3.28) and (3.39), $m_0(T_{A_0}(t)) = 0$ for each $t \geq 0$ (here, $T_{A_0} = \lim_{k \to \infty} T_{A_{n_k}}$). A combination of (3.21) and

$$\inf\{s \geq 0 : y_0(s) < 0\} = \inf\{s \geq 0 : m_0(s) > 0\}$$

enables us to conclude that $\sigma = \inf\{s \geq 0 : m_0(s) > 0\}$, whence $T_{A_0}(t) \leq \sigma$ for each $t \geq 0$. Since the pointwise limit of nondecreasing functions is nondecreasing, we infer $T_{A_0}(t) = t \wedge \sigma$. The limit does not depend on $(n_k)_{k \geq 1}$, and (3.41) follows.

We are ready to prove (3.40). To this end, we use Proposition 4.3.9 having $\lim_{n \to \infty} y_n = y_0$ on $D([0, \infty))$ and (3.41) as a starting point. In the notation of Proposition 4.3.9, $f_n = y_n$ and $g_n = T_{A_n}$ for $n \in \mathbb{N}_0$ and particularly $g_0(t) = t \wedge \sigma$ for $t \geq 0$. If some $t \in [0, \sigma)$ is a point of discontinuity of y_0, then $u \wedge \sigma \neq t$ for $u \neq t$. Equality (3.21) ensures that y_0 does not have a positive jump at σ. Thus, y_0 is continuous at σ, for it does not have negative jumps (assumption (VI)). With these at hand, (3.40) is secured by Proposition 4.3.9.

The proof of Theorem 3.2.2 is complete. □

Sketch of Proof of Corollary 3.2.1 The relation $x_n \Longrightarrow x_0$ as $n \to \infty$ follows from Theorem 3.2.2 and the Skorokhod representation theorem (Theorem 4.1.2). Relation (3.22) can be proved similarly. The assumption that y_0 and F_0 are a.s. continuous simplifies the argument significantly making some quite technical passages in the proof of Theorem 3.2.2 almost trivial. □

We proceed by giving a couple of applications of Theorem 3.2.2 and Corollary 3.2.1.

Example 3.2.1 Let $a > 0$ and $(\delta_n)_{n \geq 1}$ be a sequence of positive numbers which converges to 0 as $n \to \infty$. For each $n \in \mathbb{N}$, consider a continuous process x_n which moves like a Brownian motion before hitting $-\delta_n$, then switches a regime and moves up with the constant speed a until hitting 0, then switches a regime and moves again, independently of the past, like a Brownian motion until hitting $-\delta_n$ and so on. By the strong Markov property of a Brownian motion, the distribution of x_n coincides with the distribution of $\mathcal{G}_{\delta_n}(y_n, F_n(\cdot/\varrho_n))$, where, for each $n \in \mathbb{N}$, $\varrho_n = 1$, $F_n(t) = at$ for $t \geq 0$, $y_n = W$, and W is a Brownian motion.

Now we discuss the limit processes. Plainly,

$$y_0 = W, \quad F_0(t) = at, \quad F_0(F_0^{-1}(t)) = t, \quad t \geq 0,$$

and $\widetilde{\Gamma}(y_0, F_0) = \Gamma(y_0) = \Gamma(W) = W + m_0$ is the reflected Brownian motion W^{refl}; see Sect. 2.1.1. Moreover, the process m_0 defined by

$$m_0(t) = \max_{s \in [0,t]} ((W(s))_-), \quad t \geq 0$$

is the local time of W^{refl} at 0. Also,

$$C_1(t) = t + F_0^{-1}(m_0(t)) = t + a^{-1} m_0(t), \quad t \geq 0.$$

The process $W^{\text{refl}}(C_1^{-1})$ is called a *sticky reflected Brownian motion*.

Let $(a_n)_{n \geq 1}$ be a sequence of positive numbers. Observe that a solution to the switch problem with $\varrho_n = 1$ and $F_n(t) = a_n t$ coincides with a solution to the switch problem with $\varrho_n = a_n^{-1}$ and $F_n(t) = t$. If $\lim_{n \to \infty} a_n = +\infty$, that is, the regime below 0 (mode B), pushes up strongly, then, for any sequence of gaps $(\delta_n)_{n \geq 1}$ satisfying $\lim_{n \to \infty} \delta_n = 0$, the limit process is a (usual) reflected Brownian motion without any delay.

Example 3.2.2 Let $(\delta_n)_{n \geq 1}$ and $(\varrho_n)_{n \geq 1}$ be sequences of nonnegative and positive numbers, respectively, and $(y_n)_{n \geq 0}$ and $(F_n)_{n \geq 0}$ two sequences of stochastic processes satisfying the following conditions:

(A) $(y_n, F_n) \Longrightarrow (y_0, F_0)$ as $n \to \infty$ in the product topology on $(D([0, \infty)))^2$.
(B) $\lim_{n \to \infty} \delta_n = 0$ and $\lim_{n \to \infty} \varrho_n = \varrho_0$ for some $\varrho_0 \in [0, \infty]$.
(C) F_0 is a strictly increasing subordinator.
(D) y_0 is a spectrally positive Lévy process (without negative jumps).
(E) y_0 and F_0 are independent.

3.2 Functional Limit Theorems for Random Walks with Reflection 105

Assume that y_0 is not a subordinator. Then the process $T_0 := m_0^{-1}$ is a (possibly killed) subordinator; see Theorem 1 on p. 18 in [10]. A point t is not a point of growth of m_0 if, and only if, $m_0(t)$ is a point of jump of T_0. It is known that the subordinator T_0 is a.s. continuous at each fixed (nonrandom) point of $[0, \infty)$. Since the processes F_0 and T_0 are independent, and the set of jumps of F_0 is at most countable, assumption (VII) of Theorem 3.2.2 holds a.s. If y_0 is a subordinator, then $m_0(t) = 0$ for $t \geq 0$, and assumption (VII) holds trivially. Thus, by a standard application of the Skorokhod representation theorem (Theorem 4.1.2),

$$x_n := \mathcal{G}_{\delta_n}(y_n, F_n(\cdot/\varrho_n)) \implies \left(\widetilde{\Gamma}(y_0, F_0)\right)(C_{\varrho_0}^{-1}), \quad n \to \infty \tag{3.42}$$

on $D([0, \infty))$.

As a concrete example, consider the following sequence of storage processes (a.k.a. Moran's model of a dam, or a waiting-time process in the $M/G/1$ queue with delay at 0); see [63]. Assume that, for each $n \in \mathbb{N}$, a continuous-time Markov process $(x_n(t))_{t \geq 0}$ evolves as follows:

(1) If $x_n(t) > 0$, then x_n has a negative drift of rate r_n and jumps up according to a Poisson clock of intensity λ_n, with $\lim_{n \to \infty} \lambda_n = \infty$.
(2) If $x_n(t) = 0$, then x_n stays at 0 for a random period of time having the exponential distribution of mean $1/\hat{\lambda}_n$ and then jumps up.
(3) The distribution of each jump up of x_n is equal to the distribution of a positive random variable $\xi^{(n)}$. All jumps are independent and independent of the durations of intervals between the jumps.
(4) The following relations hold:

$$\lim_{n \to \infty} (\lambda_n \mathbb{E}\xi^{(n)} - r_n) = \mu \in \mathbb{R}, \quad \lim_{n \to \infty} \lambda_n \mathbb{E}(\xi^{(n)})^2 = \sigma^2 \in (0, \infty),$$

$$\lim_{n \to \infty} \lambda_n \mathbb{E}(\xi^{(n)})^2 \mathbb{1}_{\{\xi^{(n)} > \varepsilon\}} = 0 \quad \text{for all } \varepsilon > 0.$$

The last two limit relations entail $\lim_{n \to \infty} \lambda_n \mathbb{E}\xi^{(n)} = \infty$. This in combination with the first limit relation in part (4) ensures that

$$\lim_{n \to \infty} r_n = \infty, \quad \text{and} \quad \lim_{n \to \infty} \frac{\lambda_n \mathbb{E}\xi^{(n)}}{r_n} = 1. \tag{3.43}$$

For each $n \in \mathbb{N}$, let $(N_{\lambda_n}(t))_{t \geq 0}$ and $\left(\widehat{N}_{\hat{\lambda}_n}(t)\right)_{t \geq 0}$ be Poisson processes with intensities λ_n and $\hat{\lambda}_n$, respectively. Further, let $(\xi_k^{(n)})_{k \geq 1}$ and $(\widehat{\xi}_k^{(n)})_{k \geq 1}$ be two sequences of independent random variables having the same distribution as $\xi^{(n)}$. Assume that all these processes and sequences are jointly independent. For each $n \in \mathbb{N}$, put

$$y_n(t) := \sum_{k=1}^{N_{\lambda_n}(t)} \xi_k^{(n)} - r_n t, \quad \widehat{F}_n(t) := \sum_{k=1}^{\widehat{N}_{\hat{\lambda}_n}(t)} \widehat{\xi}_k^{(n)}, \quad t \geq 0$$

and observe that $x_n \stackrel{d}{=} \mathcal{G}_0(y_n, \widehat{F}_n)$ (see Remark 3.2.2 for a comment on the switch problems with the gap 0).

It is known that, under the present assumptions,

$$(y_n(t))_{t \geq 0} \Longrightarrow (\mu t + \sigma W(t))_{t \geq 0}, \quad n \to \infty$$

on $D([0, \infty))$, where W is a standard Brownian motion. For $n \in \mathbb{N}$, put $\varrho_n := \lambda_n / (\hat{\lambda}_n r_n)$. We intend to prove that

$$(F_n(t))_{t \geq 0} := \left(\widehat{F}_n(\varrho_n t) \right)_{t \geq 0} = \left(\sum_{k=1}^{\widehat{N}_{\hat{\lambda}_n}(\varrho_n t)} \widehat{\xi}_k^{(n)} \right)_{t \geq 0} \Longrightarrow I, \quad n \to \infty \quad (3.44)$$

on $D([0, \infty))$, where $I(t) := t$ for $t \geq 0$. Indeed, for each $n \in \mathbb{N}$, the process $(F_n(t))_{t \geq 0}$ is a.s. nondecreasing. Hence, it is sufficient to show that, for each fixed $t \geq 0$, $F_n(t) \stackrel{\mathbb{P}}{\to} t$ as $n \to \infty$. For each $n \in \mathbb{N}$, $(\widehat{N}_{\hat{\lambda}_n}(\varrho_n t))_{t \geq 0}$ is a Poisson process with intensity λ_n / r_n. Invoking $\lim_{n \to \infty} \lambda_n \mathbb{E}(\xi^{(n)})^2 = \sigma^2 \in (0, \infty)$ and (3.43) we infer

$$\mathbb{E} \sum_{k=1}^{\widehat{N}_{\hat{\lambda}_n}(\varrho_n t)} \widehat{\xi}_k^{(n)} = \frac{\lambda_n t}{r_n} \mathbb{E} \xi^{(n)} \to t, \quad n \to \infty,$$

$$\mathrm{Var} \left(\sum_{k=1}^{\widehat{N}_{\hat{\lambda}_n}(\varrho_n t)} \widehat{\xi}_k^{(n)} \right) = \frac{\lambda_n t}{r_n} \mathbb{E}(\xi^{(n)})^2 \to 0, \quad n \to \infty.$$

The convergence in probability is secured by these limit relations in combination with Chebyshev's inequality. This completes the proof of (3.44).

Assume that there exists a limit

$$\lim_{n \to \infty} \varrho_n = \varrho_0 \in [0, \infty].$$

3.2 Functional Limit Theorems for Random Walks with Reflection

Using $\widehat{F}_n(t) = F_n(t/\varrho_n)$ and (3.42) we conclude the following:

- If $\varrho_0 = 0$, then the weak limit of x_n is the Skorokhod reflection of $(\mu t + \sigma W(t))_{t \geq 0}$ at 0.
- If $\varrho_0 \in (0, \infty)$, then the weak limit of x_n is a sticky reflection of $(\mu t + \sigma W(t))_{t \geq 0}$ at 0.
- If $\varrho_0 = \infty$, then the weak limit of x_n is $(\mu t + \sigma W(t))_{t \geq 0}$ stopped at 0.

3.2.3 Perturbations with Regularly Varying Distribution Tails

Apart from MODELS LINDREFILL, OSCRW, and REPULS introduced in Sect. 3.1.2, we investigate in this section another model that we now define. As in the definitions of the models in Sect. 3.1.2, let $\xi_1, \xi_2, \ldots$ be independent copies of a real-valued random variable ξ, $\eta_1, \eta_2, \ldots$ independent copies of a real-valued random variable η, and $X(0)$ a nonnegative random variable, the two collections and $X(0)$ being mutually independent. Throughout the section we assume that S_ξ, S_η and other standard random walks start at 0.

MODEL REGEN (REGENERATIVE MODEL). Assuming that $\mathbb{E}\eta \in (0, \infty]$, put

$$X(n+1) := \begin{cases} X(n) + \xi_{n+1}, & \text{if } X(n) > 0, \\ \eta_{n+1}, & \text{if } X(n) \leq 0 \end{cases}$$

for $n \in \mathbb{N}_0$. A random sequence $(Z(n))_{n \geq 1}$ is called *regenerative* with epochs of regeneration $\theta_1, \theta_2, \ldots$, if the fragments $(Z(n))_{n \in [1, \theta_1)}$, $(Z(n))_{n \in [\theta_1, \theta_2)}$, $\ldots$ are independent and identically distributed. If $X(0) = 0$ a.s., then $(X(n))_{n \geq 1}$ is regenerative with epochs of regeneration given by the successive times j satisfying both $X(j-2) > 0$ and $X(j-1) \leq 0$. If $X(0) > 0$ a.s., then the sequence $(X(n))_{n \geq \theta_1}$ is regenerative, where $\theta_1 := \inf\{j \geq 1 : X(j-1) \leq 0\}$.

If the variables $X(0)$, ξ and η are integer-valued, then the random walk with reflection defined by MODEL REGEN is a random walk with membrane $A = -\mathbb{N}_0$ in the sense of Definition 3.1.1. The corresponding $\eta^{(x)}$ are given by $\eta^{(x)} = \eta - x$ for $x \in -\mathbb{N}_0$. In the general case where the random variables are real-valued the random walk with reflection given by MODEL REGEN can be thought of as a random walk with membrane $A = (-\infty, 0]$.

As in Sect. 3.2.1, for each $n \in \mathbb{N}$, let $(X^{(n)}(k))_{k \geq 0}$ be a Markov chain having the same transition probabilities as $(X(k))_{k \geq 0}$, with $X^{(n)}(0)$ being independent of both $(\xi_i)_{i \geq 1}$ and $(\eta_j)_{j \geq 1}$.

Theorem 3.2.3 *Assume that, for each $n \in \mathbb{N}$, $(X^{(n)}(k))_{k \geq 0}$ is given by one of the models: LINDREFILL, OSCRW, REPULS, or REGEN. Let $x \geq 0$ and assume that $\mathbb{E}\xi = 0$, $\text{Var}\,\xi = \sigma^2 \in (0, \infty)$, the distribution of η belongs to the domain of attraction of a spectrally* positive strictly β-stable distribution[6] with $\beta \in (0, 1)$, and that $X^{(n)}(0)/(\sigma\sqrt{n})$ converges in probability to x as $n \to \infty$. Then

$$\left(\frac{X^{(n)}(\lfloor nt \rfloor)}{\sigma\sqrt{n}}\right)_{t\geq 0} \Longrightarrow (W_x^{\text{refl}, \beta}(t))_{t\geq 0}$$

$$:= (x + W(t) + \mathcal{V}_\beta(\mathcal{V}_\beta^{\leftarrow}(\max_{s\in[0,t]}((x + W(s))_-)))_{t\geq 0}, \quad n \to \infty$$

on $D([0, \infty))$. Here, $\mathcal{V}_\beta$ is a drift-free β-stable subordinator (with no killing), which is independent of a standard Brownian motion W, and $\mathcal{V}_\beta^{\leftarrow}$ is its generalized inverse given by

$$\mathcal{V}_\beta^{\leftarrow}(s) := \inf\{t \geq 0 : \mathcal{V}_\beta(t) > s\}, \quad s \geq 0.$$

Remark 3.2.7 The distribution of $\mathcal{V}_\beta$ depends on the parameter β and the other parameter as is seen from the formula $\mathbb{E}\exp(-s\mathcal{V}_\beta(t)) = \exp(-cts^\beta)$ for $t, s \geq 0$ and some $c > 0$. We do not need to specify in Theorem 3.2.3 a particular value of c, for the distribution of $\mathcal{V}_\beta \circ \mathcal{V}_\beta^{\leftarrow}$ does not depend on c, and neither does the distribution of $W_x^{\text{refl}, \beta}$. To justify the first claim, put $\mathcal{V}_{\beta,a}(t) = \mathcal{V}_\beta(at)$ for $t \geq 0$ and any fixed $a > 0$. Then $\mathcal{V}_{\beta,a} \circ \mathcal{V}_{\beta,a}^{\leftarrow} = \mathcal{V}_\beta \circ \mathcal{V}_\beta^{\leftarrow}$ (for all ω), and the claim follows. We also note in passing, see Example 7 in [40], that

$$\mathbb{P}\{\mathcal{V}_\beta \circ \mathcal{V}_\beta^{\leftarrow}(t) \in dy\} = \frac{t^\beta \sin(\pi\beta)}{\pi} \frac{\mathbb{1}_{(t,\infty)}(y)}{(y-t)^\beta y} dy, \quad y > 0. \tag{3.45}$$

This formula provides another justification of the claim because its right-hand side does not depend on c.

Proof of Theorem 3.2.3 We are going to apply Theorem 3.2.2 with y_n defined by $y_n(t) = (X^{(n)}(0) + S_\xi(\lfloor nt \rfloor))/(\sigma\sqrt{n})$ for $n \in \mathbb{N}$ and $t \geq 0$ and $y_0 = x + W$. The distributional convergence $y_n \Longrightarrow y_0$ as $n \to \infty$ on $D([0, \infty))$ is secured by Donsker's theorem and the assumed convergence in probability of $X^{(n)}(0)/(\sigma\sqrt{n})$ to x.

Since the distribution of η belongs to the domain of attraction of a spectrally positive strictly β-stable distribution, there exists a function c regularly varying at infinity of index $1/\beta$ such that

$$\left(\frac{S_\eta(\lfloor nt \rfloor)}{c(n)}\right)_{t\geq 0} \Longrightarrow (\mathcal{V}_\beta(t))_{t\geq 0}, \quad n \to \infty \tag{3.46}$$

---

[6] In the notation of Sect. 4.4 the corresponding stable distribution is $U_\beta^{\mathfrak{s},1,0}$ for some $\mathfrak{s} > 0$. We stress that a random variable with distribution $U_\beta^{\mathfrak{s},1,0}$ is a.s. positive.

3.2 Functional Limit Theorems for Random Walks with Reflection

on $D([0, \infty))$; see Remark 4.4.1 and Theorem 4.4.2. Equivalently, there exists a function b regularly varying at infinity of index $\beta/2$ such that

$$\left(\frac{S_\eta(\lfloor b(n)t \rfloor)}{\sigma\sqrt{n}}\right)_{t \geq 0} \Longrightarrow (\mathcal{V}_\beta(t))_{t \geq 0}, \quad n \to \infty \tag{3.47}$$

on $D([0, \infty))$.

We first treat MODEL OSCRW. For each $n \in \mathbb{N}$, the corresponding $X^{(n)}$ satisfies

$$\frac{X^{(n)}(\lfloor n \cdot \rfloor)}{\sigma\sqrt{n}} \stackrel{d}{=} \mathcal{G}_0\left(y_n(\cdot), \frac{S_\eta(\lfloor n \cdot \rfloor)}{\sigma\sqrt{n}}\right) = \mathcal{G}_0\left(y_n(\cdot), \frac{S_\eta(\lfloor (n/b(n))b(n)\cdot \rfloor)}{\sigma\sqrt{n}}\right). \tag{3.48}$$

Here, the modes are switched upon entering $(-\infty, 0)$ and $[0, \infty)$. Remark 3.2.2 discusses a peculiarity of the switching models with the gap 0. Since $\beta \in (0, 1)$, we infer

$$\lim_{n \to \infty} \frac{b(n)}{n} = 0, \tag{3.49}$$

and the stated result for MODEL OSCRW follows from Theorem 3.2.2 with y_n specified above, $\delta_n = 0$, $\varrho_n = b(n)/n$ for $n \in \mathbb{N}$, $\varrho_0 = 0$, $F_n(t) = S_\eta(\lfloor b(n)t \rfloor)/(\sigma\sqrt{n})$ for $n \in \mathbb{N}$ and $t \geq 0$, and $F_0 = \mathcal{V}_\beta$.

We proceed by analyzing MODEL REGEN. Assume first that $\eta > 0$ a.s. If $X(k) \leq 0$ for some $k \in \mathbb{N}$, then the size of the next jump of X is the sum of two terms. The first term is $|X(k)|$, that is, the overshoot of X at 0 at the previous step, and the second term is η_{k+1}. For $i \in \mathbb{N}$, denote by γ_i the absolute value of the ith overshoot of X at 0 (on the way from the positive halfline to the negative one). The random variables $\gamma_2, \gamma_3, \ldots$ are independent, are identically distributed, and may depend on (ξ_i) and (η_j). The variable γ_1 is independent of $(\gamma_i)_{i \geq 2}$ and has a distribution that may be other than the distribution of γ_2. The reason is that γ_1 corresponds to a fragment of the path starting from $X(0)$ rather than η_k for some $k \in \mathbb{N}$. □

Lemma 3.2.2 *As* $n \to \infty$,

$$\frac{\gamma_2 + \ldots + \gamma_n}{S_\eta(n)} \stackrel{\mathbb{P}}{\to} 0, \quad n \to \infty.$$

Proof Put $\tau_0 := 0$ and, for $i \in \mathbb{N}$,

$$\tau_{i+1} := \inf\{k > \tau_i \; : \; S_\xi(k) < S_\xi(\tau_i)\} \quad \text{and} \quad \chi_i^* := S_\xi(\tau_{i-1}) - S_\xi(\tau_i).$$

The elements of the sequences $(\tau_i)_{i \geq 1}$ and $(\chi_i^*)_{i \geq 1}$ are called *descending ladder epochs* and *descending ladder heights* of S_ξ, respectively. The random variables $\chi_1^*, \chi_2^*, \ldots$ are independent and identically distributed. Denote by χ a random variable with the same distribution as χ_1^*. Plainly, $\chi > 0$ a.s. Further, the assumptions $\mathbb{E}\xi = 0$ and $\operatorname{Var}\xi < \infty$ entail $\mu := \mathbb{E}\chi < \infty$; see, for instance, formula (4b) in [39]. The random variables γ_2, $\gamma_3, \ldots$ are independent copies of a random variable γ with

$$\gamma \stackrel{d}{=} S_\chi(\nu_\chi(\eta)) - \eta,$$

where $S_\chi(0) = 0$ and

$$\nu_\chi(t) := \inf\{n \geq 0 : S_\chi(n) \geq t\}, \quad t \geq 0.$$

We start by showing that

$$\lim_{x \to \infty} \frac{\mathbb{P}\{\gamma > x\}}{\mathbb{P}\{\eta > x\}} = 0. \tag{3.50}$$

Denote by F_χ the distribution function of χ and U_χ the renewal function for $(S_\chi(n))_{n \geq 0}$, that is, $U_\chi(x) := \sum_{n \geq 0} \mathbb{P}\{S_\chi(n) \leq x\}$ for $x \in \mathbb{R}$. Then

$$\mathbb{P}\{S_\chi(\nu_\chi(z)) - z > x\} = \int_{[0, z]} \left(1 - F_\chi(z + x - y)\right) dU_\chi(y), \quad z, x \geq 0.$$

Further, for any $A > 0$,

$$\begin{aligned}
\mathbb{P}\{\gamma > x\} &= \mathbb{E}\int_{[0, \eta]} \left(1 - F_\chi(\eta + x - y)\right) dU_\chi(y) \\
&= \mathbb{E}\int_{[0, \eta]} (1 - F_\chi(\eta + x - y)) dU_\chi(y)(\mathbb{1}_{\{\eta \leq Ax\}} + \mathbb{1}_{\{\eta > Ax\}}) \quad (3.51) \\
&\leq (1 - F_\chi(x))\mathbb{E}U_\chi(\eta)\mathbb{1}_{\{\eta \leq Ax\}} + \mathbb{P}\{\eta > Ax\}
\end{aligned}$$

having utilized monotonicity of F_χ for the inequality. By the elementary renewal theorem (see, for instance, Theorem 4.1 on p. 54 in [60]), $\lim_{x \to \infty} x^{-1} U_\chi(x) = \mu^{-1}$ and thereupon

$$\frac{\mathbb{E}U_\chi(\eta)\mathbb{1}_{\{\eta \leq Ax\}}}{\mathbb{P}\{\eta > x\}} \sim \frac{\mathbb{E}\eta \mathbb{1}_{\{\eta \leq Ax\}}}{\mu \mathbb{P}\{\eta > x\}}, \quad x \to \infty.$$

Recalling that the distribution of η belongs to the domain of attraction of a β-stable distribution with $\beta \in (0, 1)$ and invoking Karamata's theorem (Theorem 1.6.4 in [13]) we infer

$$\frac{\mathbb{E}\eta \mathbb{1}_{\{\eta \leq Ax\}}}{\mu \mathbb{P}\{\eta > x\}} \sim \frac{\alpha A^{1-\beta}}{(1-\beta)\mu} x, \quad x \to \infty.$$

3.2 Functional Limit Theorems for Random Walks with Reflection

This yields

$$\lim_{x\to\infty}\frac{(1-F_\chi(x))\mathbb{E}U_\chi(\eta)\mathbb{1}_{\{\eta\leq Ax\}}}{\mathbb{P}\{\eta>x\}}=0, \quad (3.52)$$

because $\mathbb{E}\chi<\infty$ entails $\lim_{x\to\infty}(1-F_\chi(x))x=0$.

It follows from (3.51) and (3.52) that, for any $A>0$,

$$\limsup_{x\to\infty}\frac{\mathbb{P}\{\gamma>x\}}{\mathbb{P}\{\eta>x\}}\leq\lim_{x\to\infty}\frac{\mathbb{P}\{\eta>Ax\}}{\mathbb{P}\{\eta>x\}}=A^{-\beta}.$$

Since $A>0$ is arbitrary, we arrive at (3.50). Thus, we have proved that given $\varepsilon\in(0,1]$ there exists $x_0>0$ such that

$$\mathbb{P}\{\gamma>x\}\leq\varepsilon\mathbb{P}\{\eta>x\}$$

whenever $x\geq x_0$. Let $\hat{\eta}$ be a random variable with distribution

$$\mathbb{P}\{\hat{\eta}>x\}=\begin{cases}1, & \text{if } x<x_0,\\ \varepsilon\mathbb{P}\{\eta>x\}, & \text{if } x\geq x_0.\end{cases}$$

Then $\mathbb{P}\{\gamma>x\}\leq\mathbb{P}\{\hat{\eta}>x\}$ for $x\geq 0$ and, as a consequence, for each $n\in\mathbb{N}$,

$$\mathbb{P}\{\gamma_2+\ldots+\gamma_n>x\}\leq\mathbb{P}\{\hat{\eta}_1+\ldots+\hat{\eta}_n>x\}, \quad x\geq 0, \quad (3.53)$$

where $\hat{\eta}_1,\hat{\eta}_2,\ldots$ are independent copies of $\hat{\eta}$. Since $\mathbb{P}\{\hat{\eta}>x\}\sim\varepsilon x^{-\beta}\ell(x)$ as $x\to\infty$, we conclude that a counterpart of (3.46) holds for $(S_{\hat{\eta}}(n))_{n\geq 0}$. Its specialization to $t=1$ reads

$$\frac{S_{\hat{\eta}}(n)}{c(n)}\xrightarrow{d}\varepsilon^{1/\beta}\mathcal{V}_\beta(1), \quad n\to\infty. \quad (3.54)$$

Since $\varepsilon\in(0,1]$ is arbitrary, we deduce the claim of the lemma from (3.53), (3.54), and one-dimensional version of (3.46). □

For each $n\in\mathbb{N}$,

$$\frac{X^{(n)}(\lfloor n\cdot\rfloor)}{\sigma\sqrt{n}}\stackrel{d}{=}\mathcal{G}_0\Big(y_n(\cdot),|y_n(\kappa^{(n)})|\mathbb{1}_{\{\kappa^{(n)}\leq\lfloor(n/b(n))b(n)\cdot\rfloor\}}$$

$$+\frac{\gamma_2^{(n)}+\ldots+\gamma_{\lfloor(n/b(n))b(n)\cdot\rfloor}^{(n)}}{\sigma\sqrt{n}}+\frac{S_\eta(\lfloor(n/b(n))b(n)\cdot\rfloor)}{\sigma\sqrt{n}}\Big), \quad (3.55)$$

where $y_n(t) = (X^{(n)}(0) + S_\xi(\lfloor nt \rfloor))/(\sigma\sqrt{n})$, $\kappa^{(n)} := \inf\{s \geq 0 : y_n(s) \leq 0\}$ and $|y_n(\kappa^{(n)})|$, $\gamma_2^{(n)}$, $\gamma_3^{(n)}, \ldots$ are independent random variables. Furthermore, $\gamma_2^{(n)}$, $\gamma_3^{(n)}, \ldots$ have the same distribution as γ_2. We stress that the assumption $\eta > 0$ a.s. is of principal importance for the validity of (3.55). If $\mathbb{P}\{\eta \leq 0\} > 0$, then (3.55) fails to hold.

We intend to invoke Theorem 3.2.2 with y_n specified at the beginning of the proof, $\delta_n = 0$, $\varrho_n = b(n)/n$ for $n \in \mathbb{N}$, $\varrho_0 = 0$,

$$F_n(t) = |y_n(\kappa^{(n)})|\mathbb{1}_{\{\kappa^{(n)} \leq \lfloor b(n)t \rfloor\}} + (\gamma_2^{(n)} + \ldots + \gamma_{\lfloor b(n)t \rfloor}^{(n)} + S_\eta(\lfloor b(n)t \rfloor))/(\sigma\sqrt{n})$$

for $n \in \mathbb{N}$ and $t \geq 0$, and $F_0 = \mathcal{V}_\beta$. To this end, in view of (3.47) and a.s. nonnegativity of $|y_n(\kappa^{(n)})|$, $\gamma_2^{(n)}$, $\gamma_3^{(n)}, \ldots$, it suffices to prove that, for all $T > 0$,

$$|y_n(\kappa^{(n)})|\mathbb{1}_{\{\kappa^{(n)} \leq \lfloor b(n)T \rfloor\}} + \frac{\gamma_2^{(n)} + \ldots + \gamma_{\lfloor b(n)T \rfloor}^{(n)}}{\sqrt{n}} \xrightarrow{\mathbb{P}} 0, \quad n \to \infty. \quad (3.56)$$

By Lemma 3.2.2 and one-dimensional version of (3.47), relation (3.56), with the first summand removed, holds true. Finally, observe that

$$|y_n(\kappa^{(n)})|\mathbb{1}_{\{\kappa^{(n)} \leq \lfloor b(n)T \rfloor\}} \leq \sqrt{b(n)/n} \max_{1 \leq i \leq \lfloor b(n)T \rfloor} |\xi_i|/(\sigma\sqrt{b(n)}) \to 0, \quad n \to \infty \quad \text{a.s.}$$

Here, the first factor converges to 0 according to (3.49), and the second factor converges to 0 a.s. because $\mathbb{E}\xi^2 < \infty$. Thus, relation (3.56) has been proved.

Although the proof in the case $\eta > 0$ a.s. is complete, we give one more auxiliary result under the positivity assumption, which will be used in our treatment of the case $\mathbb{P}\{\eta \leq 0\} > 0$. Recall that a sequence of random variables $(\zeta_n)_{n \geq 1}$ is said to be bounded in probability if given $\varepsilon > 0$ there exists $M > 0$ such that

$$\sup_{n \geq 1} \mathbb{P}\{|\zeta_n| > M\} \leq \varepsilon.$$

Lemma 3.2.3 *Fix any $T > 0$. Let $\eta > 0$ a.s. and b be as in (3.47). Then the sequence $\left(\sum_{i=0}^{\lfloor nT \rfloor - 1} \mathbb{1}_{\{X^{(n)}(i) \leq 0\}}/b(n)\right)_{n \geq 1}$ is bounded in probability.*

Proof For each $n \in \mathbb{N}$, put

$$N_n^>(k) := \sum_{i=0}^{k-1} \mathbb{1}_{\{X^{(n)}(i) \leq 0\}}, \quad k \in \mathbb{N},$$

3.2 Functional Limit Theorems for Random Walks with Reflection

where the upper index " $>$ " indicates that $\eta > 0$ a.s. Without loss of generality we can and do assume[7] that equality (3.55) holds true a.s. rather than in distribution. Keeping the notation for $X^{(n)}$ unchanged we infer

$$X^{(n)}(k) = X^{(n)}(0) + S_\xi(k - N_n^>(k)) + |y_n(\kappa^{(n)})|\mathbb{1}_{\{\kappa^{(n)} \leq N_n^>(k)\}} + \gamma_2^{(n)} + \cdots + \gamma_{N_n^>(k)}^{(n)}$$
$$+ S_\eta(N_n^>(k)), \quad k \in \mathbb{N}.$$

Both $S_\xi(\lfloor n\cdot\rfloor)/(\sigma\sqrt{n})$ and $X^{(n)}(\lfloor n\cdot\rfloor)/(\sigma\sqrt{n})$ converge in distribution on $D([0,\infty))$ to W and $W_x^{\text{refl},\beta}$ by Donsker's theorem and the already proved part of the theorem for MODEL REGEN, respectively. Fix any $T > 0$. It follows from the discussion in Example 3.2.2 that $W_x^{\text{refl},\beta}$ is a.s. continuous at T. Since the supremum functional is continuous in the J_1-topology (Proposition 4.3.11), we conclude that both $(\max_{t\in[0,T]} X^{(n)}(\lfloor nt\rfloor))/(\sigma\sqrt{n})$ and $(\max_{t\in[0,T]} S_\xi(\lfloor nt\rfloor))/(\sigma\sqrt{n})$ converge in distribution. These facts in combination with a.s. nonnegativity of $|y_n(\kappa^{(n)})|\mathbb{1}_{\{\kappa^{(n)}\leq N_n^>(\lfloor n\cdot\rfloor)\}} + \gamma_2^{(n)} + \cdots + \gamma_{N_n^>(\lfloor n\cdot\rfloor)}^{(n)}$ ensure that the sequence $(S_\eta(N_n^>(\lfloor nT\rfloor))/\sqrt{n})_{n\geq 1}$ is bounded in probability. Using the fact that S_η is a.s. increasing we obtain that, for all $x, y > 0$,

$$\mathbb{P}\{N_n^>(\lfloor nT\rfloor) > xb_n\} \leq \mathbb{P}\{S_\eta(\lfloor xb(n)\rfloor) \leq y\sqrt{n}\} + \mathbb{P}\{S_\eta(N_n^>(\lfloor nT\rfloor)) > y\sqrt{n}\}.$$

In view of (3.47) and $\mathcal{V}_\beta(x) \stackrel{d}{=} x^{1/\beta}\mathcal{V}_\beta(1)$, we infer

$$\limsup_{n\to\infty} \mathbb{P}\{N_n^>(\lfloor nT\rfloor) > xb_n\} \leq \mathbb{P}\{\mathcal{V}_\beta(1) \leq \sigma^{-1}yx^{-1/\beta}\}$$
$$+ \sup_{n\geq 1} \mathbb{P}\{S_\eta(N_n^>(\lfloor nT\rfloor)) > y\sqrt{n}\}.$$

Given $\varepsilon > 0$ pick $y = y_\varepsilon > 0$ such that the second term on the right-hand side does not exceed $\varepsilon/2$ and then pick $x = x_\varepsilon > 0$ such that the first term on the right-hand side does not exceed $\varepsilon/2$. The claim follows. □

Now we discuss MODEL REGEN under the assumption that $\mathbb{P}\{\eta \leq 0\} > 0$. For each $n \in \mathbb{N}$, put

$$\lambda_n(0) := 0, \quad \lambda_n(k) := k - \sum_{i=0}^{k-1} \mathbb{1}_{\{X^{(n)}(i)\leq 0, X^{(n)}(i+1)\leq 0\}}, \quad k \in \mathbb{N},$$

so that $\lambda_n(k)$ is the number of elements out of $\{X^{(n)}(1), \ldots, X^{(n)}(k)\}$ obtained by a jump given by either ξ_i for some $i \in \mathbb{N}$ or *positive* η_j for some $j \in \mathbb{N}$. For each $n \in \mathbb{N}$, we

---

[7] The assumption is only made within the present proof.

extend λ_n by linearity, thereby obtaining a function of nonnegative argument. Put

$$\lambda_n^-(t) = \inf\{s \geq 0 : \lambda_n(s) \geq t\}, \quad t \geq 0.$$

Further, for $n \in \mathbb{N}$, put

$$\widetilde{X}^{(n)}(t) := X^{(n)}(\lambda_n^-(t)) = X^{(n)}(\lambda_n^-(\lfloor t \rfloor)), \quad t \geq 0. \tag{3.57}$$

The sequence $\widetilde{X}^{(n)}$ is constructed from $X^{(n)}$ by deleting positions obtained by a jump given by some nonpositive η_j. Put $\rho := \inf\{k \in \mathbb{N} : \eta_k > 0\}$. A crucial observation is that $\widetilde{X}^{(n)}$ is a Markov chain having the same transition probabilities as in MODEL REGEN, with the random variable $\widetilde{\eta} := \eta_\rho$ replacing η.

Lemma 3.2.4 *The distribution of $\widetilde{\eta}$ belongs to the domain of attraction of a spectrally positive strictly β-stable distribution with $\beta \in (0, 1)$.*

Proof By the assumption of the theorem $\mathbb{P}\{\eta > x\} \sim x^{-\beta}\ell(x)$ as $x \to \infty$ for some ℓ slowly varying at infinity. The variable $\widetilde{\eta}$ is a.s. positive. Since $\mathbb{P}\{\widetilde{\eta} > x\} = \mathbb{P}\{\eta > x | \eta > 0\} \sim (1/\mathbb{P}\{\eta > 0\})x^{-\beta}\ell(x)$ as $x \to \infty$, we arrive at the claim (see Sect. 4.4.1 for details if needed). □

It follows from Lemma 3.2.4 and the previous part of the proof for MODEL REGEN with $\eta > 0$ a.s. that

$$\left(\frac{\widetilde{X}^{(n)}(\lfloor nt \rfloor)}{\sigma\sqrt{n}}\right)_{t \geq 0} \implies (W_x^{\text{refl}, \beta}(t))_{t \geq 0}, \quad n \to \infty$$

on $D([0, \infty))$. To pass from this limit relation to that for $X^{(n)}$ we use representation (3.57) and Proposition 4.3.10 with $g_n = \lambda_n^-$, $f_n(\cdot) = X^{(n)}(\lfloor n \cdot \rfloor)/(\sigma\sqrt{n})$ for $n \in \mathbb{N}$. Further, in the notation of Proposition 4.3.10, $f_n(u_k^{(n)}-)$ is a *positive* value of $X^{(n)}(l-1)/(\sigma\sqrt{n})$ such that $X^{(n)}(l) \leq 0$ (if $X^{(n)}(0) = 0$, then $f_n((\min_k u_k^{(n)})-) = 0$ and we put $l = 1$); $u_k^{(n)}$ is then l/n and $v_k^{(n)}$ is j/n, where j is the minimal value exceeding l such that $X^{(n)}(j) > 0$. Observe that $[u_k^{(n)}, v_k^{(n)}) = \emptyset$ if $j = l + 1$ or, in other words, if the η used to produce $X^{(n)}(l+1)$ is positive. For each $n \in \mathbb{N}$, put

$$N_n(k) := \sum_{i=0}^{k-1} \mathbb{1}_{\{X^{(n)}(i) \leq 0\}}, \quad \widetilde{N}_n(k) := \sum_{i=0}^{k-1} \mathbb{1}_{\{\widetilde{X}^{(n)}(i) \leq 0\}}, \quad k \in \mathbb{N}.$$

According to the aforementioned proposition and Lemma 4.1.3 in combination with the Skorokhod representation theorem (Theorem 4.1.2), it suffices to show that, for all $T > 0$,

3.2 Functional Limit Theorems for Random Walks with Reflection

$$\sup_{t \in [0,T]} \left| \frac{\lambda_n^-(\lfloor nt \rfloor)}{n} - t \right| \overset{\mathbb{P}}{\to} 0, \quad n \to \infty \tag{3.58}$$

and

$$\frac{\max_{1 \le i \le \lfloor nT \rfloor} |\xi_i| + (\max_{1 \le k \le N_n(\lfloor nT \rfloor)} (-\eta_k)) \vee 0}{\sqrt{n}} \overset{\mathbb{P}}{\to} 0, \quad n \to \infty. \tag{3.59}$$

The form of relation (3.59) is derived from (4.5), and now we explain how to obtain it. Assume that the interval $[u_k^{(n)}, v_k^{(n)})$ is nonempty and have a nonempty intersection with $[0, T]$. Then, for $s \in [u_k^{(n)}, v_k^{(n)}) \cap [0, T]$, $\sigma\sqrt{n}(f_n(s) - f_n(u_k^{(n)}-))$ is equal to some negative ξ_i plus some negative η_j (if $X^{(n)}(0) = 0$, then it is equal to just some negative η_j). The number of negative η_j which may appear here does not exceed $N_n(\lfloor nT \rfloor)$. Hence, $\sigma\sqrt{n} \sup_{s \in [u_k^{(n)}, v_k^{(n)}) \cap [0, T]} |f_n(s) - f_n(u_k^{(n)}-)|$ does not exceed some ξ_i plus $(\max_{1 \le j \le N_n(\lfloor nT \rfloor)} (-\eta_j)) \vee 0$. Finally,

$$\sigma\sqrt{n} \sup_{k \ge 1} \sup_{s \in [u_k^{(n)}, v_k^{(n)}) \cap [0, T]} |f_n(s) - f_n(u_k^{(n)}-)|$$
$$\le \max_{1 \le i \le \lfloor nT \rfloor} |\xi_i| + (\max_{1 \le j \le N_n(\lfloor nT \rfloor)} (-\eta_j)) \vee 0,$$

thereby justifying the form of (3.59).

Proof of (3.59) The assumption $\mathbb{E}\xi^2 < \infty$ ensures that the first summand in (3.59) converges to 0 a.s. as $n \to \infty$. Left with analyzing the second summand we first need a lemma.

Lemma 3.2.5 *Assume that* $\mathbb{P}\{\eta \le 0\} > 0$. *Then the sequence* $(N_n(\lfloor nT \rfloor)/b(n))_{n \ge 1}$ *is bounded in probability.*

Proof Corresponding to $X^{(n)}(0) = 0$ (if this holds) and to each $k \in \mathbb{N}$ satisfying both $X^{(n)}(k-1) > 0$ and $X^{(n)}(k) \le 0$ is a (possibly empty) bunch of negative values of the η. Denote by $\vartheta_1, \vartheta_2, \ldots$ the numbers of elements in successive bunches, so that ϑ_1 corresponds to $k = 0$ (if $X^{(n)}(0) = 0$) or a minimal k for which the two inequalities hold, a minimal k or the second minimal k, etc. The variables $\vartheta_1, \vartheta_2, \ldots$ are independent copies of a random variable ϑ having a geometric distribution with success probability $\mathbb{P}\{\eta > 0\}$ starting at 0, that is,

$$\mathbb{P}\{\vartheta = i\} = \mathbb{P}\{\eta > 0\}(\mathbb{P}\{\eta \le 0\})^i, \quad i \in \mathbb{N}_0.$$

Here is a basic inequality for the subsequent proof: for each $n \in \mathbb{N}$ and each $k \in \mathbb{N}$, $N_n(k) \leq \sum_{i=1}^{\widetilde{N}_n(k)} (\vartheta_i + 1)$ a.s. In particular, for all $x, y > 0$,

$$\mathbb{P}\{N_n(\lfloor nT \rfloor) > b(n)x\} \leq \mathbb{P}\left\{ \sum_{i=1}^{\lfloor b(n)y \rfloor} (\vartheta_i + 1) > b(n)x \right\} + \mathbb{P}\{\widetilde{N}_n(\lfloor nT \rfloor) > b(n)y\}.$$

Since $\widetilde{X}^{(n)}$ is a counterpart of $X^{(n)}$, with $\widetilde{\eta} > 0$ a.s. replacing η, the sequence $(\widetilde{N}_n(\lfloor nT \rfloor)/b(n))_{n \geq 1}$ is bounded in probability by Lemmas 3.2.3 and 3.2.4. Hence, given $\varepsilon > 0$ there exists $y = y_\varepsilon$ such that $\sup_{n \geq 1} \mathbb{P}\{\widetilde{N}_n(\lfloor nT \rfloor) > b(n)y\} \leq \varepsilon$. Put $x = 2y_\varepsilon \mathbb{E}(\vartheta + 1)$. Using the weak law of large numbers we conclude that $\limsup_{n \to \infty} \mathbb{P}\{N_n(\lfloor nT \rfloor) > b(n)x\} \leq \varepsilon$. The proof of Lemma 3.2.5 is complete. □

The conclusion of Lemma 3.2.5 is equivalent to the statement: for any sequence of positive numbers $(v(n))_{n \geq 1}$ satisfying $\lim_{n \to \infty} v(n) = \infty$,

$$\frac{N_n(\lfloor nT \rfloor)}{b(n)v(n)} \overset{\mathbb{P}}{\to} 0, \quad n \to \infty.$$

Thus, (3.59) follows if we can show that, for all $M > 0$ and some diverging sequence $(v(n))$,

$$\frac{(\max_{0 \leq k \leq \lfloor Mb(n)v(n) \rfloor} (-\eta_k)) \vee 0}{\sqrt{n}} \overset{\mathbb{P}}{\to} 0, \quad n \to \infty.$$

For a proof, observe that $\lim_{x \to +\infty} (\mathbb{P}\{-\eta > x\}/\mathbb{P}\{\eta > x\}) = 0$. Hence, according to (4.13), with η replacing τ, $\lim_{n \to \infty} n\mathbb{P}\{-\eta > \delta c(n)\} = 0$ for all $\delta > 0$ or equivalently $\lim_{n \to \infty} b(n)\mathbb{P}\{-\eta > \varepsilon\sqrt{n}\} = 0$ for all $\varepsilon > 0$. To complete the proof of the claimed limit relation, pick any diverging $(v(n))$ satisfying

$$\lim_{n \to \infty} b(n)v(n)\mathbb{P}\{-\eta > \varepsilon\sqrt{n}\} = 0$$

for all $\varepsilon > 0$.

Proof of (3.58) According to Proposition 4.3.14, it is sufficient to show that, for all $T > 0$,

$$\sup_{t \in [0,T]} \left| \frac{\lambda_n(\lfloor nt \rfloor)}{n} - t \right| \overset{\mathbb{P}}{\to} 0, \quad n \to \infty.$$

3.2 Functional Limit Theorems for Random Walks with Reflection

Observe that $|k - \lambda_n(k)| \leq N_n(k)$ for $n, k \in \mathbb{N}$. In view of a.s. monotonicity of $k \mapsto N_n(k)$, the desired convergence follows from

$$\frac{N_n(\lfloor nT \rfloor)}{n} \xrightarrow{\mathbb{P}} 0, \quad n \to \infty,$$

which, in turn, is secured by boundedness in probability of $(N_n(\lfloor nT \rfloor)/b(n))_{n \geq 1}$ and (3.49). This completes the proof for MODEL REGEN.

The argument for treating MODELS LINDREFILL and REPULS is similar to the one used for MODEL REGEN. We omit details. □

3.2.4 Perturbations with Slowly Varying Distribution Tails

The purpose of the present section is to demonstrate that the distribution tails of perturbations which are heavier than those in the models investigated in Sect. 3.2.3 lead to another type of distributional asymptotic behavior. Informally, with high probability, random walks with reflection exhibit extremely large jumps upwards upon crossing 0. As a consequence, Donsker's scaling limit does not exist. To reveal the idea, we only discuss MODEL OSCRW with a fixed initial condition. Other models of random walks with reflection and triangular arrays could have been also investigated upon necessary adjustments.

Theorem 3.2.4 *Assume that* $\mathbb{E}\xi = 0$, $\operatorname{Var}\xi = \sigma^2 \in (0, \infty)$ *and* η *is a positive random variable such that* $x \mapsto \mathbb{P}\{\eta > x\}$ *is slowly varying*[8] *at infinity. Let* X *be defined by* MODEL OSCRW. *Then, for all* $t_0 > 0$,

$$\max_{t \in [0, t_0]} \frac{X(\lfloor nt \rfloor)}{\sqrt{n}} \xrightarrow{\mathbb{P}} +\infty, \quad n \to \infty. \tag{3.60}$$

In particular, the sequence of processes $(X(\lfloor nt \rfloor)/\sqrt{n})_{t \geq 0}$ *does not converge in distribution as* $n \to \infty$ *on* $D([0, \infty))$.

Remark 3.2.8 Suppose that the assumptions of Theorem 3.2.4 hold. Put $h(x) := 1/\mathbb{P}\{\eta > x\}$ for $x \geq 0$. Finding a (precise) scaling limit of X remains a challenge. To solve this problem, it is likely one has to exploit the result due to Kasahara (Theorem 2.1 in [79]):

$$\left(\frac{h(S_\eta(\lfloor nt \rfloor))}{n}\right)_{t \geq 0} \Longrightarrow \text{Extr}, \quad n \to \infty \tag{3.61}$$

---

[8] This means that $\lim_{x \to +\infty}(\mathbb{P}\{\eta > sx\}/\mathbb{P}\{\eta > x\}) = 1$ for each $s > 0$.

on $D([0,\infty))$, where $\text{Extr} := (\text{Extr}(t))_{t\geq 0}$ is an extremal process. It is defined by $\text{Extr}(t) = \max_{s\in[0,t]}(Z(s) - Z(s-))$ for $t \geq 0$, where $(Z(s))_{s\geq 0}$ is a Lévy process with $\mathbb{E}\exp(isZ(1)) = \exp(-\pi|s|)$ for $s \in \mathbb{R}$. Noting that $1/\text{Extr}(1)$ has an exponential distribution of unit mean and specializing (3.61) to the case $t = 1$, we arrive at the one-dimensional result due to Darling (Theorem 4.1 in [34]).

We stress that neither (3.61) nor its one-dimensional version is used in our proof of Theorem 3.2.4.

Proof of Theorem 3.2.4 Starting with a counterpart of (3.48) and then using a counterpart of (3.15) we obtain, for each $n \in \mathbb{N}$,

$$X(\lfloor n\cdot\rfloor) \stackrel{d}{=} \mathcal{G}_0\big(X(0) + S_\xi(\lfloor n\cdot\rfloor), S_\eta(\lfloor n\cdot\rfloor)\big) = X(0) + S_\xi(T_A(\lfloor n\cdot\rfloor)) + S_\eta(T_B(\lfloor n\cdot\rfloor)), \tag{3.62}$$

where, recalling that $X(0) \geq 0$ a.s. is the standing assumption,

$$T_A(k) = 1 + \sum_{i=1}^{k-1} \mathbb{1}_{\{X(i)\geq 0\}} \quad \text{and} \quad T_B(k) = \sum_{i=1}^{k-1} \mathbb{1}_{\{X(i)<0\}}$$

for $k \in \mathbb{N}$. For any measurable functions $u, v : [0,\infty) \to \mathbb{R}$ and any fixed $t_1 > 0$,

$$\sup_{t\in[0,t_1]}(u(t) + v(t)) \geq \sup_{t\in[0,t_1]} u(t) - \sup_{t\in[0,t_1]} |v(t)|.$$

Since $t \mapsto S_\eta(T_B(\lfloor nt\rfloor))$ is a.s. nondecreasing, to prove (3.60), it is enough to show[9] that, for all $t_0 > 0$,

$$\text{the sequence } \left(\frac{\max_{t\in[0,t_0]} |S_\xi(T_A(\lfloor nt\rfloor))|}{\sigma\sqrt{n}}\right)_{n\geq 1} \text{ is bounded in probability} \tag{3.63}$$

and

$$\frac{S_\eta(T_B(n))}{\sqrt{n}} \xrightarrow{\mathbb{P}} +\infty, \quad n \to \infty. \tag{3.64}$$

---

[9] Actually, $\dfrac{\max_{t\in[0,t_0]}|S_\xi(T_A(\lfloor nt\rfloor))|}{\sigma\sqrt{n}} \xrightarrow{d} \max_{t\in[0,t_0]}|W(t)|$ as $n \to \infty$, where $(W(t))_{t\geq 0}$ is a standard Brownian motion. However, we do not need such a precision here.

3.2 Functional Limit Theorems for Random Walks with Reflection

Proof of (3.63) As a consequence of Donsker's theorem (Theorem 4.3.1) applied to S_ξ and Proposition 4.3.11 in combination with the Skorokhod representation theorem (Theorem 4.1.2), $\max_{t \in [0,t_0]} |S_\xi(\lfloor nt \rfloor)|/(\sigma\sqrt{n}) \overset{d}{\to} \max_{0 \leq t \leq t_0} |W(t)|$ as $n \to \infty$. Since $T_A(k) \in [0,k]$ for $k \in \mathbb{N}$, we infer

$$\max_{t \in [0,t_0]} |S_\xi(T_A(\lfloor nt \rfloor))| \leq \max_{t \in [0,t_0]} |S_\xi(\lfloor nt \rfloor)| \quad \text{a.s.},$$

and (3.63) follows.

As a preparation for a proof of (3.64), we show that

$$\frac{T_B(n)}{n} \overset{\mathbb{P}}{\to} 0, \quad n \to \infty. \tag{3.65}$$

As has been discussed in Remark 3.2.2, Lemma 3.2.1 remains valid for the switch models with the gap 0, whence, for each $n \in \mathbb{N}$,

$$S_\eta(T_B(n) - 1) \leq m(T_A(n)) \leq S_\eta(T_B(n)) + \max_{1 \leq i \leq n} |\xi_i|, \tag{3.66}$$

where $m(k) := \max_{0 \leq i \leq k}((X(0) + S_\xi(i))_-)$ for $k \in \mathbb{N}_0$. Another application of Donsker's theorem and Proposition 4.3.11 yields

$$\frac{m(n)}{\sigma\sqrt{n}} \overset{d}{\to} \max_{s \in [0,1]}(-W(s)), \quad n \to \infty. \tag{3.67}$$

Using the left-hand inequality in (3.66) in combination with $m(T_A(n)) \leq m(n)$ a.s. which is secured by monotonicity of m and the inequality $T_A(n) \leq n$ a.s. we obtain, for each $n \in \mathbb{N}$,

$$\frac{T_B(n) - 1}{\sigma\sqrt{n}} \frac{S_\eta(T_B(n) - 1)}{T_B(n) - 1} \leq \frac{m(n)}{\sigma\sqrt{n}} \quad \text{a.s.} \tag{3.68}$$

The assumption $\mathbb{E}\xi = 0$ entails that the standard random walk oscillates. In particular, $\liminf_{n \to \infty} S_\xi(n) = -\infty$ a.s. One can check that this implies[10]

$$\lim_{n \to \infty} T_B(n) = \infty \quad \text{a.s.}$$

---

[10] If the reader finds this argument too vague, one can assume that the opposite situation prevails, that is, $\lim_{n \to \infty} T_B(n) < \infty$ a.s. Relation (3.65) then holds trivially.

Since $\mathbb{E}\eta = \infty$, an application of the strong law of large numbers for standard random walks yields

$$\lim_{n\to\infty} (S_\eta(n)/n) = +\infty \quad \text{a.s.}$$

and thereupon

$$\lim_{n\to\infty} (S_\eta(T_B(n))/T_B(n)) = +\infty \quad \text{a.s.}$$

In view of (3.67), the left-hand side of (3.68) is stochastically bounded. This proves $T_B(n)/\sqrt{n} \overset{\mathbb{P}}{\to} 0$ as $n \to \infty$, which is a stronger version of (3.65).

Proof of (3.64) Relation (3.65) together with $T_A(k) + T_B(k) = k$ for $k \in \mathbb{N}$ entails

$$\max_{t\in[0,t_0]} \left| \frac{T_A(\lfloor nt \rfloor)}{n} - t \right| \overset{\mathbb{P}}{\to} 0, \quad n \to \infty. \tag{3.69}$$

By a standard application of the Skorokhod representation theorem, we conclude, with the help of Lemma 4.1.3, that $(S_\xi(\lfloor nt \rfloor)/(\sigma\sqrt{n}), T_A(\lfloor nt \rfloor)/n)_{t\geq 0}$ converges in distribution as $n \to \infty$ in the product topology on $(D([0,\infty)))^2$ to (W, I), where $I(t) = t$. With the help of Proposition 4.3.7 we conclude that $(S_\xi(T_A(\lfloor nt \rfloor))/(\sigma\sqrt{n}))_{t\geq 0}$ converges in distribution as $n \to \infty$ on $D([0,\infty))$ to W. Further, by Proposition 4.3.11,

$$m(T_A(n))/(\sigma\sqrt{n}) \overset{d}{\to} \max_{s\in[0,1]} (-W(s)), \quad n \to \infty.$$

The limit random variable has the same distribution as $|\mathcal{N}(0,1)|$, where $\mathcal{N}(0,1)$ is a random variable with the standard normal distribution.

Given $\varepsilon > 0$ pick $c_1 > 0$ such that $\mathbb{P}\{\sigma|\mathcal{N}(0,1)| > 2c_1\} \geq 1 - \varepsilon$. The assumption $\mathbb{E}\xi^2 < \infty$ entails $\lim_{n\to\infty} \mathbb{P}\{\max_{1\leq i \leq n} |\xi_i| > c_1\sqrt{n}\} = 0$. Using now the right-hand inequality in (3.66) we infer

$$\liminf_{n\to\infty} \mathbb{P}\{S_\eta(T_B(n)) > c_1\sqrt{n}\} \geq 1 - \varepsilon.$$

Put $v(t) := \inf\{k \in \mathbb{N}_0 : S_\eta(k) > t\}$ for $t \geq 0$. By Theorem 8.8.2 in [13], slow variation[11] of $x \mapsto \mathbb{P}\{\eta > x\}$ ensures that

$$\frac{\eta_{v(t)}}{t} \overset{\mathbb{P}}{\to} \infty, \quad t \to +\infty. \tag{3.70}$$

---

[11] This is the only place in the proof where slow variation is used.

Fix any $c > c_1$. Invoking

$$\{\eta_{\nu(c_1\sqrt{n})} > c\sqrt{n}, S_\eta(T_B(n)) > c_1\sqrt{n}\} = \{\eta_{\nu(c_1\sqrt{n})} > c\sqrt{n}, \nu(c_1\sqrt{n}) \leq T_B(n)\}$$
$$\subseteq \{\nu(c\sqrt{n}) \leq T_B(n)\} = \{S_\eta(T_B(n)) > c\sqrt{n}\}$$

and (3.70) we obtain

$$\liminf_{n\to\infty} \mathbb{P}\{S_\eta(T_B(n)) > c\sqrt{n}\}$$
$$\geq \liminf_{n\to\infty} \mathbb{P}\{\eta_{\nu(c_1\sqrt{n})} \geq c\sqrt{n},\ S_\eta(T_B(n)) > c_1\sqrt{n}\}$$
$$= \liminf_{n\to\infty} \mathbb{P}\{S_\eta(T_B(n)) > c_1\sqrt{n}\} \geq 1 - \varepsilon,$$

thereby proving (3.64). □

3.3 Random Walks with Membrane and a Skew Brownian Motion

Let $m \in \mathbb{Z}$ be fixed. Consider a random walk X with the membrane A on $\mathbb{Z}$ defined by (3.1) or (3.3), where $A = \{-m, \ldots, m\}$, $\mathbb{E}\xi = 0$, and $\text{Var}\,\xi = \sigma^2 \in (0, \infty)$. If there are no perturbations, that is, $\eta^{(x)} \stackrel{d}{=} \xi$ for $x \in A$, then Donsker's scaling limit of X is a standard Brownian motion:

$$\left(\frac{X(\lfloor nt \rfloor)}{\sqrt{n}}\right)_{t \geq 0} \Longrightarrow (\sigma W(t))_{t \geq 0}, \quad n \to \infty$$

on $D([0, \infty))$. The main result of this section provides a "close-to-optimal" set of technical assumptions ensuring that Donsker's scaling limit of X is a skew Brownian motion. Actually, we shall prove a more general result, in which jumps of a random walk from points located to the right of the membrane and to the left of the membrane may have different distributions.

We first formulate a simplified version. Recall that a probability distribution is called 1-*arithmetic*, if it is concentrated on the set of integers $\mathbb{Z}$ and not concentrated on $d\mathbb{Z}$ for any $d \geq 2$.

Theorem 3.3.1 *Let X be a Markov chain on $\mathbb{Z}$ with transition probabilities*

$$\mathbb{P}\{X(n+1) \in \cdot \mid X(n) = x\} = \begin{cases} \mathbb{P}\{x + \xi^+ \in \cdot\}, & \text{if } x > m, \\ \mathbb{P}\{x + \xi^- \in \cdot\}, & \text{if } x < -m, \\ \mathbb{P}\{x + \eta^{(x)} \in \cdot\}, & \text{if } x \in \{-m, \ldots, m\}, \end{cases}$$

where $\xi^{\pm}$ are random variables with 1-arithmetic distributions and $\eta^{(x)}$ for $x \in \{-m, \ldots, m\}$ are integer-valued random variables. Assume that:

(I) $\mathbb{E}\xi^{\pm} = 0$ and $\text{Var}\,\xi^{\pm} = v_{\pm}^2 \in (0, \infty)$.

(II) *The jumps from the membrane are integrable, that is,*

$$\mathbb{E}|\eta^{(x)}| < \infty, \quad x \in \{-m, \ldots, m\}.$$

(III) *X exits the membrane with probability one, that is,*

$$\mathbb{P}\{|X(k)| > m \text{ for some } k \geq 0 | X(0) = x\} = 1, \quad x \in \{-m, \ldots, m\}.$$

(IV) *The states $\mathbb{Z} \setminus \{-m, \ldots, m\}$ of X are connected.*

Then, for any fixed initial value $X(0) \in \mathbb{Z}$,

$$\left(\varphi_{\text{sgn}}\left(\frac{X(\lfloor nt \rfloor)}{\sqrt{n}}\right)\right)_{t \geq 0} \Longrightarrow (W_\gamma^{\text{skew}}(t))_{t \geq 0}, \quad n \to \infty$$

on $D([0, \infty))$, where W_γ^{skew} is a skew Brownian motion starting at 0 with some permeability parameter $\gamma \in (-1, 1)$, and[12]

$$\varphi_{\text{sgn}}(x) := x \cdot v_{\text{sgn}(x)}^{-1} := x\big(v_+^{-1} \mathbb{1}_{[0, \infty)}(x) + v_-^{-1} \mathbb{1}_{(-\infty, 0)}(x)\big), \quad x \in \mathbb{R}. \qquad (3.71)$$

Remark 3.3.1 Throughout this section we identify the set $\{+, -\}$ with $\{-1, +1\}$. In particular, we stipulate that $\text{sgn}(x) = +$ for $x \geq 0$ and $\text{sgn}(x) = -$ for $x < 0$.

An explicit value of the permeability parameter γ can be obtained by specializing formula (3.91).

Theorem 3.3.1 is a simplified version of Theorem 3.3.3 in Sect. 3.3.1, which is itself a specialization of a general result on scaling limits of random walks on a finite collection of integer lines $\mathbb{Z} \times \{1, \ldots, d\}$, $d \geq 1$; see Theorem 3.3.2. Even the formulation of this general result, let alone its proof, requires introducing and analyzing several additional constructions, including entrance and exit laws for embedded Markov chains (in particular, the existence of their stationary distributions) and some others. A brief proof of Theorem 3.3.1 will be given in Remark 3.3.3.

---

[12] Just in case, we note that $v_{\pm}^{-1} = 1/v_{\pm}$.

Theorem 3.3.1 allows us to find Donsker's scaling limit of $\left(\frac{X(\lfloor nt \rfloor)}{\sqrt{n}}\right)_{t\geq 0}$ itself. Indeed, since the function φ_{sgn} defined by (3.71) is continuous and strictly increasing, it possesses the unique continuous inverse function $\varphi_{\text{sgn}}^{-1}$ given by

$$\varphi_{\text{sgn}}^{-1}(x) = x \cdot v_{\text{sgn}(x)} := x\big(v_+ \mathbb{1}_{[0,\infty)}(x) + v_- \mathbb{1}_{(-\infty,0)}(x)\big), \quad x \in \mathbb{R}.$$

Corollary 3.3.1 *Under the assumptions of Theorem 3.3.1,*

$$\left(\frac{X(\lfloor nt \rfloor)}{\sqrt{n}}\right)_{t\geq 0} \implies \left(\varphi_{\text{sgn}}^{-1}(W_\gamma^{\text{skew}}(t))\right)_{t\geq 0}, \quad n \to \infty$$

on $D([0,\infty))$. The process $Y = (Y(t))_{t\geq 0} := \left(\varphi_{\text{sgn}}^{-1}(W_\gamma^{\text{skew}}(t))\right)_{t\geq 0}$ is a solution to the equation

$$Y(t) = \int_0^t \left(v_- \mathbb{1}_{\{Y(s)<0\}} + v_+ \mathbb{1}_{\{Y(s)\geq 0\}}\right) dW(s)$$

$$+ \frac{\gamma(v_+ + v_-) + v_+ - v_-}{\gamma(v_+ - v_-) + v_+ + v_-} L_0^Y(t), \quad t \geq 0, \qquad (3.72)$$

where L_0^Y is the symmetric semimartingale local time of Y at 0 defined by (2.15). This local time also satisfies

$$L_0^Y(t) = \lim_{\varepsilon \to 0+} \frac{1}{2\varepsilon} \int_0^t \left(v_-^2 \mathbb{1}_{\{-\varepsilon < Y(s) < 0\}} + v_+^2 \mathbb{1}_{\{0 \leq Y(s) < \varepsilon\}}\right) ds.$$

The limit process Y appearing in Corollary 3.3.1 is called *skew oscillating Brownian motion*.

The remainder of the section is structured as follows. We formulate general results on distributional limit behavior of locally perturbed random walks in Sect. 3.3.1. Theorem 3.3.2 is concerned with locally perturbed random walks on $\mathbb{Z} \times \{1, \ldots, d\}$, whose Donsker's scaling limit is a Walsh Brownian motion. A result on scaling limits of locally perturbed random walks with values in $\mathbb{Z}$ obtained in Theorem 3.3.3 is a nontrivial corollary of Theorem 3.3.2. We determine the parameter of the Walsh Brownian motion and the permeability parameter of the skew Brownian motion explicitly, in terms of stationary distributions of some embedded entrance and exit Markov chains. Section 3.3.2 contains illustrative examples, which show that formulas obtained in Theorems 3.3.2 and 3.3.3 are consistent with previously known results. Proofs of Theorem 3.3.2, Theorem 3.3.3, and Corollary 3.3.1 are given in Sects. 3.3.3, 3.3.4, and 3.3.5, respectively.

3.3.1 Convergence to a Walsh Brownian Motion

The model treated in Theorem 3.3.1 is a particular version of a locally perturbed random walk on $\mathcal{Z}^d := \mathbb{Z} \times \{1, \ldots, d\}$, where $d \geq 1$ is fixed, that we now introduce. Let

$$\mathcal{X} := (\mathcal{X}(k))_{k \geq 0} = (R(k), l(k))_{k \geq 0}$$

be a time-homogeneous Markov chain on the state space $\mathcal{Z}^d$. We call $R(k)$ the *radius* and $l(k)$ the *label* of $\mathcal{X}(k)$. With an abuse of terminology, the radius R may attain negative values. The chain $\mathcal{X}$ evolves according to the following set of rules:

($\mathbf{A}_1$) For all $(x, i) \in \mathcal{N}^d := \mathbb{N} \times \{1, \ldots, d\}$,

$$\mathbb{P}\{\mathcal{X}(1) = (x+y, i) \mid \mathcal{X}(0) = (x, i)\} = \mathbb{P}\{\xi^{(i)} = y\}, \quad y \in \mathbb{Z},$$

where integer-valued random variables $\xi^{(1)}, \ldots, \xi^{(d)}$ satisfy

$$\mathbb{E}\xi^{(i)} = 0, \quad v_i^2 := \operatorname{Var}\xi^{(i)} \in (0, \infty), \quad 1 \leq i \leq d,$$

and the distribution of $\xi^{(i)}$, $i = 1, \ldots, d$ is 1-arithmetic.

($\mathbf{A}_2$) For all $x \leq 0$ and $1 \leq i \leq d$,

$$\mathbb{P}\{R(1) \in \mathbb{N} \mid \mathcal{X}(0) = (x, i)\} = 1.$$

($\mathbf{A}_3$) There exists $C > 0$ such that, for all $x \leq 0$ and $1 \leq i \leq d$,

$$\mathbb{E}(R(1) \mid \mathcal{X}(0) = (x, i)) \leq C(1 + |x|).$$

($\mathbf{A}_4$) The states $\mathcal{N}^d$ of the Markov chain $\mathcal{X}$ are connected.

The qualitative behavior of $\mathcal{X}$ is as follows. Assumption ($\mathbf{A}_1$) says that for a "positive" initial point $\mathcal{X}(0) = (R(0), l(0)) = (x, i)$, with $x \in \mathbb{N}$, the radius R behaves like a standard random walk $S_{\xi^{(i)}}$ (with jumps of zero mean and finite variance) until the first exit time from $\mathbb{N}$:

$$\sigma = \sigma(R) := \inf\{k \geq 0 : R(k) \leq 0\}. \tag{3.73}$$

The label l takes the constant value i up to and including the stopping time σ:

$$l(0) = l(1) = \ldots = l(\sigma) = i.$$

Once the radius has reached a nonpositive value at time σ, the Markov chain $\mathcal{X}$, with probability one, jumps to a state with a "positive" radius. The label l may also change its value, so that

$$R(\sigma + 1) > 0, \quad l(\sigma + 1) \in \{1, \ldots, d\}.$$

We do not specify explicitly the distribution of $(R(\sigma + 1), l(\sigma + 1))$, although there is the technical Assumption (**A**$_3$) which controls the size of the radius' jump at time $\sigma + 1$.

Let $(\sigma_i)_{i \geq 1}$ denote the sequence of exit times of R from $\mathbb{N}$, that is,

$$\sigma_1 := \sigma(R) = \inf\{k \geq 0 : R(k) \leq 0\} \quad \text{and} \quad \sigma_{i+1} := \inf\{k > \sigma_i : R(k) \leq 0\} \quad (3.74)$$

for $i \in \mathbb{N}$. Also, put

$$\tau_i := \sigma_i + 1, \quad i \in \mathbb{N}. \quad (3.75)$$

Assumption (**A**$_2$) ensures that $\tau_1, \tau_2, \ldots$ are consecutive entrance times of R into $\mathbb{N}$.

With the sequences $(\sigma_i)_{i \geq 1}$, $(\tau_i)_{i \geq 1}$ we associate an embedded *exit* Markov chain $\mathcal{X}_{\text{exit}} = (\mathcal{X}_{\text{exit}}(i))_{i \geq 1} := ((R_{\text{exit}}(i), l_{\text{exit}}(i)))_{i \geq 1}$ and an embedded *entrance* Markov chain $\mathcal{X}_{\text{entr}} = (\mathcal{X}_{\text{entr}}(i))_{i \geq 1} := ((R_{\text{entr}}(i), l_{\text{entr}}(i)))_{i \geq 1}$, taking values in $\{\ldots, -2, -1, 0\} \times \{1, \ldots, d\}$ and $\mathcal{N}^d$, respectively. These are defined by

$$\mathcal{X}_{\text{exit}}(i) := \mathcal{X}(\sigma_i), \quad \mathcal{X}_{\text{entr}}(i) := \mathcal{X}(\tau_i), \quad i \in \mathbb{N}. \quad (3.76)$$

We proceed by discussing the existence of stationary distributions of the embedded Markov chains and an integrability property of these stationary distributions.

Lemma 3.3.1 *Suppose*[13] *(**A**$_1$), (**A**$_2$), and (**A**$_4$). Then the Markov chains $\mathcal{X}_{\text{exit}}$ and $\mathcal{X}_{\text{entr}}$ have unique stationary probability distributions π_{exit} and π_{entr}, respectively.*

Proof Assumption (**A**$_4$) ensures that either the Markov chain $\mathcal{X}_{\text{exit}}$ has a unique stationary probability distribution π_{exit}, or a unique invariant distribution is a zero measure. We intend to prove that the Markov chain $\mathcal{X}_{\text{exit}}$ is positive recurrent, thereby demonstrating that the first alternative prevails. To this end, we show that $\{0\} \times \{1, \ldots, d\}$ is its finite regeneration set satisfying the following property: there exists $c > 0$ such that, for all $(x, i) \in \{\ldots, -2, -1, 0\} \times \{1, \ldots, d\}$,

$$\mathbb{P}\{\mathcal{X}_{\text{exit}}(2) \in \{0\} \times \{1, \ldots, d\} \mid \mathcal{X}_{\text{exit}}(1) = (x, i)\} \geq c.$$

---

[13] Assumption (**A**$_2$) is used in our definition of $\mathcal{X}_{\text{entr}}$. It ensures that the τ's defined in (3.75) are indeed the entrance times of $\mathcal{X}$ into $\mathcal{N}^d$. No other use of (**A**$_2$) is made in the proof of Lemma 3.3.1.

By decreasing c if necessary, it is sufficient to verify that

$$\mathbb{P}\{\mathcal{X}(\sigma) = (0, i) \mid \mathcal{X}(0) = (x, i)\} \geq c \tag{3.77}$$

for all $(x, i) \in \mathcal{N}^d$.

Fix $i \in \{1, \ldots, d\}$. Denote by $H_-^{(i)}$ the (strictly) descending ladder height of a standard random walk $S_{\xi^{(i)}}$, that is, the position of $S_{\xi^{(i)}}$ at the time of its first visit to the set $\{\ldots, -2, -1\}$. Assumption $(\mathbf{A}_1)$ ensures that $\mathbb{E}|H_-^{(i)}| < \infty$; see, for instance, formula (4b) in [39]. In view of assumption $(\mathbf{A}_4)$, for all $x \in \mathbb{N}$,

$$\mathbb{P}\{\mathcal{X}(\sigma) = (0, i) \mid \mathcal{X}(0) = (x, i)\} > 0. \tag{3.78}$$

For each $y \in \mathbb{N}_0$, the probability $\mathbb{P}\{\mathcal{X}(\sigma) = (-y, i) \mid \mathcal{X}(0) = (x, i)\}$ coincides with the probability of the event that a positive overshoot at $x - 1$ of $-S_{\xi^{(i)}}$, with $S_{\xi^{(i)}}(0) = 0$, is equal to $y + 1$. Hence, according to formula (10.6) on p. 104 in [60], for each fixed $y \in \mathbb{N}_0$,

$$\lim_{x \to +\infty} \mathbb{P}\{\mathcal{X}(\sigma) = (-y, i) \mid \mathcal{X}(0) = (x, i)\} = \frac{\mathbb{P}\{H_-^{(i)} \leq -y - 1\}}{\mathbb{E}|H_-^{(i)}|}$$

and particularly, for $y = 0$,

$$\lim_{x \to +\infty} \mathbb{P}\{\mathcal{X}(\sigma) = (0, i) \mid \mathcal{X}(0) = (x, i)\} = \frac{1}{\mathbb{E}|H_-^{(i)}|} > 0. \tag{3.79}$$

Now formulas (3.78) and (3.79) entail

$$\min_{i=1,\ldots,d} \inf_{x \in \mathbb{N}} \mathbb{P}\{\mathcal{X}(\sigma) = (0, i) \mid \mathcal{X}(0) = (x, i)\} > 0,$$

thereby proving (3.77), and positive recurrence of $\mathcal{X}_{\text{exit}}$ follows.

Positive recurrence of the Markov chain $\mathcal{X}_{\text{entr}}$ follows from positive recurrence of $\mathcal{X}_{\text{exit}}$ and the equalities

$$\mathcal{X}_{\text{entr}}(i) = \mathcal{X}(\tau_i) = \mathcal{X}(\sigma_i + 1) \quad \text{and} \quad \mathcal{X}_{\text{exit}}(i) = \mathcal{X}(\sigma_i), \qquad i \in \mathbb{N}.$$

$\square$

The stationary distributions π_{exit} and π_{entr} are related as follows:

$$\begin{aligned} \pi_{\text{exit}}(A) &= \mathbb{P}^{\pi_{\text{entr}}}\{\mathcal{X}(\sigma) \in A\}, & A \subseteq \{\ldots, -2, -1, 0\} \times \{1, \ldots, d\}, \\ \pi_{\text{entr}}(\widetilde{A}) &= \mathbb{P}^{\pi_{\text{exit}}}\{\mathcal{X}(1) \in \widetilde{A}\}, & \widetilde{A} \subseteq \mathcal{N}^d. \end{aligned} \tag{3.80}$$

3.3 Random Walks with Membrane and a Skew Brownian Motion

For notational simplicity we shall write the expectation with respect to the stationary distribution π_{exit} of a function $f : \{\ldots, -2, -1, 0\} \times \{1, \ldots, d\} \to \mathbb{R}$ as

$$\mathbb{E}^{\pi_{\text{exit}}} f(R(0), l(0)) := \sum_{i \leq 0} \sum_{j=1}^{d} f(i, j) \pi_{\text{exit}}(\{i, j\}). \tag{3.81}$$

The same convention applies to the expectation with respect to π_{entr}.

Lemma 3.3.2 *Suppose* $(\mathbf{A}_1)$–$(\mathbf{A}_4)$. *Then*

$$\mathbb{E}^{\pi_{\text{exit}}} |R(0)| = \sum_{x \geq 0} \sum_{j=1}^{d} x \, \pi_{\text{exit}}(\{-x, j\}) < \infty, \tag{3.82}$$

$$\mathbb{E}^{\pi_{\text{entr}}} R(0) = \sum_{x \geq 1} \sum_{j=1}^{d} x \, \pi_{\text{entr}}(\{x, j\}) < \infty. \tag{3.83}$$

Proof In view of

$$\mathbb{E}^{\pi_{\text{entr}}} R(0) = \sum_{x \geq 1} \sum_{j=1}^{d} x \, \pi_{\text{entr}}(\{x, j\}) \overset{(3.80)}{=} \sum_{x \geq 1} \sum_{j=1}^{d} x \, \mathbb{P}^{\pi_{\text{exit}}} \{\mathcal{X}(1) = \{x, j\}\}$$

$$= \mathbb{E}^{\pi_{\text{exit}}} R(1) = \sum_{x \leq 0} \sum_{j=1}^{d} \mathbb{E}^{(x,j)}(R(1)) \pi_{\text{exit}}(\{x, j\})$$

$$\overset{(\mathbf{A}_3)}{\leq} C \sum_{x \leq 0} \sum_{j=1}^{d} (1 + |x|) \pi_{\text{exit}}(\{x, j\}) = C(1 + \mathbb{E}^{\pi_{\text{exit}}} |R(0)|),$$

it suffices to check (3.82).

By the strong law of large numbers for Markov chains,

$$\lim_{n \to \infty} \frac{\sum_{k=1}^{n} |R_{\text{exit}}(k)|}{n} = \mathbb{E}^{\pi_{\text{exit}}} |R(0)| \quad \text{a.s.} \tag{3.84}$$

irrespective of the distribution of $R(0)$. We assume in the remainder of the proof that $R(0) \in \mathbb{N}$ a.s. and that $\mathbb{E} R(0) < \infty$.

Using the notation of the proof of Lemma 3.3.1, put, for $t \geq 0$,

$$\nu^{(i)}(t) := \inf\{k \in \mathbb{N} : -S_{H_{-}^{(i)}}(k) > t\}$$

and then $U^{(i)}(t) := \mathbb{E}\nu^{(i)}(t)$. It follows from the proof of Lemma 3.3.1 that $\mathbb{E}^{(x,i)}|R(\sigma)|$ is equal to -1 plus the mean of a positive overshoot at $x-1$ of $-S_{\xi^{(i)}}$. By formula[14] (2.6) on p. 83 in [60] and Wald's identity, it is also equal to

$$\mathbb{E}(-S_{H_-^{(i)}}(\nu^{(i)}(x-1))-x) = (\mathbb{E}|H_-^{(i)}|)U^{(i)}(x-1)-x.$$

By the elementary renewal theorem, see, for instance, Theorem 4.1 on p. 54 in [60], $\lim_{t\to\infty}(U^{(i)}(t)/t) = 1/\mathbb{E}|H_-^{(i)}|$. Thus, given $\varepsilon > 0$ there is $x_0 = x_0(\varepsilon) > 0$ such that $(\mathbb{E}|H_-^{(i)}|)U^{(i)}(x-1)-x \leq \varepsilon x$ whenever $x \geq x_0$. By monotonicity of $U^{(i)}$, $(\mathbb{E}|H_-^{(i)}|)U^{(i)}(x-1)-x \leq (\mathbb{E}|H_-^{(i)}|)U^{(i)}(x_0(\varepsilon)-1) =: a(\varepsilon) = a$ for $x \in [1, x_0]$. Thus, for all $(x, i) \in \mathcal{N}^d$,

$$\mathbb{E}^{(x,i)}|R(\sigma)| = \mathbb{E}^{(x,i)}|R(\sigma_1)| \leq \varepsilon x + a. \tag{3.85}$$

In what follows we fix some $\varepsilon \in (0, C^{-1})$, where C is the constant appearing in $(\mathbf{A}_3)$. For $n \in \mathbb{N}$, put

$$x_n := \mathbb{E}|R(\sigma_n)| = \mathbb{E}|R_{\text{exit}}(n)|, \quad y_n := \mathbb{E}R(\tau_n) = \mathbb{E}R_{\text{entr}}(n).$$

Formula (3.85) in combination with the assumption imposed on $R(0)$ ensures that $x_1 < \infty$. Using

$$y_{n+1} = \sum_{x \leq 0} \mathbb{E}\big(|R_{\text{entr}}(n+1)| \,\big|\, R_{\text{exit}}(n+1) = x\big)\mathbb{P}\{R_{\text{exit}}(n+1) = x\}$$

$$\overset{(\mathbf{A}_3)}{\leq} \sum_{x \leq 0} C(1+|x|)\mathbb{P}\{R_{\text{exit}}(n+1) = x\} = C(1+x_{n+1})$$

for $n \in \mathbb{N}_0$ and

$$x_{n+1} = \sum_{x \geq 1} \mathbb{E}\big[|R_{\text{exit}}(n+1)| \,\big|\, R_{\text{entr}}(n) = x\big]\mathbb{P}\{R_{\text{entr}}(n) = x\}$$

$$\overset{(3.85)}{\leq} \sum_{x \geq 1}(\varepsilon x + a)\mathbb{P}\{R_{\text{entr}}(n) = x\} = \varepsilon y_n + a$$

for $n \in \mathbb{N}$, we conclude that, for each $n \in \mathbb{N}$, x_n and y_n are finite. Furthermore,

$$x_{n+1} \leq \varepsilon C x_n + \varepsilon C + a, \quad n \in \mathbb{N}$$

---

[14] The formula stems from the fact that a positive overshoot occurs at a ladder epoch, so that the corresponding position of a standard random walk is a ladder height.

3.3 Random Walks with Membrane and a Skew Brownian Motion

and thereupon

$$\overline{K} := \sup_{n \geq 1} x_n < \infty$$

because $\varepsilon C \in (0, 1)$. By (3.84) and Fatou's lemma,

$$\mathbb{E}^{\pi_{\text{exit}}} |R(0)| = \mathbb{E}\left(\lim_{n \to \infty} \frac{\sum_{k=1}^{n} |R_{\text{exit}}(k)|}{n} \right) \leq \liminf_{n \to \infty} \mathbb{E}\left(\frac{\sum_{k=1}^{n} |R_{\text{exit}}(k)|}{n} \right) \leq \overline{K}.$$

The proof of Lemma 3.3.2 is complete. □

To formulate our result on scaling limits of $\mathcal{X}$ we need another piece of notation. For each $n \in \mathbb{N}$, let $(\mathcal{X}^{(n)}(k))_{k \geq 0} := (R^{(n)}(k), l^{(n)}(k))_{k \geq 0}$ be a Markov chain having the same transition probabilities as $(\mathcal{X}(k))_{k \geq 0}$, but possibly satisfying a different initial condition. For each $n \in \mathbb{N}$, the sequences of stopping times $(\sigma_i^{(n)})_{i \geq 1}$ and $(\tau_i^{(n)})_{i \geq 1}$ are defined by (3.74) and (3.75), with $R^{(n)}(k)$ replacing $R(k)$. Assuming $(\mathbf{A}_1)$, put, for $t \geq 0$,

$$X^{(n)}(t) := \left(\frac{R^{(n)}(\lfloor nt \rfloor)}{v_1 \sqrt{n}} \mathbb{1}_{\{l^{(n)}(\lfloor nt \rfloor) = 1\}}, \ldots, \frac{R^{(n)}(\lfloor nt \rfloor)}{v_d \sqrt{n}} \mathbb{1}_{\{l^{(n)}(\lfloor nt \rfloor) = d\}} \right)$$

and

$$L^{(n)}(t) = \frac{1}{\sqrt{n}} \sum_{i: \tau_i^{(n)} \leq \lfloor nt \rfloor} \left(\frac{R^{(n)}(\tau_i^{(n)})}{v_{l^{(n)}(\tau_i^{(n)})}} - \frac{R^{(n)}(\sigma_i^{(n)})}{v_{l^{(n)}(\sigma_i^{(n)})}} \right). \tag{3.86}$$

Let E_d be the union of nonnegative coordinate semi-axes in $\mathbb{R}^d$ as defined in Sect. 2.2.2 and $W_{\mathbf{p}}(\cdot, x)$ a Walsh Brownian motion starting at $x \in E_d$ with parameter $\mathbf{p} = (p_1, \ldots, p_d)$; see Sect. 2.2.2 for the definition. Denote by $L_0^{W_{\mathbf{p}}(\cdot, x)} := \left(L_0^{W_{\mathbf{p}}(\cdot, x)}(t) \right)_{t \geq 0}$ the local time of $W_{\mathbf{p}}(\cdot, x)$ at 0; see (2.51).

Theorem 3.3.2 *Suppose* $(\mathbf{A}_1)$–$(\mathbf{A}_4)$ *and that*

$$X^{(n)}(0) \xrightarrow{\mathbb{P}} x, \quad n \to \infty \tag{3.87}$$

for some $x \in E_d$. Then

$$(X^{(n)}(t), L^{(n)}(t))_{t \geq 0} \implies \left(W_{\mathbf{p}}(t, x), L_0^{W_{\mathbf{p}}(\cdot, x)}(t) \right)_{t \geq 0}, \quad n \to \infty$$

in the J_1-topology on $D([0,\infty), \mathbb{R}^2)$. The parameter $\mathbf{p}$ of $W_\mathbf{p}$ is given by

$$\begin{aligned}
p_k &:= \frac{\mathbb{E}^{\pi_{\text{entr}}}[R(0)v_k^{-1}\mathbb{1}_{\{l(0)=k\}}] - \mathbb{E}^{\pi_{\text{exit}}}[R(0)v_k^{-1}\mathbb{1}_{\{l(0)=k\}}]}{\mathbb{E}^{\pi_{\text{entr}}}[R(0)v_{l(0)}^{-1}] - \mathbb{E}^{\pi_{\text{exit}}}[R(0)v_{l(0)}^{-1}]} \\
&= \frac{\mathbb{E}^{\pi_{\text{entr}}}[(R(0) - R(\sigma))v_k^{-1}\mathbb{1}_{\{l(0)=k\}}]}{\mathbb{E}^{\pi_{\text{entr}}}[(R(0) - R(\sigma))v_{l(0)}^{-1}]} \\
&= \frac{\mathbb{E}^{\pi_{\text{exit}}}[R(1)v_{l(1)}^{-1}\mathbb{1}_{\{l(1)=k\}} - R(0)v_{l(0)}^{-1}\mathbb{1}_{\{l(0)=k\}}]}{\mathbb{E}^{\pi_{\text{exit}}}[R(1)v_{l(1)}^{-1} - R(0)v_{l(0)}^{-1}]}, \quad k = 1, \ldots, d.
\end{aligned} \tag{3.88}$$

We shall prove this theorem in Sect. 3.3.3.

Remark 3.3.2 Equalities (3.88) follow from (3.80) and the definition of the embedded exit and entrance Markov chains.

Let X be a random walk with a membrane $A = \{-m, \ldots, m\}$ as defined in Theorem 3.3.1. Our purpose is to identify the permeability parameter γ of the limit skew Brownian motion in Theorem 3.3.1. To this end, we introduce embedded entrance and exit Markov chains, similar to those described in Theorem 3.3.2.

Define the stopping times

$$\begin{aligned}
\widetilde{\sigma}_0 &:= 0, \\
\widetilde{\tau}_i &:= \inf\{k \geq \widetilde{\sigma}_i : |X(k)| > m\}, \quad i \in \mathbb{N}_0, \\
\widetilde{\sigma}_{i+1} &= \inf\{k > \widetilde{\tau}_i : (X(k-1) > m \text{ and } X(k) \leq m) \\
&\qquad \text{or } (X(k-1) < -m \text{ and } X(k) \geq -m)\}, \quad i \in \mathbb{N}_0.
\end{aligned} \tag{3.89}$$

The difference between definition (3.89) and definitions (3.74) and (3.75) is twofold. First, if $X(\widetilde{\sigma}_i) \in \{-m, \ldots, m\}$, the Markov chain X may stay within the membrane $\{-m, \ldots, m\}$ for more than one unit of time. Hence, in general, $\widetilde{\tau}_i \geq \widetilde{\sigma}_i + 1$ in this case. Second, when X "jumps over" the membrane, that is, when $X(\widetilde{\sigma}_i - 1) > m$ and $X(\widetilde{\sigma}_i) < -m$ or vice versa, we infer $\widetilde{\tau}_i = \widetilde{\sigma}_i$.

With this notation at hand, we restate assumptions (I)–(IV) of Theorem 3.3.1 in a form adapted to Assumptions ($\mathbf{A}_1$)–($\mathbf{A}_4$) of Theorem 3.3.2.

($\mathbf{B}_1$)   For all $y \in \mathbb{Z}$,

$$\mathbb{P}\{X(1) = x + y \mid X(0) = x\} = \begin{cases} \mathbb{P}\{\xi_+ = y\}, & \text{if } x > m, \\ \mathbb{P}\{\xi_- = y\}, & \text{if } x < -m, \end{cases}$$

3.3 Random Walks with Membrane and a Skew Brownian Motion

where integer-valued random variables ξ_- and ξ_+ satisfy

$$\mathbb{E}\xi_\pm = 0, \quad v_\pm^2 = \text{Var}\,\xi_\pm \in (0, \infty),$$

and their distributions are 1-arithmetic.

(**B$_2$**)   For all $x \in \{-m, \ldots, m\}$,

$$\mathbb{E}\big(|X(\widetilde{\tau}_0)|\big|X(0) = x\big) < \infty.$$

(**B$_3$**)   For all $x \in \{-m, \ldots, m\}$,

$$\mathbb{P}\{\widetilde{\tau}_0 < \infty | X(0) = x\} = 1.$$

(**B$_4$**)   The states $\mathbb{Z} \setminus \{-m, \ldots, m\}$ of the Markov chain X are connected.

Remark 3.3.3 A condition

$$\max_{|x| \leq m} \mathbb{E}\big(|X(1)| \,\big|\, X(0) = x\big) < \infty$$

is equivalent to condition (II) in Theorem 3.3.1. Also, it entails Assumption (**B$_2$**). Indeed, for all $|x| \leq m$,

$$\mathbb{E}\big(|X(\widetilde{\tau}_0)| \,\big|\, X(0) = x\big) = \sum_{|y| \leq m}\sum_{k \geq 1} \mathbb{P}\{\widetilde{\tau}_0 = k, X(k-1) = y \mid X(0) = x\}$$

$$\times \mathbb{E}\big(|X(k)| \,\big|\, \widetilde{\tau}_0 = k, X(k-1) = y\big)$$

$$= \sum_{|y| \leq m}\sum_{k \geq 1} \mathbb{P}\{\widetilde{\tau}_0 = k, X(k-1) = y \mid X(0) = x\}$$

$$\times \mathbb{E}\big(|X(1)| \,\big|\, \widetilde{\tau}_0 = 1, X(0) = y\big)$$

$$= \sum_{|y| \leq m}\sum_{k \geq 1} \mathbb{P}\{\widetilde{\tau}_0 = k, X(k-1) = y \mid X(0) = x\}$$

$$\times \frac{\mathbb{E}(|X(1)|\mathbb{1}_{\{|X(1)|>m\}} \mid X(0) = y)}{\mathbb{P}\{|X(1)| > m \mid X(0) = y\}}$$

$$\leq \sum_{|y| \leq m} \frac{\mathbb{E}[|X(1)|\mathbb{1}_{\{|X(1)|>m\}} \mid X(0) = y]}{\mathbb{P}\{|X(1)| > m \mid X(0) = y\}} < \infty.$$

In the first line we have used Assumption (**B$_3$**). In the last line we have interpreted $0/0$ as 0.

As before, we introduce embedded Markov chains $(X_{\text{exit}}(i))_{i\geq 1} := (X(\widetilde{\sigma}_i))_{i\geq 1}$ and $(X_{\text{entr}}(i))_{i\geq 0} := (X(\widetilde{\tau}_i))_{i\geq 0}$. Similarly to Lemmas 3.3.1 and 3.3.2, it can be shown that X_{exit} and X_{entr} have unique stationary distributions $\widetilde{\pi}_{\text{exit}}$ and $\widetilde{\pi}_{\text{entr}}$, respectively, and that these are integrable.

For each $n \in \mathbb{N}$, let $(X^{(n)}(k))_{k\geq 0}$ be a Markov chain having the same transition probabilities as $(X(k))_{k\geq 0}$, but possibly satisfying a different initial condition.

Theorem 3.3.3 *Suppose* (**B**$_1$)–(**B**$_4$) *and that*

$$\varphi_{\text{sgn}}\left(\frac{X^{(n)}(0)}{\sqrt{n}}\right) \xrightarrow{\mathbb{P}} x, \quad n \to \infty \qquad (3.90)$$

for some $x \in \mathbb{R}$, where φ_{sgn} is as defined in (3.71). Then

$$\left(\varphi_{\text{sgn}}\left(\frac{X^{(n)}(\lfloor nt \rfloor)}{\sqrt{n}}\right)\right)_{t\geq 0} \Longrightarrow \left(W^{\text{skew}}_\gamma(t, x)\right)_{t\geq 0}, \quad n \to \infty$$

on $D([0, \infty))$, where $(W^{\text{skew}}_\gamma(t, x))_{t\geq 0}$ is a skew Brownian motion starting at x with the permeability parameter

$$\gamma = \frac{\mathbb{E}^{\widetilde{\pi}_{entr}}\varphi_{\text{sgn}}(X(0) - X(\widetilde{\sigma}_1))}{\mathbb{E}^{\widetilde{\pi}_{entr}}|\varphi_{\text{sgn}}(X(0) - X(\widetilde{\sigma}_1))|} = \frac{\mathbb{E}^{\widetilde{\pi}_{entr}}(X(0) - X(\widetilde{\sigma}_1))v^{-1}_{\text{sgn}(X(0)-X(\widetilde{\sigma}_1))}}{\mathbb{E}^{\widetilde{\pi}_{entr}}|X(0) - X(\widetilde{\sigma}_1)|v^{-1}_{\text{sgn}(X(0)-X(\widetilde{\sigma}_1))}}$$

$$= \frac{\mathbb{E}^{\widetilde{\pi}_{entr}}(X(0) - X(\widetilde{\sigma}_1))v^{-1}_{\text{sgn}\, X(0)}}{\mathbb{E}^{\widetilde{\pi}_{entr}}|X(0) - X(\widetilde{\sigma}_1)|v^{-1}_{\text{sgn}\, X(0)}}. \qquad (3.91)$$

Remark 3.3.4 The last equality in (3.91) follows from the fact that $\mathbb{P}^{\widetilde{\pi}_{\text{entr}}}\{|X(0)| > m\} = \widetilde{\pi}_{\text{entr}}(\{x \in \mathbb{Z} : |x| > m\}) = 1$. Together with the definition of $\widetilde{\sigma}_1$ it implies that $\text{sgn}\, X(0) = \text{sgn}(X(0) - X(\widetilde{\sigma}_1))$ under $\widetilde{\pi}_{\text{entr}}$.

3.3.2 Examples

Example 3.3.1 We work in the setting of Theorem 3.3.3. Assume that the original unperturbed random walk acting on both sides beyond the membrane is a simple symmetric random walk. Thus, $\xi_+ \stackrel{d}{=} \xi_- \stackrel{d}{=} \xi$, where $\mathbb{P}\{\xi = \pm 1\} = \frac{1}{2}$. As a consequence, $v^2_\pm = 1$ and $\varphi_{\text{sgn}}(x) = x$.

3.3 Random Walks with Membrane and a Skew Brownian Motion

CASE $m = 0$ in which the membrane consists of the single point $\{0\}$. Since $X(\widetilde{\sigma}_k) = 0$ for each $k \in \mathbb{N}$, the stationary distribution $\widetilde{\pi}_{\text{exit}}$ is given by $\widetilde{\pi}_{\text{exit}}(\{0\}) = 1$. Let η be a random variable having the same distribution as a generic jump from 0, that is,

$$\mathbb{P}\{X(1) = j \mid X(0) = 0\} = \mathbb{P}\{\eta = j\}, \quad j \in \mathbb{Z}.$$

Assumption (**B**$_3$) simply means that $\mathbb{P}\{\eta = 0\} < 1$. Hence, the entrance stationary distribution coincides with the distribution of X at the time of exit from 0, that is,

$$\widetilde{\pi}_{\text{entr}}(\{j\}) = \mathbb{P}\{X(\widetilde{\tau}_0) = j \mid X(0) = 0\} = \frac{\mathbb{P}\{\eta = j\}}{\mathbb{P}\{\eta \neq 0\}}, \quad j \neq 0.$$

Thus, according to (3.91) the permeability parameter γ of the limit skew Brownian motion is

$$\gamma = \frac{\mathbb{E}^{\widetilde{\pi}_{\text{entr}}} \varphi_{\text{sgn}}(X(0) - X(\widetilde{\sigma}_1))}{\mathbb{E}^{\widetilde{\pi}_{\text{entr}}} |\varphi_{\text{sgn}}(X(0) - X(\widetilde{\sigma}_1))|} = \frac{\mathbb{E}^{\widetilde{\pi}_{\text{entr}}}(X(0) - X(\widetilde{\sigma}_1))}{\mathbb{E}^{\widetilde{\pi}_{\text{entr}}} |X(0) - X(\widetilde{\sigma}_1)|}$$

$$= \frac{\mathbb{E}^{\widetilde{\pi}_{\text{entr}}} X(0)}{\mathbb{E}^{\widetilde{\pi}_{\text{entr}}} |X(0)|} = \frac{\sum_{j \in \mathbb{Z}} j \frac{\mathbb{P}\{\eta = j\}}{\mathbb{P}\{\eta \neq 0\}}}{\sum_{j \in \mathbb{Z}} |j| \frac{\mathbb{P}\{\eta = j\}}{\mathbb{P}\{\eta \neq 0\}}} = \frac{\mathbb{E}\eta}{\mathbb{E}|\eta|}. \quad (3.92)$$

In particular, if

$$\mathbb{P}\{\eta = 1\} = \mathbb{P}\{X(1) = 1 \mid X(0) = 0\} = p$$

and

$$\mathbb{P}\{\eta = -1\} = \mathbb{P}\{X(1) = -1 \mid X(0) = 0\} = 1 - p$$

for some $p \in (0, 1)$, then $|\eta| = 1$ a.s. and

$$\gamma = 2p - 1. \quad (3.93)$$

Formulas (3.92) and (3.93) appeared for the first time in [65] (formula (3.92) was only announced there).

CASE $m \in \mathbb{N}$. Assume that X exits from $\{-m, \ldots, m\}$ to the neighboring points $m + 1$ or $-m - 1$ only. Then the entrance Markov chain X_{entr} and the distribution $\widetilde{\pi}_{\text{entr}}$ are supported by the two-point set $\{-m - 1, m + 1\}$.

Put

$$\alpha := \mathbb{P}\{X(\widetilde{\tau}_0) = m + 1 \mid X(0) = -m\} \quad \text{and} \quad \beta := \mathbb{P}\{X(\widetilde{\tau}_0) = -m - 1 \mid X(0) = m\}.$$

The transition probabilities of the entrance Markov chain X_{entr} are given by

$$\mathbb{P}\{X_{\text{entr}}(1) = -m-1 \mid X_{\text{entr}}(0) = -m-1\} = 1-\alpha,$$
$$\mathbb{P}\{X_{\text{entr}}(1) = m+1 \mid X_{\text{entr}}(0) = -m-1\} = \alpha,$$
$$\mathbb{P}\{X_{\text{entr}}(1) = -m-1 \mid X_{\text{entr}}(0) = m+1\} = \beta,$$
$$\mathbb{P}\{X_{\text{entr}}(1) = m+1 \mid X_{\text{entr}}(0) = m+1\} = 1-\beta.$$

The stationary distribution $\widetilde{\pi}_{\text{entr}}$ is given by

$$\widetilde{\pi}_{\text{entr}}(\{m+1\}) = \frac{\alpha}{\alpha+\beta}, \quad \widetilde{\pi}_{\text{entr}}(\{-m-1\}) = \frac{\beta}{\alpha+\beta}.$$

Observe that $X(\widetilde{\sigma}_1) = m$ if $X(0) = m+1$ and $X(\widetilde{\sigma}_1) = -m$ if $X(0) = -m-1$. Hence, $X(0) - X(\widetilde{\sigma}_1) = \operatorname{sgn} X(0)$ provided that $|X(0)| = m+1$, and the permeability parameter γ is equal to

$$\gamma = \widetilde{\pi}_{\text{entr}}(\{m+1\}) - \widetilde{\pi}_{\text{entr}}(\{-m-1\}) = \frac{\alpha-\beta}{\alpha+\beta}.$$

This result was originally obtained in [127].

Example 3.3.2 We keep working in the setting of Theorem 3.3.3. Let $m \in \mathbb{N}$ and assume that $\mathbb{P}\{\xi_+ \geq -1\} = \mathbb{P}\{\xi_- \leq 1\} = 1$. Then the exit Markov chain X_{exit} and the distribution $\widetilde{\pi}_{\text{exit}}$ are supported by the two-point set $\{-m, m\}$.

Recall that $\widetilde{\tau}_0 = \inf\{k \geq 0 \colon |X(k)| > m\}$ and put

$$\alpha := \mathbb{P}\{X(\widetilde{\tau}_0) > m \mid X(0) = -m\}, \quad \beta := \mathbb{P}\{X(\widetilde{\tau}_0) < -m \mid X(0) = m\}.$$

Since $X(\widetilde{\sigma}_1) = m$ if $X(0) > m$ and $X(\widetilde{\sigma}_1) = -m$ if $X(0) < m$, the transition probabilities of the exit Markov chain X_{exit} are given by

$$\mathbb{P}\{X_{\text{exit}}(2) = -m \mid X_{\text{exit}}(1) = -m\} = 1-\alpha,$$
$$\mathbb{P}\{X_{\text{exit}}(2) = m \mid X_{\text{exit}}(1) = -m\} = \alpha,$$
$$\mathbb{P}\{X_{\text{exit}}(2) = -m \mid X_{\text{exit}}(1) = m\} = \beta,$$
$$\mathbb{P}\{X_{\text{exit}}(2) = m \mid X_{\text{exit}}(1) = m\} = 1-\beta.$$

The stationary exit distribution $\widetilde{\pi}_{\text{exit}}$ is given by

$$\pi_+ := \widetilde{\pi}_{\text{exit}}(\{m\}) = \frac{\alpha}{\alpha+\beta}, \quad \pi_- := \widetilde{\pi}_{\text{exit}}(\{-m\}) = \frac{\beta}{\alpha+\beta},$$

3.3 Random Walks with Membrane and a Skew Brownian Motion

and the stationary entrance distribution $\widetilde{\pi}_{\mathrm{entr}}$ is given by

$$\pi_{\mathrm{entr}}(\{x\}) = \pi_-\mathbb{P}\{X(\widetilde{\tau}_0) = x \mid X(0) = -m\} + \pi_+\mathbb{P}\{X(\widetilde{\tau}_0) = x \mid X(0) = m\},$$

$$|x| > m.$$

Observe that $X(0) - X(\widetilde{\sigma}_1) = X(0) - m \operatorname{sgn} X(0)$ provided that $|X(0)| > m$. Hence, the permeability parameter γ is equal to

$$\gamma = \frac{\mathbb{E}^{\widetilde{\pi}_{\mathrm{entr}}}(X(0) - m \operatorname{sgn} X(0)) v_{\operatorname{sgn} X(0)}^{-1}}{\mathbb{E}^{\widetilde{\pi}_{\mathrm{entr}}}|X(0) - m \operatorname{sgn} X(0)| v_{\operatorname{sgn} X(0)}^{-1}}.$$

In particular, if $v_+ = v_-$, then

$$\gamma = \frac{\mathbb{E}^{\widetilde{\pi}_{\mathrm{entr}}}(X(0) - m \operatorname{sgn} X(0))}{\mathbb{E}^{\widetilde{\pi}_{\mathrm{entr}}}|X(0) - m \operatorname{sgn} X(0)|}.$$

Example 3.3.3 We now switch to the setting of Theorem 3.3.2. Fix $d \in \mathbb{N}$ and let $\mathcal{X}$ be a locally perturbed random walk on $\mathcal{Z}^d$, as defined at the beginning of Sect. 3.3.1. Assume that

$$\mathbb{P}\{\xi^{(i)} = \pm 1\} = 1/2, \quad \mathbb{P}\{\mathcal{X}(0) = (1, j) \mid \mathcal{X}(0) = (0, i)\} = q_{i,j}, \quad 1 \le i, j \le d,$$

where $Q = (q_{i,j})_{1 \le i, j \le d}$ is an irreducible stochastic matrix.

The embedded chains $\mathcal{X}_{\mathrm{exit}} = (\mathcal{X}_{\mathrm{exit}}(k))_{k \ge 1} = (\mathcal{X}(\sigma_k))_{k \ge 1}$ and $\mathcal{X}_{\mathrm{entr}} = (\mathcal{X}_{\mathrm{entr}}(k))_{k \ge 1} = (\mathcal{X}(\tau_k))_{k \ge 1}$ have the state spaces $\{(0, i)\}_{1 \le i \le d}$ and $\{(1, i)\}_{1 \le i \le d}$, respectively. Also, these satisfy

$$\mathbb{P}\{\mathcal{X}_{\mathrm{entr}}(k+1) = (1, j) \mid \mathcal{X}_{\mathrm{entr}}(k) = (1, i)\} = q_{i,j}$$

and

$$\mathcal{X}(\sigma_{k+1}) = (0, i) \iff \mathcal{X}(\tau_k) = (1, i), \quad k \in \mathbb{N}, \quad 1 \le i \le d.$$

Let $\pi = (\pi_i)_{1 \le i \le d}$ be a unique stationary distribution corresponding to the matrix Q. Then the stationary distribution π_{entr} is given by

$$\pi_{\mathrm{entr}}(\{1, i\}) = \pi_i, \quad 1 \le i \le d.$$

Since $v_i = 1$, $1 \le i \le d$, the second equality in (3.88) implies that the parameter $\mathbf{p} = (p_1, \ldots, p_d)$ of the limit Walsh Brownian motion $W_{\mathbf{p}}$ in Theorem 3.3.2 coincides with π.

If $q_{i,j} = q_j$, that is, if the choice of the next label is independent of the previous one, then $\pi_i = q_i$, $1 \le i \le d$. Another particular case has been investigated in [47]. In that paper, $q_{i,j}$ depends on $(j - i) \bmod d$. It turns out that the stationary distribution π is then uniform on $\{1, \ldots, d\}$.

Example 3.3.4 Following [116], let $(Y(k))_{k \geq 0}$ be a time-homogeneous Markov chain defined by

$$Y(k+1) := \begin{cases} Y(k) + \xi_{k+1}^+, & \text{if } Y(k) > 0 \text{ and } Y(k) + \xi_{k+1}^+ > 0, \\ Y(k) + \xi_{k+1}^-, & \text{if } Y(k) < 0 \text{ and } Y(k) + \xi_{k+1}^- < 0, \\ \eta_{k+1}, & \text{if } Y(k) = 0, \\ 0, & \text{otherwise} \end{cases}$$

for $k \in \mathbb{N}_0$. Here, $(\xi_k^+)_{k \geq 1}$, $(\xi_k^-)_{k \geq 1}$, and $(\eta_k)_{k \geq 1}$ are mutually independent (also independent of $Y(0)$) sequences of independent copies of random variables ξ^+, ξ^-, and η, respectively. Moreover, ξ^- and ξ^+ are random variables with zero mean, variances $v_\pm^2 = \operatorname{Var} \xi^\pm \in (0, \infty)$, and 1-arithmetic distributions, and η is an integer-valued random variable with $\mathbb{E}|\eta| \in (0, \infty)$.

Since Y is forced to hit 0 at any attempt to jump over the origin, it does not satisfy the setting of Theorem 3.3.3. Nevertheless, our theory applies, and now we explain details. Let $\mathcal{X} = (\mathcal{X}(k))_{k \geq 0} := ((R(k), l(k)))_{k \geq 0}$ be an auxiliary Markov chain with the state space $\mathbb{Z} \times \{-1, +1\}$ defined as follows. Put

$$R(k+1) = \begin{cases} R(k) + \xi_{k+1}^+, & \text{if } l(k) = +1 \text{ and } R(k) > 0, \\ R(k) - \xi_{k+1}^-, & \text{if } l(k) = -1 \text{ and } R(k) > 0, \\ |\eta_k|, & \text{if } R(k) \leq 0 \end{cases}$$

for $k \in \mathbb{N}_0$ and

$$l(k+1) = \begin{cases} l(k), & \text{if } R(k) > 0, \\ +1, & \text{if } R(k) \leq 0 \text{ and } \eta_k \geq 0, \\ -1, & \text{if } R(k) \leq 0 \text{ and } \eta_k < 0 \end{cases}$$

for $k \in \mathbb{N}_0$, with the initial conditions

$$R(0) = |Y(0)|, \quad l(0) = \begin{cases} +1, & \text{if } Y(0) \geq 0, \\ -1, & \text{if } Y(0) < 0. \end{cases}$$

For $k \in \mathbb{N}_0$, the equality $Y(k+1) = l(k+1)R(k+1)$ holds in each of the following three cases:

$$\left\{Y(k) > 0 \text{ and } Y(k) + \xi_{k+1}^+ > 0\right\} \text{ or}$$

$$\left\{Y(k) < 0 \text{ and } Y(k) + \xi_{k+1}^- < 0\right\} \text{ or } \left\{Y(k) = 0\right\}.$$

Otherwise,

$$|Y(k+1) - l(k+1)R(k+1)| \leq |\xi_{k+1}^-| \vee |\xi_{k+1}^+|.$$

Since the second moment of $\xi^\pm$ is finite, we infer

$$\frac{1}{\sqrt{n}} \max_{0 \leq k \leq \lfloor nt \rfloor} \left(|\xi_{k+1}^-| \vee |\xi_{k+1}^+|\right) \xrightarrow{\mathbb{P}} 0, \quad n \to \infty$$

for each $t \geq 0$. Hence, by Slutsky's lemma, the weak limit of $\left(\frac{Y(\lfloor nt \rfloor)}{\sqrt{n}}\right)_{t \geq 0}$ as $n \to \infty$ exists if, and only if, the weak limit of $\left(\frac{l(\lfloor nt \rfloor)R(\lfloor nt \rfloor)}{\sqrt{n}}\right)_{t \geq 0}$ as $n \to \infty$ exists. Moreover, if exist, these have to be the same.

We apply Theorem 3.3.2 to derive the limit behavior of $\mathcal{X}$. The stationary distribution π_{entr} of the Markov chain $\mathcal{X}_{\text{entr}}$ is given by

$$\pi_{\text{entr}}(x, +1) = \mathbb{P}\{\eta = x \mid \eta \neq 0\},$$
$$\pi_{\text{entr}}(x, -1) = \mathbb{P}\{\eta = -x \mid \eta \neq 0\}, \quad x \in \mathbb{N}.$$

Recall the notation $\sigma = \inf\{k \geq 0 : R(k) \leq 0\}$ and let

$$f(x, \pm) := \mathbb{E}(R(\sigma) \mid (R(0), l(0)) = (x, \pm)), \quad x \in \mathbb{N}$$

denote the mean of the overshoot at 0. Hence, by formula (3.88),

$$p_+ = \frac{\mathbb{E}^{\pi_{\text{entr}}}(R(0) - R(\sigma))v_+^{-1}\mathbb{1}_{\{l(0)=+\}}}{\mathbb{E}^{\pi_{\text{entr}}}(R(0) - R(\sigma))v_{l(0)}^{-1}}$$

$$= \frac{\mathbb{E}(\eta - f(\eta, +))v_+^{-1}\mathbb{1}_{\{\eta > 0\}}}{\mathbb{E}(\eta - f(\eta, +))v_+^{-1}\mathbb{1}_{\{\eta > 0\}} + \mathbb{E}(-\eta - f(-\eta, -))v_-^{-1}\mathbb{1}_{\{\eta < 0\}}},$$

$$p_- = 1 - p_+.$$

According to Theorems 3.3.2 and 2.2.4, the weak limit of $\left(\varphi_{\text{sgn}}(Y(\lfloor nt \rfloor)/\sqrt{n})\right)_{t \geq 0}$, where φ_{sgn} is as defined in (3.71), is a skew Brownian motion with the permeability parameter $\gamma = p_+ - p_-$.

We intend to compare our formula for γ with the formula obtained in Theorem 1.1 of [116]. As a preparation, let $H^{(+)}$ denote the strictly descending ladder height of a standard random walk S_{ξ^+} with $S_{\xi^+}(0) = 0$ and $H^{(-)}$ denote the strictly ascending ladder height of a standard random walk S_{ξ^-} with $S_{\xi^-}(0) = 0$. Put

$$U^{(\pm)}(t) := 1 + \sum_{k \geq 1} \mathbb{P}\{\mp S_{H^{(\pm)}}(k) \leq t\}, \quad t \geq 0.$$

Using the calculation preceding formula (3.85) we infer

$$f(x, \pm) = x - (\mathbb{E}|H^{(\pm)}|)U^{(\pm)}(x-1), \quad x \in \mathbb{N}$$

and thereupon

$$\gamma = p_+ - p_- = \frac{\mathbb{E}(\eta - f(\eta, +))v_+^{-1}\mathbb{1}_{\{\eta > 0\}} - \mathbb{E}(-\eta - f(-\eta, -))v_-^{-1}\mathbb{1}_{\{\eta < 0\}}}{\mathbb{E}(\eta - f(\eta, +))v_+^{-1}\mathbb{1}_{\{\eta > 0\}} + \mathbb{E}(-\eta - f(-\eta, -))v_-^{-1}\mathbb{1}_{\{\eta < 0\}}}$$

$$= \frac{v_+^{-1}(\mathbb{E}|H^{(+)}|)\mathbb{E}U^{(+)}(\eta - 1)\mathbb{1}_{\{\eta \geq 1\}} - v_-^{-1}(\mathbb{E}H^{(-)})\mathbb{E}U^{(-)}(-\eta - 1)\mathbb{1}_{\{\eta \leq -1\}}}{v_+^{-1}(\mathbb{E}|H^{(+)}|)\mathbb{E}U^{(+)}(\eta - 1)\mathbb{1}_{\{\eta \geq 1\}} + v_-^{-1}(\mathbb{E}H^{(-)})\mathbb{E}U^{(-)}(-\eta - 1)\mathbb{1}_{\{\eta \leq -1\}}}.$$

Here is the formula given in [116]:

$$\gamma = \frac{v_+^{-1}(\mathbb{E}|H^{(+)}|)\mathbb{E}U^{(+)}(\eta)\mathbb{1}_{\{\eta \geq 1\}} - v_-^{-1}(\mathbb{E}H^{(-)})\mathbb{E}U^{(-)}(-\eta)\mathbb{1}_{\{\eta \leq -1\}}}{v_+^{-1}(\mathbb{E}|H^{(+)}|)\mathbb{E}U^{(+)}(\eta)\mathbb{1}_{\{\eta \geq 1\}} + v_-^{-1}(\mathbb{E}H^{(-)})\mathbb{E}U^{(-)}(-\eta)\mathbb{1}_{\{\eta \leq -1\}}},$$

that is, the argument of $U^{(\pm)}$ is $x-1$ in our formula, whereas it is x in the formula of [116]. We think that Lemma 2.1 in [116] suggests that $h(x)$ which is our $U^{(+)}(x)$ and $h'(x)$ which is our $U^{(-)}(x)$ should actually read $h(x-1)$ and $h'(x-1)$. With this amendment, our formula for γ and that in [116] do agree.

In the present example the distributions of ξ^+ and ξ^- are 1-arithmetic. As a consequence, $U^{(\pm)}(x-) = U^{(\pm)}(x-1)$ and the formula for γ can be written in an equivalent form:

$$\gamma = \frac{v_+^{-1}(\mathbb{E}|H^{(+)}|)\mathbb{E}U^{(+)}(\eta-)\mathbb{1}_{\{\eta > 0\}} - v_-^{-1}(\mathbb{E}H^{(-)})\mathbb{E}U^{(-)}((-\eta)-)\mathbb{1}_{\{\eta < 0\}}}{v_+^{-1}(\mathbb{E}|H^{(+)}|)\mathbb{E}U^{(+)}(\eta-)\mathbb{1}_{\{\eta > 0\}} + v_-^{-1}(\mathbb{E}H^{(-)})\mathbb{E}U^{(-)}((-\eta)-)\mathbb{1}_{\{\eta < 0\}}}.$$

We believe (but omit any further details) that the latter formula remains valid in a more general situation where $\xi^{\pm}$ and/or η are no longer integer-valued.

3.3 Random Walks with Membrane and a Skew Brownian Motion

Example 3.3.5 In addition to the conditions of Theorem 3.3.1, assume that

$$\mathbb{E}(X(1) \mid X(0) = x) = 0, \quad |x| \leq m. \tag{3.94}$$

According to Corollary 3.3.1, the weak limit of $\left(\frac{X(\lfloor nt \rfloor)}{\sqrt{n}}\right)_{t \geq 0}$ is a skew oscillating Brownian motion satisfying (3.72). On the other hand, in view of $\mathbb{E}\xi^{\pm} = 0$ and (3.94), the Markov chain X is a martingale (with respect to the natural filtration). It can be checked that Y, being the scaling limit of X, is also a martingale. As a consequence, the local time L_0^Y in (3.72) is identically zero, and the weak limit of $\left(\frac{X(\lfloor nt \rfloor)}{\sqrt{n}}\right)_{t \geq 0}$ is an oscillating Brownian motion that solves the stochastic differential equation

$$dY(t) = (v_- \mathbb{1}_{\{Y(t) < 0\}} + v_+ \mathbb{1}_{\{Y(t) \geq 0\}}) \, dW(t).$$

The latter convergence result was obtained in Corollary 8.4 of [66] in the case $m = 0$ of a one-point membrane.

3.3.3 Proof of Theorem 3.3.2

We shall only prove the theorem in the case where all $\mathcal{X}^{(n)}$ are independent of n, that is, we shall treat a single Markov chain rather than triangular arrays. A complete proof can be found in the paper [120]. The only aspect requiring an additional work in the general case is the law of large numbers for triangular arrays of Markov chains, which share the same transition probabilities but possibly satisfy different initial conditions. We shall indicate the relevant point in the proof but omit details, as they are standard.

Our proof is divided into six steps.

STEP 1. First we introduce several objects associated with the Markov chain $(\mathcal{X}(k))_{k \geq 0} = ((R(k), l(k)))_{k \geq 0}$. We say that the radial part R is in the *normal mode* at time k if $R(k) > 0$ and that it is in the *critical mode* at time k, otherwise. Denote by

$$T_{\text{norm}}(0) := 0, \quad T_{\text{norm}}(k) := \sum_{i=0}^{k-1} \mathbb{1}_{\{R(i) \in \mathbb{N}\}}, \quad k \in \mathbb{N} \tag{3.95}$$

the number of jumps in the normal mode up to time k, and by

$$T_{\text{crit}}(k) := \sum_{i=0}^{k-1} \mathbb{1}_{\{R(i) \in -\mathbb{N}_0\}} = k - T_{\text{norm}}(k) \tag{3.96}$$

the number of jumps in the critical mode up to time k. Here, $-\mathbb{N}_0$ is the set of nonpositive integers. Also, we define the left-continuous inverses (first-passage time processes)

$$T^-_{\text{norm}}(i) := \min\{k \geq 0 : T_{\text{norm}}(k) \geq i\},$$
$$T^-_{\text{crit}}(i) := \min\{k \geq 0 : T_{\text{crit}}(k) \geq i\}, \quad i \in \mathbb{N}.$$

For $i \in \mathbb{N}$, the stopping times σ_i (the ith exit of R from $\mathbb{N}$) and τ_i (the ith entrance of R into $\mathbb{N}$), defined in (3.74) and (3.75), can be represented as follows:

$$\sigma_i = T^-_{\text{crit}}(i) - 1 \quad \text{and} \quad \tau_i = T^-_{\text{crit}}(i), \quad i \in \mathbb{N}.$$

Define a random sequence $(V(i))_{i \geq 0}$ by putting $V(0) := 0$ and

$$V(i) - V(i-1) := \frac{R(T^-_{\text{norm}}(i))}{v_{l(T^-_{\text{norm}}(i))}} - \frac{R(T^-_{\text{norm}}(i) - 1)}{v_{l(T^-_{\text{norm}}(i) - 1)}}$$

$$= \frac{R(T^-_{\text{norm}}(i)) - R(T^-_{\text{norm}}(i) - 1)}{v_{l(T^{-1}_{\text{norm}}(i) - 1)}}$$

$$= \frac{R(T^-_{\text{norm}}(i)) - R(T^-_{\text{norm}}(i) - 1)}{v_{l(T^{-1}_{\text{norm}}(i))}}, \quad i \in \mathbb{N}.$$

We have used the fact that the label l takes a constant value as long as R is in the normal mode. Similarly, put $U(0) := 0$ and

$$U(i) - U(i-1) := \frac{R(T^-_{\text{crit}}(i))}{v_{l(T^-_{\text{crit}}(i))}} - \frac{R(T^-_{\text{crit}}(i) - 1)}{v_{l(T^-_{\text{crit}}(i) - 1)}} = \frac{R(\tau_i)}{v_{l(\tau_i)}} - \frac{R(\sigma_i)}{v_{l(\sigma_i)}}, \quad i \in \mathbb{N}. \quad (3.97)$$

Observe that $T^-_{\text{norm}}(T_{\text{norm}}(k)) = k$ on the event $\{R(k-1) > 0\}$ and $T^-_{\text{crit}}(T_{\text{crit}}(k)) = k$ on the complementary event $\{R(k-1) \leq 0\}$. Therefore,

$$\frac{R(k)}{v_{l(k)}} - \frac{R(k-1)}{v_{l(k-1)}} = \Big(V(T_{\text{norm}}(k)) - V(T_{\text{norm}}(k) - 1)\Big)\mathbb{1}_{\{R(k-1) > 0\}}$$

$$+ \Big(U(T_{\text{crit}}(k)) - U(T_{\text{crit}}(k) - 1)\Big)\mathbb{1}_{\{R(k-1) \leq 0\}}.$$

Fix $k \in \mathbb{N}$ and note that the set $\{1, 2, \ldots, k\}$ is a disjoint union of

$$\{T^-_{\text{crit}}(1), T^-_{\text{crit}}(2), \ldots, T^-_{\text{crit}}(T_{\text{crit}}(k))\} \quad \text{and}$$

$$\{T^-_{\text{norm}}(1), T^-_{\text{norm}}(2), \ldots, T^-_{\text{norm}}(T_{\text{norm}}(k))\}.$$

3.3 Random Walks with Membrane and a Skew Brownian Motion

Hence, for $k \in \mathbb{N}_0$,

$$\sum_{i=1}^{T_{\text{norm}}(k)} (V(i) - V(i-1)) + \sum_{i=1}^{T_{\text{crit}}(k)} (U(i) - U(i-1))$$

$$= \sum_{i=1}^{T_{\text{norm}}(k)} \left(\frac{R(T_{\text{norm}}^-(i))}{v_{l(T_{\text{norm}}^-(i))}} - \frac{R(T_{\text{norm}}^-(i) - 1)}{v_{l(T_{\text{norm}}^-(i) - 1)}} \right)$$

$$+ \sum_{i=1}^{T_{\text{crit}}(k)} \left(\frac{R(T_{\text{crit}}^-(i))}{v_{l(T_{\text{crit}}^-(i))}} - \frac{R(T_{\text{crit}}^-(i) - 1)}{v_{l(T_{\text{crit}}^-(i) - 1)}} \right)$$

$$= \sum_{i=1}^{k} \left(\frac{R(i)}{v_{l(i)}} - \frac{R(i-1)}{v_{l(i-1)}} \right) = \frac{R(k)}{v_{l(k)}} - \frac{R(0)}{v_{l(0)}}$$

and thereupon

$$\frac{R(k)}{v_{l(k)}} = \frac{R(0)}{v_{l(0)}} + V(T_{\text{norm}}(k)) + U(T_{\text{crit}}(k)), \quad k \in \mathbb{N}_0. \tag{3.98}$$

For each $i \in \mathbb{N}$, $T_{\text{norm}}^-(i) - 1$ are stopping times with respect to the filtration generated by $\mathcal{X}$. This in combination with $\mathbb{E}\xi^{(i)} = 0$ for $1 \le i \le d$ enables us to conclude that the sequence $(V(k))_{k \ge 0}$ is a martingale with respect to the filtration generated by the consecutively stopped $\mathcal{X}$. The embedded entrance and exit Markov chains introduced in (3.76) are given by

$$\left(R(T_{\text{crit}}^-(i)), l(T_{\text{crit}}^-(i)) \right)_{i \ge 1} \quad \text{and} \quad \left(R(T_{\text{crit}}^-(i) - 1), l(T_{\text{crit}}^-(i) - 1) \right)_{i \ge 1},$$

respectively. For $t \ge 0$, define

$$R(t) := R(\lfloor t \rfloor), \quad V(t) := V(\lfloor t \rfloor), \quad U(t) := U(\lfloor t \rfloor), \quad l(t) := l(\lfloor t \rfloor), \tag{3.99}$$

extend T_{norm} to the function on $[0, \infty)$ by linear interpolation, and put

$$T_{\text{crit}}(t) := t - T_{\text{norm}}(t), \quad t \ge 0.$$

Note that T_{norm} and T_{crit} are globally Lipschitz continuous with the Lipschitz constants 1. A continuous-time counterpart of (3.98) reads

$$\frac{R(t)}{v_{l(t)}} = \frac{R(0)}{v_{l(0)}} + V(T_{\text{norm}}(t)) + U(T_{\text{crit}}(t)), \quad t \ge 0.$$

Introducing similar notation for processes $\mathcal{X}^{(n)}$ we obtain an analogue of (3.98):

$$\frac{R^{(n)}(nt)}{v_{l^{(n)}(nt)}\sqrt{n}} = \frac{R^{(n)}(0)}{v_{l^{(n)}(0)}\sqrt{n}} + \frac{V^{(n)}(T_{\mathrm{norm}}^{(n)}(nt))}{\sqrt{n}} + \frac{U^{(n)}(T_{\mathrm{crit}}^{(n)}(nt))}{\sqrt{n}}. \quad (3.100)$$

Observe that, by construction,

$$\frac{U^{(n)}(T_{\mathrm{crit}}^{(n)}(nt))}{\sqrt{n}} = L^{(n)}(t), \quad t \geq 0, \quad (3.101)$$

where $L^{(n)}$ is as defined in (3.86).

STEP 2 is concerned with proving a lemma.

Lemma 3.3.3 *Under the assumptions of Theorem 3.3.2,*

$$\left(\frac{V^{(n)}(nt)}{\sqrt{n}}, \frac{U^{(n)}(\sqrt{n}t)}{\sqrt{n}}\right)_{t\geq 0} \Longrightarrow \left(W(t), \mu t\right)_{t\geq 0}, \quad n \to \infty \quad (3.102)$$

in the J_1-topology on $D([0, \infty), \mathbb{R}^2)$, where W is a standard Brownian motion,

$$\begin{aligned}\mu &:= \mathbb{E}^{\pi_{\mathrm{entr}}}\left(\frac{R(0)}{v_{l(0)}}\right) - \mathbb{E}^{\pi_{\mathrm{exit}}}\left(\frac{R(0)}{v_{l(0)}}\right) = \mathbb{E}^{\pi_{\mathrm{exit}}}\left(\frac{R(1)}{v_{l(1)}} - \frac{R(0)}{v_{l(0)}}\right) \\ &= \mathbb{E}^{\pi_{\mathrm{entr}}}\left(\frac{R(0)}{v_{l(0)}} - \frac{R(\sigma)}{v_{l(\sigma)}}\right) = \mathbb{E}^{\pi_{\mathrm{entr}}}\left(\frac{R(0) - R(\sigma)}{v_{l(0)}}\right) \in (0, \infty)\end{aligned} \quad (3.103)$$

and σ is as defined in (3.73).

Proof of Lemma 3.3.3 The second coordinate of the limit in (3.102) is nonrandom. Hence, according to Lemma 4.1.3, it is sufficient to prove the convergence of each coordinate separately.

In order to prove

$$\left(\frac{V^{(n)}(nt)}{\sqrt{n}}\right)_{t\geq 0} \Longrightarrow (W(t))_{t\geq 0}, \quad n \to \infty \quad (3.104)$$

on $D([0, \infty))$, we intend to apply a functional central limit theorem for martingale differences (Theorem 18.2 in [12]). By Assumption $(\mathbf{A}_1)$, $\mathrm{Var}\,\xi^{(j)} = v_j^2 \in (0, \infty)$ for

3.3 Random Walks with Membrane and a Skew Brownian Motion

$1 \leq j \leq d$. This ensures that the following Lindeberg condition holds true: for each $t \geq 0$ and all $\varepsilon > 0$,

$$\frac{1}{n} \sum_{i=1}^{\lfloor nt \rfloor} \mathbb{E}\Big((V^{(n)}(i) - V^{(n)}(i-1))^2 \mathbb{1}_{\{|V^{(n)}(i) - V^{(n)}(i-1)| \geq \sqrt{n}\varepsilon\}}\Big)$$

$$\leq t \sum_{j=1}^{d} \mathbb{E}\bigg[\frac{(\xi^{(j)})^2}{v_j^2} \mathbb{1}_{\{|\xi^{(j)}| \geq v_j \sqrt{n}\varepsilon\}}\bigg] \to 0, \quad n \to \infty.$$

Furthermore, for each $t \geq 0$,

$$\frac{1}{n} \sum_{i=1}^{\lfloor nt \rfloor} \mathbb{E}(V^{(n)}(i) - V^{(n)}(i-1))^2 = \frac{\lfloor nt \rfloor}{n} \to t, \quad n \to \infty.$$

Thus, (3.104) does indeed hold, by Theorem 18.2 in [12].

Now we prove the convergence of the second coordinates to the linear function $t \mapsto \mu t$. While doing so, we assume that $\mathcal{X}^{(n)}(k) = \mathcal{X}(k)$ for all $k \in \mathbb{N}_0$ and all $n \in \mathbb{N}$. Representation (3.97) together with the strong law of large numbers for Markov chains guarantees that the limit relation

$$\frac{U(\sqrt{n}t)}{\sqrt{n}} = \frac{1}{\sqrt{n}} \sum_{k=1}^{\lfloor \sqrt{n}t \rfloor} \Big(U(k) - U(k-1)\Big)$$

$$= \frac{\lfloor \sqrt{n}t \rfloor}{\sqrt{n}} \frac{1}{\lfloor \sqrt{n}t \rfloor} \sum_{k=1}^{\lfloor \sqrt{n}t \rfloor} \bigg(\frac{R(\tau_k)}{v_{l(\tau_k)}} - \frac{R(\sigma_k)}{v_{l(\sigma_k)}}\bigg) \to \mu t, \quad n \to \infty \quad (3.105)$$

holds a.s., for each fixed $t \geq 0$. Since $R(\tau_k) \geq 0$ and $R(\sigma_k) \leq 0$, the function $t \mapsto U(t)$ is a.s. nondecreasing. Using also that the limit is continuous, we conclude that (3.105) holds a.s. locally uniformly on $[0, \infty)$. The proof of Lemma 3.3.3 is complete.[15] $\square$

STEP 3. The purpose of this step is to prove Proposition 3.3.1. The result settles joint distributional convergence of the radius process, the corresponding local time process, $U^{(n)}$ and $T_{\text{crit}}^{(n)}$, properly scaled, and particularly demonstrates that the scaling limit of the radius process is a reflected Brownian motion. For each $n \in \mathbb{N}$, let $L^{(n)}$ be as defined in (3.86).

---

[15] The assumption $\mathcal{X}^{(n)} = \mathcal{X}$ has only been used when obtaining (3.105). A proof that does not require such an assumption can be found in [120].

Proposition 3.3.1 *Under the assumptions of Theorem 3.3.2,*

$$\left(\frac{R^{(n)}(nt)}{v_{l^{(n)}(nt)}\sqrt{n}}, L^{(n)}(t), \frac{U^{(n)}(\sqrt{n}t)}{\sqrt{n}}, \frac{T^{(n)}_{\mathrm{crit}}(nt)}{\sqrt{n}}\right)_{t\geq 0}$$

$$\Longrightarrow \left(|\|x\| + W(t)|, \nu(t), \mu t, \frac{\nu(t)}{\mu}\right)_{t\geq 0}, \quad n\to\infty \quad (3.106)$$

in the J_1-topology on $D([0,\infty), \mathbb{R}^4)$, where $(\nu(t))_{t\geq 0} := \left(L_0^{|\|x\|+W|}(t)\right)_{t\geq 0}$ is the symmetric semimartingale local time at 0 of the reflected Brownian motion $(|\|x\|+W(t)|)_{t\geq 0}$ and[16] $x\in E^d$ is the same as in (3.87).

Proof We first prove the convergence of pairs

$$\left(\frac{R^{(n)}(nt)}{v_{l^{(n)}(nt)}\sqrt{n}}, L^{(n)}(t)\right)_{t\geq 0} \Longrightarrow \left(|\|x\|+W(t)|, L_0^{|\|x\|+W|}(t)\right)_{t\geq 0}, \quad n\to\infty \quad (3.107)$$

in the product topology on $(D([0,\infty), \mathbb{R}))^2$.

For $n\in\mathbb{N}$ and $t\geq 0$, put

$$y_n(t) := \frac{R^{(n)}(0)}{v_{l^{(n)}(0)}\sqrt{n}} + \frac{V^{(n)}(nt)}{\sqrt{n}}, \quad F_n(t) := \frac{U^{(n)}(\sqrt{n}t)}{\sqrt{n}}$$

and

$$x_n(t) := \frac{R^{(n)}(nt)}{v_{l^{(n)}(nt)}\sqrt{n}}.$$

The basic observation for the subsequent proof is that $x_n(t) = \mathcal{G}_0(y_n(t), F_n(\sqrt{n}t))$ for $n\in\mathbb{N}$ and $t\geq 0$, that is, x_n is a solution to the switch problem with the noise y_n, the regulator $F_n(\sqrt{n}\cdot)$, and the gap 0; see Definition 3.2.1. As discussed in Remark 3.2.2, the theory developed in Sect. 3.2.2 for the switch problems with nonzero gaps applies equally well to the switch problems with the gap 0 provided these are well-defined. In the present context the modes A_n and B_n are collections of time intervals, in which $R^{(n)}(n\cdot)$ stays in $\mathbb{N}$ and $-\mathbb{N}_0$, respectively. For $n\in\mathbb{N}$ and $t\geq 0$, put

$$T_{A_n}(t) = \frac{T^{(n)}_{\mathrm{norm}}(nt)}{n} \quad \text{and} \quad T_{B_n}(t) = \frac{T^{(n)}_{\mathrm{crit}}(nt)}{n}. \quad (3.108)$$

Then equality (3.100) is a counterpart of formula (3.15).

---

[16] Just in case we note that x is a d-dimensional vector with one nonnegative coordinate x_0, say and $d-1$ zero coordinates. Hence, $\|x\| = x_0$.

3.3 Random Walks with Membrane and a Skew Brownian Motion

We intend to apply Corollary 3.2.1 with $\varrho_n = n^{-1/2}$ and $\delta_n = 0$ for $n \in \mathbb{N}$, $\varrho_0 = 0$, and y_n, F_n and x_n defined in the previous paragraph. By Lemma 3.3.3, $(y_n, F_n) \Longrightarrow (y_0, F_0)$ in the J_1-topology on $D([0, \infty), \mathbb{R}^2)$, where $y_0(t) = \|x\| + W(t)$ and $F_0(t) = \mu t$ for $t \geq 0$. Condition (VII) of Theorem 3.2.2 trivially holds true, for F_0 is a continuous function. The other conditions (II)–(VI) of Theorem 3.2.2 also hold. Since both y_0 and F_0 are a.s. continuous, an application of Corollary 3.2.1 yields $(x_n, F_n(\varrho_n^{-1} T_{B_n})) \Longrightarrow (x_0, m_0(C_0^{-1}))$ as $n \to \infty$ in the product topology on $(D([0, \infty), \mathbb{R}))^2$. Here, $C_0^{-1}(t) = t$ for $t \geq 0$ and $x_0 = \widetilde{\Gamma}(y_0, F_0) = y_0 + F_0 \circ F_0^{\leftarrow} \circ m_0 = y_0 + m_0$, where $m_0(t) = \sup_{s \in [0,t]} ((y_0(s))_-)$ for $t \geq 0$. Thus, recalling from (3.98) and (3.108) that $L^{(n)} = F_n(\varrho_n^{-1} T_{B_n})$ the latter limit relation reads

$$\left(\frac{R^{(n)}(nt)}{v_{l^{(n)}(nt)} \sqrt{n}}, L^{(n)}(t) \right)_{t \geq 0}$$

$$\Longrightarrow \left(\|x\| + W(t) + \max_{s \in [0,t]} ((\|x\| + W(s))_-), \max_{s \in [0,t]} ((\|x\| + W(s))_-) \right)_{t \geq 0}, \quad n \to \infty$$

in the product topology on $(D([0, \infty), \mathbb{R}))^2$. Using the first equality in (2.17), (2.19), and the remark following formula (2.19), we conclude that the process on the right-hand side has the same distribution as

$$(|\|x\| + W(t)|, L_0^{|\|x\|+W|}(t))_{t \geq 0}.$$

Thus, (3.107) holds true.

To prove (3.106), we first observe that the distributional convergence of the first three coordinates is secured by Lemmas 3.3.3 and 4.1.3. By Proposition 4.3.14 (II),

$$\left(\frac{R^{(n)}(nt)}{v_{l^{(n)}(nt)} \sqrt{n}}, L^{(n)}(t), F_n(t), F_n^{\leftarrow}(t) \right)_{t \geq 0}$$

$$\Longrightarrow \left(|\|x\| + W(t)|, v(t), \mu t, \mu^{-1} t \right)_{t \geq 0}, \quad n \to \infty$$

in the product topology on $(D([0, \infty), \mathbb{R}))^4$. Since the limit is continuous with probability one, Proposition 4.3.9 ensures that

$$\left(\frac{R^{(n)}(nt)}{v_{l^{(n)}(nt)} \sqrt{n}}, L^{(n)}(t), F_n(t), F_n^{\leftarrow}(L_n(t)) \right)_{t \geq 0}$$

$$\Longrightarrow \left(|\|x\| + W(t)|, v(t), \mu t, \mu^{-1} v(t) \right)_{t \geq 0}, \quad n \to \infty$$

in the product topology on $(D([0, \infty), \mathbb{R}))^4$. Note that

$$F_n^{\leftarrow}(L_n(t)) = F_n^{\leftarrow}\left(F_n\left(\frac{T_{\text{crit}}^{(n)}(nt))}{\sqrt{n}}\right)\right), \quad n \in \mathbb{N}, \ t \geq 0.$$

By Proposition 4.3.9, for all $T > 0$,

$$\sup_{t \in [0,T]} |(F_n^{\leftarrow} \circ F_n)(t) - t| \xrightarrow{\mathbb{P}} 0, \quad n \to \infty,$$

which together with the a.s. monotonicity of $t \mapsto F_n^{\leftarrow} \circ F_n(t)$ entails, for all $T > 0$,

$$\sup_{t \in [0,T]} \left|F_n^{\leftarrow}\left(F_n\left(\frac{T_{\text{crit}}^{(n)}(nt))}{\sqrt{n}}\right)\right) - \frac{T_{\text{crit}}^{(n)}(nt))}{\sqrt{n}}\right| \xrightarrow{\mathbb{P}} 0, \quad n \to \infty.$$

This proves that (3.106) holds in the product topology. Since the limit process is a.s. continuous, it also holds in the J_1-topology on $D([0, \infty), \mathbb{R}^4)$. The proof of Proposition 3.3.1 is complete. $\square$

STEP 4. We have to introduce another piece of notation. First assume for simplicity that the initial values $\mathcal{X}^{(n)}(0) = \mathcal{X}(0)$ do not depend on n. Consider projections of $\mathcal{X}$ on the line $\mathbb{Z} \times \{j\}$, $j = 1, \ldots, d$:

$$R_j(k) := R(k)\mathbb{1}_{\{l(k)=j\}}, \quad R(k) := R_1(k) + \ldots + R_d(k).$$

Similarly to (3.95) we define

$$T_{\text{norm},j}(k) := \sum_{i=0}^{k-1} \mathbb{1}_{\{R_j(i) \in \mathbb{N}\}}, \quad k \in \mathbb{N}$$

the number of positive jumps on the jth ray up to time k, and the inverse function

$$T_{\text{norm},j}^{-}(i) := \inf\{k \geq 0 : T_{\text{norm},j}(k) \geq i\}, \quad i \in \mathbb{N}_0.$$

For $j = 1, \ldots, d$, put $V_j(0) = U_j(0) := 0$,

$$V_j(i) - V_j(i-1) := \frac{R_j(T_{\text{norm},j}^{-}(i))}{v_{l(T_{\text{norm},j}^{-}(i))}} - \frac{R_j(T_{\text{norm},j}^{-}(i) - 1)}{v_{l(T_{\text{norm},j}^{-1}(i)-1)}}$$

$$= \frac{R_j(T_{\text{norm},j}^{-}(i)) - R_j(T_{\text{norm},j}^{-}(i) - 1)}{v_j}, \quad i \in \mathbb{N}$$

3.3 Random Walks with Membrane and a Skew Brownian Motion

and

$$U_j(i) - U_j(i-1) = \frac{R(\tau_i)}{v_j} \mathbb{1}_{\{l(\tau_i)=j\}} - \frac{R(\sigma_i)}{v_j} \mathbb{1}_{\{l(\sigma_i)=j\}}, \quad i \in \mathbb{N}.$$

It follows from the assumptions of the theorem that $(V_j(i))_{i\geq 0}$, $j = 1, \ldots, d$ are independent standard random walks, with jumps distributed like $\xi^{(j)}/v_j$, $1 \leq j \leq d$. Similarly to the derivation of (3.98), we write, for $k \in \mathbb{N}$,

$$\frac{R_j(k)}{v_j} - \frac{R_j(0)}{v_j} = \sum_{i=1}^{k} \left(\frac{R_j(i)}{v_j} - \frac{R_j(i-1)}{v_j} \right)$$

$$= \sum_{i=1}^{T_{\text{norm}}(k)} \left(\frac{R_j(T_{\text{norm}}^-(i))}{v_j} - \frac{R_j(T_{\text{norm}}^-(i)-1)}{v_j} \right)$$

$$+ \sum_{i=1}^{T_{\text{crit}}(k)} \left(\frac{R(T_{\text{crit}}^-(i))}{v_j} \mathbb{1}_{\{l(T_{\text{crit}}^-(i))=j\}} - \frac{R(T_{\text{crit}}^-(i)-1)}{v_j} \mathbb{1}_{\{l(T_{\text{crit}}^-(i)-1)=j\}} \right)$$

$$= \sum_{i=1}^{T_{\text{norm},j}(k)} \left(\frac{R_j(T_{\text{norm},j}^-(i))}{v_j} - \frac{R_j(T_{\text{norm},j}^-(i)-1)}{v_j} \right) + \sum_{i=1}^{T_{\text{crit}}(k)} (U_j(i) - U_j(i-1)),$$

where for the last equality we have used that R_j does not increase whenever T_{norm} increases while $T_{\text{norm},j}$ remains unchanged. Thus, we obtain a version of (3.98):

$$\frac{R_j(k)}{v_j} = \frac{R_j(0)}{v_j} + V_j(T_{\text{norm},j}(k)) + U_j(T_{\text{crit}}(k)), \quad k \in \mathbb{N}_0. \tag{3.109}$$

It is important for what follows that the function $k \mapsto U_j(T_{\text{crit}}(k))$ is nondecreasing, for each fixed $j = 1, \ldots, d$. Moreover, the following decompositions hold true:

$$V(T_{\text{norm}}(k)) = V_1(T_{\text{norm},1}(k)) + \ldots + V_d(T_{\text{norm},d}(k)), \tag{3.110}$$

$$U(T_{\text{crit}}(k)) = U_1(T_{\text{crit}}(k)) + \ldots + U_d(T_{\text{crit}}(k)). \tag{3.111}$$

If the initial value $\mathcal{X}^{(n)}(0)$ depends on n, we argue similarly and define the sequences $R_j^{(n)}$, $V_j^{(n)}$, $U_j^{(n)}$, etc. Also, we define a continuous-time extension of $T_{\text{norm},j}^{(n)}$ to the nonnegative halfline $[0, \infty)$ by linearly interpolating the points $T_{\text{norm},j}^{(n)}(i)$, $i \in \mathbb{N}_0$.

STEP 5. To prove the convergence stated in Theorem 3.3.2, we shall utilize a martingale characterization of the Walsh Brownian motion, as given in Theorem 2.2.5. First, we show that the sequence of distributions of

$$\left(\left(\frac{R_j^{(n)}(nt)}{\sqrt{n}}, \frac{V_j^{(n)}(T_{\text{norm},j}^{(n)}(nt))}{\sqrt{n}}, \frac{U_j^{(n)}(T_{\text{crit}}^{(n)}(nt))}{\sqrt{n}}\right)_{1\le j\le d}, \frac{T_{\text{crit}}^{(n)}(nt)}{\sqrt{n}}\right)_{t\ge 0}, \quad n\in\mathbb{N} \quad (3.112)$$

is tight on $D([0,\infty), \mathbb{R}^{3d+1})$, and that each its limit point is a.s. continuous.

In view of Proposition 3.3.1, $(U^{(n)}(T_{\text{crit}}^{(n)}(nt))/\sqrt{n})_{t\ge 0}$ converges in distribution as $n\to\infty$ to a continuous nondecreasing process (the local time). Equation (3.111) implies that the modulus of continuity of each nondecreasing process $t\mapsto U_j^{(n)}(T_{\text{crit}}^{(n)}(nt))/\sqrt{n}$ is dominated by the modulus of continuity of $t\mapsto U^{(n)}(T_{\text{crit}}^{(n)}(nt))/\sqrt{n}$, as follows from the inequality

$$\frac{U_j^{(n)}(T_{\text{crit}}^{(n)}(nt))}{\sqrt{n}} - \frac{U_j^{(n)}(T_{\text{crit}}^{(n)}(ns))}{\sqrt{n}} \le \frac{U^{(n)}(T_{\text{crit}}^{(n)}(nt))}{\sqrt{n}} - \frac{U^{(n)}(T_{\text{crit}}^{(n)}(ns))}{\sqrt{n}}, \quad 0\le s\le t.$$

Hence the sequence of distributions of $t\mapsto (U_j^{(n)}(T_{\text{crit}}^{(n)}(nt))/\sqrt{n})_{1\le j\le d}$, $n\in\mathbb{N}$, is tight on $D([0,\infty), \mathbb{R}^d)$, each limit point $\widehat{U}^{(\infty)} = (\widehat{U}_1^{(\infty)}(t), \ldots, \widehat{U}_d^{(\infty)}(t))_{t\ge 0}$ is a.s. continuous, and each coordinate $t\mapsto \widehat{U}_j^\infty(t)$ is a.s. nondecreasing.

By Donsker's theorem,

$$\left(\frac{V_1^{(n)}(nt)}{\sqrt{n}}, \ldots, \frac{V_d^{(n)}(nt)}{\sqrt{n}}\right)_{t\ge 0} \Longrightarrow \left(W_1(t), \ldots, W_d(t)\right)_{t\ge 0}, \quad n\to\infty$$

in the J_1-topology on $D([0,\infty), \mathbb{R}^d)$, where $W_1, \ldots, W_d$ are independent Brownian motions, possibly with different variances. By the Lipschitz continuity of $t\mapsto T_{\text{norm},j}^{(n)}(t)$,

$$0 \le T_{\text{norm},j}^{(n)}(t) - T_{\text{norm},j}^{(n)}(s) \le t - s, \quad 0\le s\le t.$$

3.3 Random Walks with Membrane and a Skew Brownian Motion

As a consequence, the sequence of distributions of $t \mapsto T^{(n)}_{\text{norm},j}(nt)/n$, $n \in \mathbb{N}$ is tight on $D([0,\infty))$. Therefore, the sequence of distributions of

$$\left(\frac{V^{(n)}_1(T^{(n)}_{\text{norm},1}(nt))}{\sqrt{n}}, \ldots, \frac{V^{(n)}_d(T^{(n)}_{\text{norm},d}(nt))}{\sqrt{n}}\right)_{t \geq 0}, \quad n \in \mathbb{N}$$

is tight on $D([0,\infty), \mathbb{R}^d)$, and each limit point $M^{(\infty)} = (M^{(\infty)}_1(t), \ldots, M^{(\infty)}_d(t))_{t \geq 0}$ is an a.s. continuous process. Tightness of the sequence of distributions $(R^{(n)}_j(nt)/\sqrt{n})_{t \geq 0}$, $n \in \mathbb{N}$ follows from (3.109) and tightness of the sequence of distributions of $(T^{(n)}_{\text{crit}}(nt)/\sqrt{n})_{t \geq 0}$, $n \in \mathbb{N}$ follows from Proposition 3.3.1. Since all weak limits $M^{(\infty)}_j$ and $\widehat{U}^{(\infty)}_j$ are a.s. continuous, so are weak limits $R^{(\infty)}_j$, and the sequence of distributions of the processes in (3.112) is tight on $D([0,\infty), \mathbb{R}^{3d})$.

By Proposition 3.3.1, $(R^{(n)}(nt)/(v_{l^{(n)}(nt)}\sqrt{n}))_{t \geq 0}$ converges in distribution as $n \to \infty$ to $(\|x\| + W(t) + \max_{s \in [0,t]}((\|x\| + W(s))_-))_{t \geq 0}$, a reflected Brownian motion, and $(U^{(n)}(T^{(n)}_{\text{crit}}(nt))/\sqrt{n})_{t \geq 0}$ converges in distribution to its local time at 0 that we denoted by $v = (v(t))_{t \geq 0}$. In view of representation (3.100), Proposition 3.3.1 ensures that $(V^{(n)}(T^{(n)}_{\text{norm}}(nt))/\sqrt{n})_{t \geq 0}$ converges in distribution to a Brownian motion W. Furthermore, the three aforementioned distributional convergences are joint. Hence, each weak limit

$$((R^{(\infty)}_j(t), M^{(\infty)}_j(t), \widehat{U}^{(\infty)}_j(t))_{1 \leq j \leq d}, \mu^{-1}v(t))_{t \geq 0} \tag{3.113}$$

of (3.112) is a continuous process with the following three properties:

(**P**$_1$) For each $t \geq 0$,

$$\sum_{j=1}^d M^{(\infty)}_j(t) = W(t), \tag{3.114}$$

$$\sum_{j=1}^d \widehat{U}^{(\infty)}_j(t) = v(t).$$

(**P**$_2$) For each $1 \leq j \leq d$ and $t \geq 0$,

$$X^{(\infty)}_j(t) := \frac{R^{(\infty)}_j(t)}{v_j} = x_j + M^{(\infty)}_j(t) + \widehat{U}^{(\infty)}_j(t),$$

where $X^{(\infty)}_j$ is a weak limit point of the jth coordinate of $(X^{(n)})_{n \geq 1}$, and x_j is the jth coordinate of x.

(**P**$_3$) For each $j = 1, \ldots, d$, $\widehat{U}_j^{(\infty)}(0) = 0$ and the process $(\widehat{U}_j^{(\infty)}(t))_{t \geq 0}$ is a.s. nondecreasing.

Since the process $\left(\sum_{j=1}^d X_j^{(\infty)}(t)\right)_{t \geq 0}$ is a reflected Brownian motion whose local time at 0 is v, the following equalities hold a.s.

$$\int_{[0,\infty)} \mathbb{1}\left\{\sum_{j=1}^d X_j^{(\infty)}(t) > 0\right\} dv(t) = 0 \quad \text{and} \quad \int_0^\infty \mathbb{1}\left\{\sum_{j=1}^d X_j^{(\infty)}(t) = 0\right\} dt = 0; \quad (3.115)$$

see the discussion at the end of Sect. 2.1.1. Also, note that $X_i^{(\infty)}(t) X_j^{(\infty)}(t) = 0$ for each $i \neq j$ and each $t \geq 0$. Since

$$\sum_{j=1}^d X_j^{(\infty)}(t) \geq 0, \quad t \geq 0 \quad \text{a.s.,}$$

we conclude that $X_j^{(\infty)}(t) \geq 0$ for all $t \geq 0$ and $j = 1, \ldots, d$ with probability one.

STEP 6. It remains to demonstrate that each weak limit $(X_j^{(\infty)}(t))_{1 \leq j \leq d, t \geq 0}$ is a Walsh Brownian motion with parameters $p_1, \ldots, p_d$ and the starting point x. Without loss of generality, we can and do assume that (a) the sequence given in (3.112) itself converges in distribution as $n \to \infty$ in the J_1-topology on $D([0, \infty), \mathbb{R}^{3d+1})$ to

$$((X_j^{(\infty)}(t), M_j^{(\infty)}(t), \widehat{U}_j^{(\infty)}(t))_{1 \leq j \leq d}, \mu^{-1} v(t))_{t \geq 0};$$

and that (b) this convergence holds jointly with

$$\left(\frac{U^{(n)}(\sqrt{n}t)}{\sqrt{n}}\right)_{t \geq 0} \Longrightarrow (\mu t)_{t \geq 0}, \quad n \to \infty$$

in the J_1-topology on $D([0, \infty))$; see Lemma 4.1.3. We shall now use the Skorokhod representation theorem (Theorem 4.1.2). Retaining the same notation for versions that converge a.s. we can and do assume that

$$\left(\left(\frac{R_j^{(n)}(nt)}{v_j \sqrt{n}}, \frac{V_j^{(n)}(T_{\text{norm},j}^{(n)}(nt))}{\sqrt{n}}, \frac{U_j^{(n)}(T_{\text{crit}}^{(n)}(nt))}{\sqrt{n}}\right)_{1 \leq j \leq d}, \frac{U^{(n)}(\sqrt{n}t)}{\sqrt{n}}, \frac{T_{\text{crit}}^{(n)}(nt)}{\sqrt{n}}\right)_{t \geq 0}$$

$$\to ((X_j^{(\infty)}(t), M_j^{(\infty)}(t), \widehat{U}_j^{(\infty)}(t))_{1 \leq j \leq d}, \mu t, \mu^{-1} v(t))_{t \geq 0} \quad n \to \infty \text{ a.s.}$$

in the J_1-topology on $D([0, \infty), \mathbb{R}^{3d+2})$.

3.3 Random Walks with Membrane and a Skew Brownian Motion

We intend to check the conditions of Theorem 2.2.5 which provides a martingale characterization of the Walsh Brownian motion. Taking into account (3.115), we need to show that:

(a) For any $j = 1, \ldots, d$ and $t \geq 0$

$$\frac{\widehat{U}_j^{(\infty)}(t)}{v(t)} = \frac{\widehat{U}_j^{(\infty)}(t)}{\sum_{l=1}^d \widehat{U}_l^{(\infty)}(t)} = p_j \quad \text{a.s.}, \tag{3.116}$$

where p_j are as given in (3.88).

(b) For each $j = 1, \ldots, d$, the process $M_j^{(\infty)}$ is a continuous square integrable martingale with respect to the natural filtration $(\mathcal{F}_t^{(\infty)})_{t \geq 0}$ generated by the process

$$(R_j^{(\infty)}(t), M_j^{(\infty)}(t), \widehat{U}_j^{(\infty)}(t))_{1 \leq j \leq d, t \geq 0},$$

having the predictable quadratic variation

$$\langle M_j^{(\infty)}, M_j^{(\infty)} \rangle(t) = \int_0^t \mathbb{1}_{\{X_j^{(\infty)}(s) > 0\}} \, ds. \tag{3.117}$$

Proof of (3.116) Fix some $j = 1, \ldots, d$. Analogously to the proof of (3.105), applying the strong law of large numbers for Markov chains and using monotonicity of the processes involved yield

$$\left(\frac{U_j^{(n)}(\sqrt{n}t)}{\sqrt{n}} \right)_{t \geq 0} \to (p_j \mu t)_{t \geq 0}, \quad n \to \infty$$

a.s. locally uniformly in t, where p_j is as defined in (3.88). By Proposition 4.3.7, this in combination with the assumed convergence $\lim_{n \to \infty} (T_{\text{crit}}^{(n)}(n \cdot)/\sqrt{n}) = \mu^{-1} v(\cdot)$ a.s. in the J_1-topology on $D([0, \infty), \mathbb{R})$ entails

$$\left(\frac{U_j^{(n)}(T_{\text{crit}}^{(n)}(nt))}{\sqrt{n}} \right)_{t \geq 0} = \left(\frac{U_j^{(n)}\left(\sqrt{n} \frac{T_{\text{crit}}^{(n)}(nt)}{\sqrt{n}}\right)}{\sqrt{n}} \right)_{t \geq 0} \to (p_j v(t))_{t \geq 0}, \quad n \to \infty$$

in the J_1-topology on $D([0, \infty), \mathbb{R})$. On the other hand,

$$\lim_{n \to \infty} \frac{U_j^{(n)}(T_{\text{crit}}^{(n)}(n \cdot))}{\sqrt{n}} = \widehat{U}_j^{(\infty)}(\cdot) \quad \text{a.s.}$$

in the J_1-topology on $D([0, \infty), \mathbb{R})$, whence $\widehat{U}_j^{(\infty)}(\cdot) = p_j v(\cdot)$ a.s.

Proof of (3.117) Fix some $j = 1, \ldots, d$. It can be checked that W defined in (3.114) is a $(\mathcal{F}_t^{(\infty)})_{t \geq 0}$-Brownian motion. Hence, (3.117) follows if we can show that

$$M_j^{(\infty)}(t) = \int_0^t \mathbb{1}_{\{X_j^{(\infty)}(s) > 0\}} \, dW(s), \quad t \geq 0 \quad \text{a.s.} \tag{3.118}$$

Formula (3.110) ensures that

$$\frac{V_j^{(n)}(T_{\text{norm}, j}^{(n)}(nt))}{\sqrt{n}} = \int_{[0,t]} \mathbb{1}_{\{R_j^{(n)}((ns)-) > 0\}} \, d_s \frac{V^{(n)}(T_{\text{norm}}^{(n)}(ns))}{\sqrt{n}},$$

where the integral is understood as a stochastic integral with respect to the square integrable martingale $(V^{(n)}(T_{\text{norm}}^{(n)}(nt))/\sqrt{n}, \mathcal{F}_t^{(n)})_{t \geq 0}$, and

$$\mathcal{F}_t^{(n)} = \sigma\{R_j^{(n)}(ns), V_j^{(n)}(T_{\text{norm}, j}^{(n)}(ns)), U_j^{(n)}(T_{\text{crit}}^{(n)}(ns)), \, j = 1, \ldots, d, \, s \leq t\}.$$

Given next is a lemma which secures (3.118).

Lemma 3.3.4 *For $j = 1, \ldots, d$ and $t \geq 0$,*

$$\int_{[0,t]} \mathbb{1}_{\{R_j^{(n)}((ns)-) > 0\}} \, d_s \frac{V^{(n)}(T_{\text{norm}}^{(n)}(ns))}{\sqrt{n}} \xrightarrow{\mathbb{P}} \int_0^t \mathbb{1}_{\{X_j^{(\infty)}(s) > 0\}} \, dW(s), \quad n \to \infty.$$

Proof For each $t_0 > 0$,

$$\mathbb{E} \left| \int_{[0,t_0)} \mathbb{1}_{\{R_j^{(n)}((ns)-) > 0\}} \, d_s \frac{V^{(n)}(T_{\text{norm}}^{(n)}(ns))}{\sqrt{n}} \right|^2 \leq t_0,$$

$$\mathbb{E} \left| \int_0^{t_0} \mathbb{1}_{\{X_j^{(\infty)}(s) > 0\}} \, dW(s) \right|^2 \leq t_0.$$

Therefore, to prove the lemma it is sufficient to check that, for each $t_0 \in (0, t]$,

$$\int_{[t_0,t]} \mathbb{1}_{\{R_j^{(n)}((ns)-) > 0\}} \, d_s \frac{V^{(n)}(T_{\text{norm}}^{(n)}(ns))}{\sqrt{n}} \xrightarrow{\mathbb{P}} \int_{t_0}^t \mathbb{1}_{\{X_j^{(\infty)}(s) > 0\}} \, dW(s), \quad n \to \infty. \tag{3.119}$$

3.3 Random Walks with Membrane and a Skew Brownian Motion

To this end, we use Theorem on p. 32 in [151], which amounts to checking the following conditions:

(i) For each fixed $s \in [t_0, t]$,
$$\lim_{n \to \infty} \frac{V^{(n)}(T^{(n)}_{\text{norm}}(ns))}{\sqrt{n}} = W(s) \quad \text{a.s.}$$

(ii) For each[17] fixed $t \geq 0$ and all $n \in \mathbb{N}$, $\mathbb{E}(V^{(n)}(T^{(n)}_{\text{norm}}(nt))/\sqrt{n})^2 \leq t$.

(iii) For each fixed $s \in [t_0, t]$,
$$\lim_{n \to \infty} \mathbb{1}_{\{X_j^{(n)}(s-)>0\}} = \mathbb{1}_{\{X_j^{(\infty)}(s)>0\}} \quad \text{a.s.,}$$

where
$$X_j^{(n)}(t) := \frac{R_j^{(n)}(nt)}{v_j \sqrt{n}}, \quad n \in \mathbb{N},\ t \geq 0.$$

(iv) For all $\varepsilon > 0$,
$$\lim_{h \to 0+} \limsup_{n \to \infty} \sup_{|t_2 - t_1| < h} \mathbb{P}\left\{ \left| \mathbb{1}_{\{X_j^{(n)}(t_2-)>0\}} - \mathbb{1}_{\{X_j^{(n)}(t_1-)>0\}} \right| > \varepsilon \right\} = 0. \tag{3.120}$$

The a.s. convergence in part (i) follows from (3.110) and (3.114). Condition (ii) holds true by the construction of the sequence $V^{(n)}$. To prove (iii), we recall that $X^{(\infty)} = \sum_{l=1}^{d} X_l^{(\infty)}$ is a reflected Brownian motion. Fix $s > 0$ and note that $X^{(\infty)}(s) > 0$ with probability one. Hence, for almost all ω, there exists $k = k(s, \omega)$ such that $X_k^{(\infty)}(s, \omega) > 0$. Let $\lambda = \lambda(s, \omega) > 0$ be such that

$$\inf_{r \in [s-\lambda, s+\lambda]} X_k^{(\infty)}(r) > 0. \tag{3.121}$$

Then, for $i \neq k$,
$$X_i^{(\infty)}(r) = 0, \quad r \in [s - \lambda, s + \lambda]. \tag{3.122}$$

---

[17] In [151], a stronger assumption $\lim_{n \to \infty} \mathbb{E}(V^{(n)}(T^{(n)}_{\text{norm}}(ns))/\sqrt{n})^2 = s$ for each $s \in [t_0, t]$ was imposed. However, a perusal of the argument given in [151] shows that only boundedness is actually needed.

Recall that, for each $j = 1, \ldots, d$, $\lim_{n \to \infty} X_j^{(n)} = X_j^{(\infty)}$ a.s. locally uniformly and particularly uniformly on $[s - \lambda, s + \lambda]$. Thus, for n large enough,

$$X_k^{(n)}(r) > 0, \quad r \in [s - \lambda, s + \lambda] \tag{3.123}$$

and

$$X_i^{(n)}(r) = 0, \quad r \in [s - \lambda, s + \lambda] \tag{3.124}$$

whenever $i \neq k$. This proves part (iii).

Turning to part (iv) we first note that, by (3.121) and (3.122), the process $s \mapsto \mathbb{1}_{\{X^{(\infty)}(s) > 0\}}$ is continuous in probability on $[t_0, t]$. Hence, it is also uniformly continuous in probability on $[t_0, t]$, that is, for all $\varepsilon > 0$,

$$\lim_{h \to 0+} \sup_{|t_2 - t_1| < h} \mathbb{P}\{|\mathbb{1}_{\{X^{(\infty)}(t_2) > 0\}} - \mathbb{1}_{\{X^{(\infty)}(t_1) > 0\}}| > \varepsilon\} = 0$$

which is equivalent to

$$\lim_{h \to 0+} \sup_{|t_2 - t_1| < h} \mathbb{P}\{\mathbb{1}_{\{X^{(\infty)}(t_2) > 0\}} \neq \mathbb{1}_{\{X^{(\infty)}(t_1) > 0\}}\} = 0.$$

Formulas (3.121), (3.122), (3.123), and (3.124) ensure that, given $\delta > 0$ and $s \in [t_0, t]$, there exist $\lambda = \lambda(s)$ and $n_0 = n_0(s)$ such that

$$\sup_{|z - s| < \lambda(s)} \mathbb{P}\{\mathbb{1}_{\{X_j^{(n)}(z-) > 0\}} \neq \mathbb{1}_{\{X_j^{(\infty)}(z) > 0\}}\} \leq \delta$$

for all $j = 1, \ldots, d$ and all $n \geq n_0(s)$. The set $[t_0, t]$ is compact. Its infinite covering by the sets $(s - \lambda(s)/2, s + \lambda(s)/2)$ for $s \in [t_0, t]$ contains a finite subcovering $(s_p - \lambda(s_p)/2, s_p + \lambda(s_p)/2)$ for $p = 1, \ldots, N$. Hence, if t_1 and t_2 satisfy $|t_2 - t_1| < h$ with $h := (\min_{p=1,\ldots,N} \lambda(s_p))/2$, then also $t_1, t_2 \in (s_p - \lambda(s_p), s_p + \lambda(s_p))$ for some $p = p(t_1, t_2)$. Combining fragments together we infer, for all $n \geq \max\{n_0(s_1), \ldots, n_0(s_N)\}$,

$$\mathbb{P}\{\mathbb{1}_{\{X_j^{(n)}(t_1-) > 0\}} \neq \mathbb{1}_{\{X_j^{(n)}(t_2-) > 0\}}\} \leq \mathbb{P}\{\mathbb{1}_{\{X_j^{(n)}(t_1-) > 0\}} \neq \mathbb{1}_{\{X_j^{(\infty)}(t_1) > 0\}}\}$$

$$+ \mathbb{P}\{\mathbb{1}_{\{X_j^{(\infty)}(t_1) > 0\}} \neq \mathbb{1}_{\{X_j^{(\infty)}(t_2) > 0\}}\}$$

$$+ \mathbb{P}\{\mathbb{1}_{\{X_j^{(\infty)}(t_2) > 0\}} \neq \mathbb{1}_{\{X_j^{(n)}(t_2-) > 0\}}\}$$

$$\leq 2\delta + \sup_{|s_2 - s_1| < h} \mathbb{P}\{\mathbb{1}_{\{X^{(\infty)}(s_2) > 0\}} \neq \mathbb{1}_{\{X^{(\infty)}(s_1) > 0\}}\}$$

and thereupon

$$\limsup_{h\to 0+}\limsup_{n\to\infty}\sup_{|t_2-t_1|<h}\mathbb{P}\{|\mathbb{1}_{\{X_j^{(n)}(t_2-)>0\}} - \mathbb{1}_{\{X_j^{(n)}(t_1-)>0\}}| > \varepsilon\} \leq 2\delta,$$

what finishes the proof of (3.120) and (3.119). The proofs of Lemma 3.3.4 and Theorem 3.3.2 are complete. □

3.3.4 Proof of Theorem 3.3.3

It follows from Theorem 2.2.4 that if (W_-, W_+) is a Walsh Brownian motion with parameters p_- and p_+ satisfying $p_- + p_+ = 1$ and $p_\pm \geq 0$ then the process $W_+ - W_-$ is a skew Brownian motion with permeability parameter $\gamma = p_+ - p_-$. To prove Theorem 3.3.3, we intend to apply Theorem 3.3.2 with $d = 2$ to some auxiliary Markov chain $(R^{(n)}, l^{(n)})$ on $\mathbb{Z} \times \{-, +\}$ (identified with $\mathbb{Z} \times \{-1, +1\}$) associated with the initial random walk $X^{(n)}$.

We first explain informally the construction of this auxiliary Markov chain. While doing so, we skip the superscript (n). It is natural to identify the set $\{m+1, m+2, \ldots\} \subset \mathbb{Z}$ with $\mathbb{N} \times \{+\} \subset \mathbb{Z} \times \{-, +\}$ and $\{-m-1, -m-2, \ldots\} \subset \mathbb{Z}$ with $\mathbb{N} \times \{-\} \subset \mathbb{Z} \times \{-, +\}$ and to put $(R(k), l(k)) = (x, +)$ if $X(k) = x + m$ for $x \in \mathbb{N}$ and $(R(k), l(k)) = (x, -)$ if $X(k) = -x - m$ for $x \in \mathbb{N}$. Next, we define the transition probabilities from the corresponding states: for $x, y \in \mathbb{N}$ and $k \in \mathbb{N}_0$,

$$\mathbb{P}\{(R(k+1), l(k+1)) = (x+y, +) \mid (R(k), l(k)) = (x, +)\} = \mathbb{P}\{\xi_+ = y\},$$
$$\mathbb{P}\{(R(k+1), l(k+1)) = (x+y, -) \mid (R(k), l(k)) = (x, -)\} = \mathbb{P}\{-\xi_- = y\}.$$
(3.125)

It is tempting to do this for all k and apply Theorem 3.3.2 to the resulting Markov chain. However, if X jumps over the membrane $\{-m, \ldots, m\}$, then l must change the sign, which is not reflected in (3.125). To overcome this problem, we add additional steps to the Markov chain. For instance, if X goes from $x > m$ to $y < -m$, we add one additional step by postulating that first (R, l) jumps to the state $(y, +)$ and then jumps to $(m + y, -)$ at the next step. Unfortunately, Theorem 3.3.2 is still not applicable. Indeed, one of its assumptions is that a Markov chain stays in a membrane for one unit of time, whereas the X constructed so far is allowed to stay in $\{-m, \ldots, m\}$ for more than one unit of time. To cope with this issue we simply remove extra steps except the very first one by which X enters the membrane. This informal construction will be made rigorous a few lines below at STEP 1.

The whole proof is divided into four steps and is carried out in a "shuffled" order. We start by defining a Markov chain $(R(k), l(k))_{k\geq 0}$ which satisfies the assumptions of Theorem 3.3.2. In this way we obtain distributional convergence of $(l(\lfloor nt \rfloor) R(\lfloor nt \rfloor)/\sqrt{n})_{t\geq 0}$

as $n \to \infty$ to a skew Brownian motion.[18] We then add and remove some steps, thereby obtaining a Markov chain with the same transition probabilities as X. Finally, we show that the contribution of the number of added/removed steps is negligible and finalize the proof by applying Propositions 4.3.7 and 4.3.10.

STEP 1. As discussed above, we construct a Markov chain $\mathcal{X}^{(n)} = (\mathcal{X}^{(n)}(k))_{k \geq 0} = (R^{(n)}(k), l^{(n)}(k))_{k \geq 0}$ on $\mathbb{Z} \times \{-, +\}$ that satisfies the assumptions of Theorem 3.3.2. The original Markov chain $X^{(n)}$ on $\mathbb{Z}$ is then obtained from $\mathcal{X}^{(n)}$ via two transformations.

We stipulate that the transition probabilities of $\mathcal{X}^{(n)}$ are independent of n and define them as follows. If $R^{(n)}(0) = x \in \mathbb{N}$, then, for each $y \in \mathbb{Z}$, put

$$\mathbb{P}\{(R^{(n)}(1), l^{(n)}(1)) = (y, +) \mid (R^{(n)}(0), l^{(n)}(0)) = (x, +)\} := \mathbb{P}\{\xi_+ = y - x\},$$

$$\mathbb{P}\{(R^{(n)}(1), l^{(n)}(1)) = (y, -) \mid (R^{(n)}(0), l^{(n)}(0)) = (x, -)\} := \mathbb{P}\{-\xi_- = y - x\}.$$

Thus, if R is positive, then the label does not change, and the distribution of a jump is equal to the distribution of ξ_+ if the label is $+$ and that of $(-\xi_-)$ if the label is $-$.

Assume that $R^{(n)}(0) = x \in \mathbb{Z} \setminus \mathbb{N}$ which means that the Markov chain $\mathcal{X}^{(n)}$ starts from the membrane. If $R^{(n)}(0) \leq -2m - 1$, then we put

$$R^{(n)}(1) := -R^{(n)}(0) - 2m, \quad l^{(n)}(1) := -l^{(n)}(0). \tag{3.126}$$

In other words, the random walk changes its label and the radius gets reflected symmetrically with respect to the point $-m$. Notice that $R^{(n)}(1) \in \mathbb{N}$.

Recall the definition of $\widetilde{\tau}_0$ in (3.89). If $R^{(n)}(0) = x \in \{-2m, \ldots, 0\}$ (so that $x + m \in \{-m, \ldots, m\}$) and $y \in \mathbb{N}$, then we put

$$\mathbb{P}\{(R^{(n)}(1), l^{(n)}(1)) := (y, +) \mid (R^{(n)}(0), l^{(n)}(0)) = (x, +)\}$$
$$:= \mathbb{P}\{X(\widetilde{\tau}_0) = y + m \mid X(0) = x + m\},$$

$$\mathbb{P}\{(R^{(n)}(1), l^{(n)}(1)) := (y, -) \mid (R^{(n)}(0), l^{(n)}(0)) = (x, +)\}$$
$$:= \mathbb{P}\{X(\widetilde{\tau}_0) = -(y + m) \mid X(0) = x + m\},$$

$$\mathbb{P}\{(R^{(n)}(1), l^{(n)}(1)) := (y, -) \mid (R^{(n)}(0), l^{(n)}(0)) = (x, -)\}$$
$$:= \mathbb{P}\{X(\widetilde{\tau}_0) = -(y + m) \mid X(0) = -(x + m)\},$$

$$\mathbb{P}\{(R^{(n)}(1), l^{(n)}(1)) := (y, +) \mid (R^{(n)}(0), l^{(n)}(0)) = (x, -)\}$$
$$:= \mathbb{P}\{X(\widetilde{\tau}_0) = y + m \mid X(0) = -(x + m)\}.$$

---

[18] The self-explaining notation $l(k)R(k)$ means $\pm R(k)$ with the sign being $l(k)$.

3.3 Random Walks with Membrane and a Skew Brownian Motion

Define the initial condition $\mathcal{X}^{(n)}(0) = (R^{(n)}(0), l^{(n)}(0))$ as follows:

$$(R^{(n)}(0), l^{(n)}(0)) = \begin{cases} (X^{(n)}(0) - m, +), & \text{if } X^{(n)}(0) \geq -m, \\ (-X^{(n)}(0) - m, -), & \text{if } X^{(n)}(0) < -m, \end{cases}$$

where $(X^{(n)}(k))_{k \geq 0}$ is the original Markov chain. Then

$$\left(\frac{R^{(n)}(0)}{v_-\sqrt{n}} \mathbb{1}_{\{l^{(n)}(0)=-\}}, \frac{R^{(n)}(0)}{v_+\sqrt{n}} \mathbb{1}_{\{l^{(n)}(0)=+\}} \right) \xrightarrow{\mathbb{P}} (x_-, x_+), \quad n \to \infty,$$

where x is the same as in (3.90) and, as usual, x_+ and x_- denote the positive part and the negative part of x, respectively.

Theorems 3.3.2 and 2.2.4 imply that

$$\left(\varphi_{\text{sgn}}\left(\frac{l^{(n)}(nt) R^{(n)}(\lfloor nt \rfloor)}{\sqrt{n}} \right) \right)_{t \geq 0} \Longrightarrow (W_\gamma^{\text{skew}}(t, x))_{t \geq 0}, \quad n \to \infty \quad (3.127)$$

on $D([0, \infty), \mathbb{R})$, where φ_{sgn} is as defined in (3.71). According to (3.88) and Theorem 2.2.4 the permeability parameter γ is equal to

$$\gamma = p_+ - p_- = \frac{\mathbb{E}^{\pi_{\text{entr}}}(R(0) - R(\sigma))(v_+^{-1} \mathbb{1}_{\{l(0)=+\}} - v_-^{-1} \mathbb{1}_{\{l(0)=-\}})}{\mathbb{E}^{\pi_{\text{entr}}}(R(0) - R(\sigma))v_{l(0)}^{-1}}. \quad (3.128)$$

Here, the stopping time σ is as given in (3.74) and the stationary measure π_{entr} is as defined in Lemma 3.3.1. The right-hand side of (3.128) is equal to that of (3.91) because the distribution of $(X(0) - X(\widetilde{\sigma}_1))\mathbb{1}_{\{X(0) \geq 0\}}$ (respectively, the distribution of $(X(0) - X(\widetilde{\sigma}_1))\mathbb{1}_{\{X(0) < 0\}}$) under $\widetilde{\pi}_{\text{entr}}$ coincides with the distribution of $(R(0) - R(\sigma))\mathbb{1}_{\{l(0)=+\}}$ (respectively, $-(R(0) - R(\sigma))\mathbb{1}_{\{l(0)=-\}}$) under π_{entr}.

We note for later needs that (3.127) entails

$$\left(\varphi_{\text{sgn}}\left(\frac{l^{(n)}(\lfloor nt \rfloor)(R^{(n)}(\lfloor nt \rfloor) + m)}{\sqrt{n}} \right) \right)_{t \geq 0} \Longrightarrow (W_\gamma^{\text{skew}}(t, x))_{t \geq 0}, \quad n \to \infty$$

$$(3.129)$$

on $D([0, \infty), \mathbb{R})$.

STEP 2. The mapping $(R, l) \mapsto l(R + m)$ is a bijection of the sets $\mathbb{N} \times \{+\}$ and $\{m+1, m+2, \ldots\}$ and also $\mathbb{N} \times \{-\}$ and $\{\ldots, -m-2, -m-1\}$. Fix any $n \in \mathbb{N}$. The sequence

$(l^{(n)}(k)(R^{(n)}(k)+m))_{k\geq 0}$ exhibits the following behavior. Assume that the sequence $R^{(n)}$ jumps over the set $\{-2m,\ldots,0\}$ downwards, that is, for some k,

$$R^{(n)}(k+1) > 0 \quad \text{and} \quad R^{(n)}(k+2) < -2m.$$

Then, according to (3.126),

$$l^{(n)}(k)(R^{(n)}(k)+m) = l^{(n)}(k+1)(R^{(n)}(k+1)+m),$$

which means that, in this case, the sequence $(l^{(n)}(j)(R^{(n)}(j)+m))_{j\geq 1}$ spends two units of time at the same point. We intend to construct a new sequence $(\widehat{X}^{(n)}(k))_{k\geq 0}$ from the sequence $(l^{(n)}(k)(R^{(n)}(k)+m))_{k\geq 0}$, which omits one of these steps. To this end, define a random time change by setting

$$\widehat{\lambda}_n(0) := 0, \quad \widehat{\lambda}_n(k) := k - \sum_{j=1}^{k} \mathbb{1}_{\{R^{(n)}(k)<-2m,\, R^{(n)}(k+1)>0\}}, \quad k \in \mathbb{N}.$$

The time-changed sequence $(\widehat{X}^{(n)}(k))_{k\geq 0}$ is then defined by

$$\widehat{X}^{(n)}(k) := l^{(n)}(\widehat{\lambda}_n^{-}(k))(R^{(n)}(\widehat{\lambda}_n^{-}(k))+m), \quad k \in \mathbb{N}_0,$$

where $\widehat{\lambda}_n^{-}(k) := \inf\{j \geq 0 : \widehat{\lambda}_n(j) \geq k\}$. Observe that $\widehat{X}^{(n)}$ is a Markov chain with transition probabilities

$$\mathbb{P}\{\widehat{X}^{(n)}(1) = j \mid \widehat{X}^{(n)}(0) = i\} = \begin{cases} \mathbb{P}\{X(1) = j \mid X(0) = i\}, & \text{if } |i| > m, \\ \mathbb{P}\{X(\widetilde{\tau}_0) = j \mid X(0) = i\}, & \text{if } |i| \leq m, \end{cases}$$

where $(X(k))_{k\geq 0}$ is the original random walk with membrane, and $\widetilde{\tau}_0$ is as defined in (3.89).

Now we show that the contribution of the number of omitted steps is negligible as $n \to \infty$ in the sense that

$$\left(\varphi_{\text{sgn}}\left(\frac{\widehat{X}^{(n)}(\lfloor nt \rfloor)}{\sqrt{n}}\right)\right)_{t\geq 0} \Longrightarrow (W_\gamma^{\text{skew}}(t,x))_{t\geq 0}, \quad n \to \infty \quad (3.130)$$

on $D([0,\infty), \mathbb{R})$. Indeed, with $L^{(n)}$ defined in (3.86),

$$\lfloor nt \rfloor - \widehat{\lambda}_n(\lfloor nt \rfloor) \leq \sqrt{n} L^{(n)}(t) \max(v_-, v_+), \quad t > 0. \quad (3.131)$$

3.3 Random Walks with Membrane and a Skew Brownian Motion

According to Theorem 3.3.2, for each $t \geq 0$, $L^{(n)}(t)$ converges in distribution to an a.s. finite random variable. Hence, for each $t \geq 0$,

$$\left|\frac{\widehat{\lambda}_n(\lfloor nt \rfloor)}{n} - t\right| \leq \frac{L^{(n)}(t)\max(v_-, v_+)}{\sqrt{n}} + \frac{1}{n} \xrightarrow{\mathbb{P}} 0, \quad n \to \infty. \tag{3.132}$$

By monotonicity of $t \mapsto \widehat{\lambda}_n(\lfloor nt \rfloor)$, this convergence is locally uniform. As a consequence, for all $T > 0$,

$$\sup_{t \in [0,T]} \left|\frac{\widehat{\lambda}_n(\lfloor nt \rfloor)}{n} - t\right| \xrightarrow{\mathbb{P}} 0, \quad n \to \infty$$

and, by Proposition 4.3.14,

$$\sup_{t \in [0,T]} \left|\frac{\widehat{\lambda}_n^-(\lfloor nt \rfloor)}{n} - t\right| \xrightarrow{\mathbb{P}} 0, \quad n \to \infty.$$

With this at hand, relation (3.130) follows from (3.129), Lemma 4.1.3, Proposition 4.3.7, and the Skorokhod representation theorem (Theorem 4.1.2).

STEP 3. Now we consider the original Markov chain $(X^{(n)}(k))_{k \geq 0}$ and define a random time change that removes jumps of $X^{(n)}$ inside the set $\{-m, \ldots, m\}$ after entering this set. Put

$$\lambda_n(0) = 0, \quad \lambda_n(k) := k - \sum_{j=1}^{k} \mathbb{1}_{\{|X^{(n)}(j)| \leq m, \, |X^{(n)}(j-1)| \leq m\}}, \quad k \in \mathbb{N}$$

and then $\lambda_n^-(k) := \inf\{j \geq 0 \colon \lambda_n(j) \geq k\}$ for $k \in \mathbb{N}_0$. It can be checked that

$$\left(\widehat{X}^{(n)}(k)\right)_{k \geq 0} \stackrel{d}{=} \left(X^{(n)}(\lambda_n^-(k))\right)_{k \geq 0}.$$

From now on, without loss of generality, we assume that, for each $n \in \mathbb{N}$,

$$\left(\frac{\widehat{X}^{(n)}(\lfloor nt \rfloor)}{\sqrt{n}}\right)_{t \geq 0} = \left(\frac{X^{(n)}(\lambda_n^-(\lfloor nt \rfloor))}{\sqrt{n}}\right)_{t \geq 0}.$$

STEP 4. To prove the distributional convergence of $\left(X^{(n)}(\lambda_n^-(\lfloor nt \rfloor))/\sqrt{n}\right)_{t \geq 0}$ we intend to use Proposition 4.3.10 in combination with the Skorokhod representation theorem.

Limit relation (4.5) in Proposition 4.3.10 holds true because the corresponding supremum in (4.5) does not exceed $2m/\sqrt{n}$. Hence, it remains to prove that, for all $T > 0$,

$$\sup_{t \in [0,T]} \left| \frac{\lambda_n^-(\lfloor nt \rfloor)}{n} - t \right| \xrightarrow{\mathbb{P}} 0, \quad n \to \infty. \tag{3.133}$$

The variable

$$\widehat{N}^{(n)}(k) = \sum_{j=0}^{k} \mathbb{1}_{\{|\widehat{X}^{(n)}(j)| \leq m\}}$$

counts the number of visits of $\widehat{X}^{(n)}$ to the set $\{-m, \ldots, m\}$ up to time k, for $k \in \mathbb{N}_0$. Define the stopping times by

$$\widehat{\sigma}_0^{(n)} := 0,$$
$$\widehat{\tau}_k^{(n)} := \inf\{j \geq \widehat{\sigma}_k^{(n)} : |X^{(n)}(j)| > m\}, \quad k \in \mathbb{N}_0,$$
$$\widehat{\sigma}_{k+1}^{(n)} = \inf\{j > \widehat{\tau}_k^{(n)} : |X^{(n)}(j)| \leq m\}, \quad k \in \mathbb{N}_0$$

and put $\zeta_k^{(n)} := \widehat{\tau}_k^{(n)} - \widehat{\sigma}_k^{(n)}$ for $k \in \mathbb{N}_0$. The variables $\zeta_1^{(n)} - 1$, $\zeta_2^{(n)} - 1, \ldots$ count the numbers of consecutive jumps inside the membrane. The variable $\zeta_0^{(n)} - 1$ does so provided that $|X^{(n)}(0)| \leq m$. Since

$$k - \lambda_n(k) \leq \sum_{j=0}^{\widehat{N}^{(n)}(k)} \zeta_j^{(n)}, \quad k \in \mathbb{N}_0,$$

we infer

$$\left| t - \frac{\lambda_n(\lfloor nt \rfloor)}{n} \right| \leq \frac{1}{n} + \frac{1}{n} \sum_{j=0}^{\widehat{N}^{(n)}(\lfloor nt \rfloor)} \zeta_j^{(n)}, \quad t \geq 0.$$

Observe that

$$\frac{\widehat{N}^{(n)}(\lfloor nt \rfloor)}{\sqrt{n}} \leq L^{(n)}(t) \max(v_-, v_+),$$

where $L^{(n)}$ is as defined in (3.86); see also (3.131). Similarly to (3.132), for each $t \geq 0$,

$$\frac{\widehat{N}^{(n)}(\lfloor nt \rfloor)}{n} \xrightarrow{\mathbb{P}} 0, \quad n \to \infty. \tag{3.134}$$

3.3 Random Walks with Membrane and a Skew Brownian Motion

For $|x| \leq m$, put $f(x) := \mathbb{E}[\zeta_0^{(n)} | X^{(n)}(0) = x]$. The function f is bounded and does not depend on n. We shall use a decomposition

$$\sum_{j=0}^{\widehat{N}^{(n)}(\lfloor nt \rfloor)} \zeta_j^{(n)} = \sum_{j=0}^{\widehat{N}^{(n)}(\lfloor nt \rfloor)} \left(\zeta_j^{(n)} - f(X^{(n)}(\widehat{\sigma}_j^{(n)})) \right) + \sum_{j=0}^{\widehat{N}^{(n)}(\lfloor nt \rfloor)} f(X^{(n)}(\widehat{\sigma}_j^{(n)})).$$

Relation (3.134) and boundedness of f ensure that, for each $t \geq 0$,

$$\frac{1}{n} \sum_{j=0}^{\widehat{N}^{(n)}(\lfloor nt \rfloor)} f(X^{(n)}(\widehat{\sigma}_j^n)) \xrightarrow{\mathbb{P}} 0, \quad n \to \infty.$$

The random sequence

$$\sum_{j=0}^{k} \left(\zeta_j^{(n)} - f(X^{(n)}(\widehat{\sigma}_j^{(n)})) \right), \quad k \geq 0$$

is a martingale. Doob's inequality entails, for each $\ell \in \mathbb{N}_0$,

$$\mathbb{E} \max_{0 \leq k \leq \ell} \left(\sum_{j=0}^{k} \left(\zeta_j^{(n)} - f(X^{(n)}(\widehat{\sigma}_j^{(n)})) \right) \right)^2$$

$$\leq 4\mathbb{E} \sum_{j=0}^{\ell} \left(\zeta_j^{(n)} - f(X^{(n)}(\widehat{\sigma}_j^{(n)})) \right)^2$$

$$\leq 4(\ell + 1) \max_{|x| \leq m} \mathbb{E}\left[\left(\zeta_0^{(n)} - f(X^{(n)}(\widehat{\sigma}_0^n)) \right)^2 \big| X^{(n)}(0) = x \right]$$

$$\leq C(\ell + 1) \tag{3.135}$$

for[19] some $C > 0$. Hence, for all $\delta > 0$ and each $t \geq 0$,

$$\frac{1}{n} \left| \sum_{j=0}^{\widehat{N}^{(n)}(\lfloor nt \rfloor)} \left(\zeta_j^{(n)} - f(X^{(n)}(\widehat{\sigma}_j^{(n)})) \right) \right|$$

$$\leq \frac{1}{n} \mathbb{1}_{\{\widehat{N}^{(n)}(\lfloor nt \rfloor) \leq n\delta\}} \left| \sum_{j=0}^{\widehat{N}^{(n)}(\lfloor nt \rfloor)} \left(\zeta_j^{(n)} - f(X^{(n)}(\widehat{\sigma}_j^{(n)})) \right) \right|$$

---

[19] As follows from the general theory of Markov chains with finite state spaces, the conditional distribution of $\zeta_0^{(n)}$ has some finite exponential moments, hence a finite second moment.

$$+ \frac{1}{n}\mathbb{1}_{\{\widehat{N}^{(n)}(\lfloor nt \rfloor) > n\delta\}} \bigg| \sum_{j=0}^{\widehat{N}^{(n)}(\lfloor nt \rfloor)} (\zeta_j^{(n)} - f(X^{(n)}(\widehat{\sigma}_j^{(n)}))) \bigg|$$

$$\leq \frac{1}{n} \max_{0 \leq k \leq \lfloor n\delta \rfloor} \bigg| \sum_{j=0}^{k} (\zeta_j^{(n)} - f(X^{(n)}(\widehat{\sigma}_j^{(n)}))) \bigg|$$

$$+ \frac{1}{n}\mathbb{1}_{\{\widehat{N}^{(n)}(\lfloor nt \rfloor) > n\delta\}} \bigg| \sum_{j=0}^{\widehat{N}^{(n)}(\lfloor nt \rfloor)} (\zeta_j^{(n)} - f(X^{(n)}(\widehat{\sigma}_j^{(n)}))) \bigg|.$$

The first term in the latter formula converges to zero in L_2 in view of (3.135), and the second term converges to zero in probability according to (3.134). Combining all the fragments together we obtain

$$\bigg| \frac{\lambda_n(\lfloor nt \rfloor)}{n} - t \bigg| \xrightarrow{\mathbb{P}} 0, \quad n \to \infty.$$

By monotonicity, for all $T > 0$,

$$\sup_{t \in [0,T]} \bigg| \frac{\lambda_n(\lfloor nt \rfloor)}{n} - t \bigg| \xrightarrow{\mathbb{P}} 0, \quad n \to \infty.$$

Invoking Proposition 4.3.14 we arrive at (3.133) and the desired limit relation

$$\left(\frac{X^{(n)}(\lfloor nt \rfloor)}{\sqrt{n}} \right)_{t \geq 0} \Longrightarrow (W_\gamma^{\text{skew}}(t,x))_{t \geq 0}, \quad n \to \infty$$

on $D([0,\infty), \mathbb{R})$. The proof of Theorem 3.3.3 is complete.

3.3.5 Proof Corollary 3.3.1

Representation (3.72) for the limit process is the only thing that needs to be proved.

Recall stochastic differential equation (2.44) and apply the Itô-Tanaka formula with the symmetric semimartingale local time (see, for instance, Exercise 1.25 on p. 234 in [140]) to W_γ^{skew} with $W_\gamma^{\text{skew}}(0) = 0$:

$$Y(t) = \varphi_{\text{sgn}}^{-1}(W_\gamma^{\text{skew}}(t)) = \int_0^t (v_+ \mathbb{1}_{\{W_\gamma^{\text{skew}}(s) > 0\}} + v_- \mathbb{1}_{\{W_\gamma^{\text{skew}}(s) < 0\}}) \, dW(s)$$

$$+ \frac{1}{2}\gamma(v_+ + v_-)L_0^{W_\gamma^{\text{skew}}}(t) + \frac{1}{2}(v_+ - v_-)L_0^{W_\gamma^{\text{skew}}}(t)$$

3.3 Random Walks with Membrane and a Skew Brownian Motion

$$= \int_0^t (v_+ \mathbb{1}_{\{Y(s)>0\}} + v_- \mathbb{1}_{\{Y(s)<0\}})) \, dW(s)$$
$$+ \frac{1}{2}\gamma(v_+ + v_-)L_0^{W_\gamma^{\text{skew}}}(t) + \frac{1}{2}(v_+ - v_-)L_0^{W_\gamma^{\text{skew}}}(t) \tag{3.136}$$

and

$$|Y(t)| = |\varphi_{\text{sgn}}^{-1}(W_\gamma^{\text{skew}}(t))| = \int_0^t (v_+ \mathbb{1}_{\{W_\gamma^{\text{skew}}(s)>0\}} + v_- \mathbb{1}_{\{W_\gamma^{\text{skew}}(s)<0\}}) \, dW(s)$$
$$+ \frac{1}{2}\gamma(v_+ - v_-)L_0^{W_\gamma^{\text{skew}}}(t) + \frac{1}{2}(v_+ + v_-)L_0^{W_\gamma^{\text{skew}}}(t). \tag{3.137}$$

On the other hand, the process Y itself is a continuous semimartingale. Hence, the Itô-Tanaka formula applied to the process Y yields

$$|Y(t)| = \int_0^t (\mathbb{1}_{\{Y(s)>0\}} - \mathbb{1}_{\{Y(s)<0\}}) \, dY(s) + L_0^Y(t) \tag{3.138}$$
$$= \int_0^t (v_+ \mathbb{1}_{\{W_\gamma^{\text{skew}}(s)>0\}} + v_- \mathbb{1}_{\{W_\gamma^{\text{skew}}(s)<0\}}) \, dW(s) + L_0^Y(t),$$

where $L_0^Y(t)$ is the symmetric semimartingale local time defined in (2.15), that is,

$$L_0^Y(t) = \lim_{\varepsilon \to 0+} \frac{1}{2\varepsilon} \int_{[0,t]} \mathbb{1}_{\{|Y(s)|<\varepsilon\}} d\langle Y, Y \rangle(s)$$
$$= \lim_{\varepsilon \to 0+} \frac{1}{2\varepsilon} \int_0^t (v_+^2 \mathbb{1}_{\{0 \le Y(s)<\varepsilon\}} + v_-^2 \mathbb{1}_{\{-\varepsilon<Y(s)<0\}}) \, ds.$$

Comparing the representations (3.137) and (3.138) we conclude that

$$L_0^Y(t) = \frac{1}{2}(\gamma(v_+ - v_-) + (v_+ + v_-))L_0^{W_\gamma^{\text{skew}}}(t).$$

Substituting this formula into (3.136) and recalling that Y spends zero time at zero we obtain (3.72).

3.4 Limit Theorems for Heavy-Tailed Random Walks with Membrane at 0

3.4.1 Main Result

Let ξ be a random variable with zero mean and a 1-arithmetic distribution. The latter means that the distribution of ξ is concentrated on the set of integers $\mathbb{Z}$ and not concentrated on $d\mathbb{Z}$ for any $d \geq 2$. It is known (see, for instance, Theorem 8 on p. 23 in [153]) that S_ξ visits every integer point, and particularly 0, infinitely often a.s. Let $(X(n))_{n \geq 0}$ be a standard random walk perturbed at 0, that is, a Markov chain with transition probabilities

$$\mathbb{P}\{X(n+1) = j \mid X(n) = i\} = \begin{cases} \mathbb{P}\{\xi = j - i\}, & \text{if } i \neq 0; \\ \mathbb{P}\{\eta = j\}, & \text{if } i = 0, \end{cases}$$

where η is an integer-valued random variable with $\mathbb{P}\{\eta = 0\} < 1$.

In addition to the conditions imposed on the distribution of ξ above we assume that the distribution of ξ belongs to the domain of attraction of a symmetric α-stable distribution with $\alpha \in (1, 2)$. Thus, $\mathbb{E}\xi = 0$ and in the setting of Theorem 4.4.2 $\tau = \xi$, $\gamma = \alpha$ and $\beta = 0$. An application of the aforementioned theorem yields

$$\left(\frac{S_\xi(\lfloor vt \rfloor)}{a(v)}\right)_{t \geq 0} \implies (U_\alpha(t))_{t \geq 0}, \quad v \to \infty \tag{3.139}$$

on $D([0, \infty)) = D([0, \infty), \mathbb{R})$, where a is any positive function satisfying

$$\lim_{x \to \infty} x\mathbb{P}\{|\xi| > a(x)\} = (1 - \alpha)/(\Gamma(2 - \alpha)\cos(\pi\alpha/2)).$$

Here, as in Sect. 2.3.1, $U_\alpha = (U_\alpha(t))$ is a symmetric α-stable Lévy process with the characteristic function $\mathbb{E}\exp(izU_\alpha(t)) = \exp(-t|z|^\alpha)$ for $z \in \mathbb{R}$ and $t \geq 0$. Also, we assume that either $\mathbb{E}|\eta| < \infty$ or the distribution of η belongs to the domain of attraction of a β-stable distribution with $\beta \in (0, 1)$. The latter means (see Theorem 4.4.1) that the function $x \mapsto \mathbb{P}\{|\eta| > x\}$ is regularly varying at $+\infty$ of index $-\beta$ and that

$$\mathbb{P}\{\eta > x\} \sim c_+ \mathbb{P}\{|\eta| > x\} \quad \text{and} \quad \mathbb{P}\{-\eta > x\} \sim c_- \mathbb{P}\{|\eta| > x\}, \quad x \to \infty \tag{3.140}$$

for some nonnegative c_+ and c_- summing up to one.[20] Furthermore, by Theorem 4.4.2,

$$\left(\frac{S_\eta(\lfloor vt \rfloor)}{c(v)}\right)_{t \geq 0} \implies (\mathcal{S}_\beta(t))_{t \geq 0}, \quad v \to \infty$$

---

[20] In the notation of Theorem 4.4.1 $c_+ = (1 + \beta)/2$. However, β in Theorem 4.4.1, being a skewness parameter, should not be confused with (the negative of) the index of regular variation of the distribution of $|\eta|$.

on $D([0, \infty))$. Here, c is any positive function satisfying

$$\lim_{x \to \infty} x\mathbb{P}\{|\eta| > c(x)\} = (1 - \beta)/(\Gamma(2 - \beta)\cos(\pi\beta/2))$$

and $\mathcal{S}_\beta$ is a β-stable Lévy process with the characteristic function

$$\mathbb{E}\exp(iz\mathcal{S}_\beta(t)) = \exp(-t|z|^\beta(1 - i(c_+ - c_-)\tan(\pi\beta/2)\operatorname{sign} z), \quad z \in \mathbb{R}$$

with the constants c_+ and c_- as in (3.140). According to Remark 4.4.1, the functions a and c are regularly varying at ∞ of indices $1/\alpha$ and $1/\beta$, respectively. We conclude that $\lim_{v \to \infty}(a(v)/c(v)) = 0$ because $\alpha > \beta$.

For each $v > 0$, let $(X_v(n))_{n \geq 0}$ be a Markov chain having the same transition probabilities as $(X(n))_{n \geq 0}$ but possibly satisfying a different initial condition. We are ready to state the main result of the section. The function a appearing in the theorem is the same as in (3.139).

Theorem 3.4.1 *Let $x \in \mathbb{R}$ and assume that $X_v(0)/a(v)$ converges in probability to x as $v \to \infty$.*

(a) If the distribution of η belongs to the domain of attraction of a β-stable distribution with $\beta < \alpha - 1$, then

$$\left(\frac{X_v(\lfloor vt \rfloor)}{a(v)}\right)_{t \geq 0} \Longrightarrow \left(U_{\alpha,\beta}(x, t)\right)_{t \geq 0}, \quad v \to \infty$$

on $D([0, \infty))$, where $(U_{\alpha,\beta}(x, t))_{t \geq 0}$ is a skew α-stable Lévy process with parameter β starting from x.

(b) If $\mathbb{E}|\eta| < \infty$ or the distribution of η belongs to the domain of attraction of a β-stable distribution with $\beta > \alpha - 1$, then

$$\left(\frac{X_v(\lfloor vt \rfloor)}{a(v)}\right)_{t \geq 0} \Longrightarrow \left(x + U_\alpha(t)\right)_{t \geq 0}, \quad v \to \infty \qquad (3.141)$$

on $D([0, \infty))$, where U_α is a symmetric α-stable Lévy process.[21]

---

[21] We attract the reader's attention to the fact that $U_\alpha(0) = 0$ a.s.

Our proof of Theorem 3.4.1 exploits a resolvent approach and bears a similarity to the proof of Theorem 2.3.1. In the cited result, the skew α-stable Lévy process was constructed as a scaling limit of small perturbations at 0 of a symmetric α-stable process. The main achievement of Theorem 3.4.1 is a new construction of a skew stable Lévy process as a scaling limit of locally perturbed standard random walks. On the technical side, a passage from continuous-time processes to random sequences requires at places additional nontrivial arguments. Last but not least, part (b) of Theorem 3.4.1 is a discrete-time counterpart of part (b) of Theorem 2.3.1.

Here is a counterpart of Remark 2.3.1 which dealt with weak convergence of continuous-time processes. When the perturbations have finite nonzero means, there is an essential difference between the case where $\mathbb{E}\xi^2 = \sigma^2 \in (0, \infty)$ and the case where the distribution of ξ belongs to the domain of attraction of an α-stable distribution with $\alpha \in (1, 2)$ (then automatically $\sigma^2 = \infty$). In the latter case, according to Theorem 3.4.1(b) the perturbations have no effect asymptotically, and the scaling limit of locally perturbed standard random walks is, up to a shift, the same as the scaling limit of the unperturbed random walks. In the former case, according to Theorem 3.3.1, the scaling limit of locally perturbed standard random walks is a skew Brownian motion, rather than a Brownian motion (the scaling limit of the unperturbed random walks).

3.4.2 Proof of Theorem 3.4.1(a)

For each $v > 0$, define the process $\mathfrak{X}_v = (\mathfrak{X}_v(t))_{t \geq 0}$ by $\mathfrak{X}_v(t) := X_v(\lfloor vt \rfloor)/a(v)$ for $t \geq 0$. We intend to Poissonize the $\mathfrak{X}_v$. To this end, let $(N(t))_{t \geq 0}$ denote a Poisson process on $[0, \infty)$ of unit intensity, which is independent of $((X_v(k))_{k \geq 0})_{v > 0}$. For each $v > 0$, define now $\widetilde{\mathfrak{X}}_v := (\widetilde{\mathfrak{X}}_v(t))_{t \geq 0}$, a Poissonized version of $\mathfrak{X}_v$, by

$$\widetilde{\mathfrak{X}}_v(t) := \frac{X_v(N(vt))}{a(v)}, \quad t \geq 0.$$

The Poissonized version $\widetilde{\mathfrak{X}}_v$ is a continuous-time Markov chain. The sizes of its jumps are the same as those of $\mathfrak{X}_v$, but unlike in $\mathfrak{X}_v$, the jumps occur at random epochs given by the successive positions of a standard random walk with exponentially distributed increments of mean $1/v$. The process $\widetilde{\mathfrak{X}}_v$ is an instance of the holding and jumping process discussed in Example 4.6.3. The main reason behind using the Poissonization in the present setting is availability of formula (4.29).

For each $T > 0$,

$$\lim_{v \to \infty} \sup_{t \in [0, T]} |v^{-1} N(vt) - t| = 0 \quad \text{a.s.} \tag{3.142}$$

3.4 Limit Theorems for Heavy-Tailed Random Walks with Membrane at 0

Since the limit function is nonrandom, continuous, and increasing, Remark 4.3.1 and Proposition 4.3.7 tell us that the weak limits of $\mathfrak{X}_v$ and $\widetilde{\mathfrak{X}}_v$ are the same, provided these exist. In particular, it is enough to prove that

$$\widetilde{\mathfrak{X}}_v \implies U_{\alpha,\beta}, \quad v \to \infty \qquad (3.143)$$

on $D([0, \infty))$. For later use, we note that, according to Proposition 4.3.7 and Remark 4.3.1, relations (3.139) and (3.142) entail

$$\widetilde{S}_v \implies U_\alpha, \quad v \to \infty \qquad (3.144)$$

on $D([0, \infty))$, where $\widetilde{S}_v := (S_\xi(N(vt))/a(v))_{t \geq 0}$.

We intend to prove (3.143) with the help of Theorem 4.5.2. Since $U_{\alpha,\beta}$ and $\widetilde{\mathfrak{X}}_v, v > 0$ are strong Markov processes, invoking Lemma 4.5.1 (with $x^* = 0$) yields, for $\lambda > 0$,

$$R_\lambda^{U_{\alpha,\beta}} f(x) = V_\lambda^{U_{\alpha,\beta}} f(x) + \mathbb{E}^x e^{-\lambda \sigma(U_{\alpha,\beta})} R_\lambda^{U_{\alpha,\beta}} f(0), \quad x \in \mathbb{R}$$

and

$$R_\lambda^{\widetilde{\mathfrak{X}}_v} f(l/a(v)) = V_\lambda^{\widetilde{\mathfrak{X}}_v} f(l/a(v)) + \mathbb{E}^{l/a(v)} e^{-\lambda \sigma(\widetilde{\mathfrak{X}}_v)} R_\lambda^{\widetilde{\mathfrak{X}}_v} f(0), \quad l \in \mathbb{Z}.$$

By Theorem 4.5.2, (3.143) follows if we can show that, for each $f \in C_0(\mathbb{R})$ and $\lambda > 0$,

$$\lim_{v \to \infty} \sup_{l \in \mathbb{Z}} |V_\lambda^{\widetilde{\mathfrak{X}}_v} f(l/a(v)) - V_\lambda^{U_{\alpha,\beta}} f(l/a(v))| = 0, \qquad (3.145)$$

$$\lim_{v \to \infty} \sup_{l \in \mathbb{Z}} \left| \mathbb{E}^{l/a(v)} e^{-\lambda \sigma(\widetilde{\mathfrak{X}}_v)} - \mathbb{E}^{l/a(v)} e^{-\lambda \sigma(U_{\alpha,\beta})} \right| = 0 \qquad (3.146)$$

and

$$\lim_{v \to \infty} \left| R_\lambda^{\widetilde{\mathfrak{X}}_v} f(0) - R_\lambda^{U_{\alpha,\beta}} f(0) \right| = 0. \qquad (3.147)$$

Observe that, for each $v > 0$ and each $x \in \mathbb{R}$,

$$\mathbb{P}\{(X_v(k)\mathbb{1}_{\{k \leq \sigma(X_v)\}})_{k \geq 0} \in \cdot \mid X_v(0) = x\}$$
$$= \mathbb{P}\{(S_\xi(k)\mathbb{1}_{\{k \leq \sigma(S_\xi)\}})_{k \geq 0} \in \cdot \mid S_\xi(0) = x\}.$$

This implies that

$$V_\lambda^{\widetilde{\mathfrak{X}}_v} = V_\lambda^{\widetilde{S}_v}, \quad \lambda > 0 \qquad (3.148)$$

and, for each $l \in \mathbb{N}$,

$$\mathbb{E}^{l/a(v)} e^{-\lambda \sigma(\widetilde{\mathfrak{X}}_v)} = \mathbb{E}^{l/a(v)} e^{-\lambda \sigma(\widetilde{S}_v)} = \mathbb{E}^l e^{-(\lambda/v)\sigma(S_\xi \circ N)}, \quad \lambda \geq 0.$$

Here, $\circ$ denotes composition, and the last equality follows by a direct computation. Also, for each $x \in \mathbb{R}$,

$$\mathbb{P}\{(U_{\alpha,\beta}(t)\mathbb{1}_{\{t<\sigma(U_{\alpha,\beta})\}})_{t\geq 0} \in \cdot \mid U_{\alpha,\beta}(0) = x\}$$
$$= \mathbb{P}\{(U_\alpha(t)\mathbb{1}_{\{t<\sigma(U_\alpha)\}})_{t\geq 0} \in \cdot \mid U_\alpha(0) = x\}.$$

This entails $V_\lambda^{U_{\alpha,\beta}} = V_\lambda^{U_\alpha}$, $\lambda > 0$ and, for each $x \in \mathbb{R}$, $\mathbb{E}^x e^{-\lambda \sigma(U_{\alpha,\beta})} = \mathbb{E}^x e^{-\lambda \sigma(U_\alpha)}$, $\lambda \geq 0$. As a consequence, (3.145) and (3.146) are equivalent to

$$\lim_{v \to \infty} \sup_{l \in \mathbb{Z}} |V_\lambda^{\widetilde{S}_v} f(l/a(v)) - V_\lambda^{U_\alpha} f(l/a(v))| = 0 \quad (3.149)$$

and

$$\lim_{v \to \infty} \sup_{l \in \mathbb{Z}} \left| \mathbb{E}^{l/a(v)} e^{-\lambda \sigma(\widetilde{S}_v)} - \mathbb{E}^{l/a(v)} e^{-\lambda \sigma(U_\alpha)} \right|$$
$$= \lim_{v \to \infty} \sup_{l \in \mathbb{Z}} \left| \mathbb{E}^l e^{-(\lambda/v)\sigma(S_\xi \circ N)} - \mathbb{E}^{l/a(v)} e^{-\lambda \sigma(U_\alpha)} \right| = 0. \quad (3.150)$$

Another application of Lemma 4.5.1 (with $x^* = 0$) to strong Markov processes U_α and $\widetilde{S}_v$, $v > 0$ enables us to conclude that, for $\lambda > 0$,

$$R_\lambda^{U_\alpha} f(x) = V_\lambda^{U_\alpha} f(x) + \mathbb{E}^x e^{-\lambda \sigma(U_\alpha)} R_\lambda^{U_\alpha} f(0), \quad x \in \mathbb{R}$$

and

$$R_\lambda^{\widetilde{S}_v} f(l/a(v)) = V_\lambda^{\widetilde{S}_v} f(l/a(v)) + \mathbb{E}^{l/a(v)} e^{-\lambda \sigma(\widetilde{S}_v)} R_\lambda^{\widetilde{S}_v} f(0), \quad l \in \mathbb{Z}.$$

Thus, if we can prove (3.150) and

$$\lim_{v \to \infty} \sup_{x \in \mathbb{R}} |R_\lambda^{\widetilde{S}_v} f(x) - R_\lambda^{U_\alpha} f(x)| =$$
$$\lim_{v \to \infty} \sup_{x \in \mathbb{R}} \left| \int_0^\infty \Big(\mathbb{E} f(x + S_\xi(N(vt))/a(v)) - \mathbb{E} f(x + U_\alpha(t)) \Big) e^{-\lambda t} dt \right| = 0 \quad (3.151)$$

for $f \in C_0(\mathbb{R})$, then (3.149) holds. Once this is done, the only remaining thing is to check (3.147).

3.4 Limit Theorems for Heavy-Tailed Random Walks with Membrane at 0

Proof of (3.151) Note that each $f \in C_0(\mathbb{R})$ is uniformly continuous and put, for $\gamma > 0$, $\omega_f(\gamma) := \sup_{x,y \in \mathbb{R},\, |x-y| \leq \gamma} |f(x) - f(y)|$, that is, ω_f is the modulus of continuity of f. Let $(v_k)_{k \geq 1}$ be any sequence of positive numbers satisfying $\lim_{k \to \infty} v_k = \infty$. Using (3.144) together with the Skorokhod representation theorem (Theorem 4.1.2) we conclude that there exist $(\hat{S}_{v_k})_{k \geq 1}$, versions of $(\widetilde{S}_{v_k})_{k \geq 1}$, and $\hat{U}_\alpha$, a version of U_α, such that

$$\lim_{k \to \infty} \hat{S}_{v_k}(t) = \hat{U}_\alpha(t) \quad \text{a.s.}$$

on $D([0, \infty))$. In particular, this entails the a.s. convergence for almost all $t \geq 0$ with respect to Lebesgue measure. Hence,

$$\limsup_{k \to \infty} \sup_{x \in \mathbb{R}} \int_0^\infty \left| \mathbb{E} f(x + S_\xi(N(v_k t))/a(v_k)) - \mathbb{E} f(x + U_\alpha(t)) \right| e^{-\lambda t} \, dt$$

$$\leq \limsup_{k \to \infty} \sup_{x \in \mathbb{R}} \int_0^\infty \mathbb{E} \left| f(x + \hat{S}_{v_k}(t)) - f(x + \hat{U}_\alpha(t)) \right| e^{-\lambda t} \, dt$$

$$\leq \lim_{k \to \infty} \int_0^\infty \mathbb{E} \left[\left(\omega_f\left(\hat{S}_{v_k}(t) - \hat{U}_\alpha(t) \right) \right) \wedge (2\|f\|) \right] e^{-\lambda t} \, dt = 0,$$

where the last equality is justified by the Lebesgue dominated convergence theorem. Since the diverging sequence $(v_k)_{k \geq 1}$ is arbitrary, the proof of (3.151) is complete.

To prove (3.150) we first derive in Corollary 3.4.1 a formula for $\mathbb{E}^l e^{-\lambda \sigma(S_\xi \circ N)}$. As a preparation, we prove an auxiliary result.

Lemma 3.4.1 *Let $Y := (Y(k))_{k \geq 0}$ be a Markov chain on a finite or countable set G. For $x^* \in G$, put $\sigma := \sigma_{x^*} := \inf\{k \in \mathbb{N}_0 : Y(k) = x^*\}$. Then*

$$\mathbb{E}^x s^\sigma = \frac{u_s(x, x^*)}{u_s(x^*, x^*)}, \quad x \in G, \quad |s| < 1, \tag{3.152}$$

where $u_s(x, x^*) = \sum_{k \geq 0} s^k \mathbb{P}\{Y(k) = x^* \mid Y(0) = x\}$ for $x \in G$. In particular, if $x^* = 0$, $Y(k) = S_\tau(k)$ for $k \in \mathbb{N}_0$ and $(Y(k))_{k \geq 0}$ lives on the lattice $G = b\mathbb{Z}$ for some $b > 0$, then

$$\mathbb{E}^x s^\sigma = \frac{u_s(-x)}{u_s(0)}, \quad x \in b\mathbb{Z}, \quad |s| < 1, \tag{3.153}$$

where

$$u_s(x) = u_s(x, 0) = \sum_{k \geq 0} s^k \mathbb{P}\{S_\tau(k) = 0 \mid S_\tau(0) = x\}$$

$$= \mathbb{1}_{\{0\}}(x) + \sum_{k \geq 1} s^k \mathbb{P}\{S_\tau(k) = -x\}.$$

Alternatively,

$$u_s(-x) = \frac{1}{2\pi} \int_{-\pi}^{\pi} \frac{e^{ix\theta}}{1 - s\mathbb{E}e^{i\theta\tau}} d\theta, \quad x \in b\mathbb{Z}, \; |s| < 1. \tag{3.154}$$

Proof Denote by R_s the resolvent of Y, so that

$$R_s f(x) = \sum_{k \geq 0} s^k \mathbb{E}^x f(Y(k)), \quad x \in G, \; |s| < 1$$

for bounded functions $f : G \to \mathbb{R}$. The R_s satisfies a formula similar to (4.18)

$$R_s f(x) = \sum_{k=0}^{\sigma-1} s^k \mathbb{E}^x f(Y(k)) + \mathbb{E}^x s^\sigma R_s f(x^*), \quad x \in G, \; |s| < 1.$$

Put $f(x) = \mathbb{1}_{\{x^*\}}(x)$. It follows from the penultimate centered formula that $R_s f(x) = u_s(x, x^*)$ for $x \in G$ and $|s| < 1$. On the other hand, the first summand on the right-hand side of the last centered formula vanishes, whence $R_s f(x) = \mathbb{E}^x s^\sigma R_s f(x^*) = \mathbb{E}^x s^\sigma u_s(x^*, x^*)$ for $x \in G$ and $|s| < 1$. This proves (3.152). Formula (3.153) is just a specialization of (3.152). To prove (3.154), write with the help of Fubini's theorem

$$\sum_{x \in b\mathbb{Z}} u_s(-x) e^{ix\theta} = 1 + \sum_{x \in b\mathbb{Z}} \sum_{k \geq 1} s^k \mathbb{P}\{S_\tau(k) = x\} e^{ix\theta}$$

$$= 1 + \sum_{k \geq 1} s^k \sum_{x \in b\mathbb{Z}} \mathbb{P}\{S_\tau(k) = x\} e^{ix\theta} = 1 + \sum_{k \geq 1} s^k (\mathbb{E}e^{i\theta\tau})^k = \frac{1}{1 - s\mathbb{E}e^{i\theta\tau}}$$

for $|s| < 1$ and $\theta \in \mathbb{R}$. With this at hand, (3.154) is an immediate consequence of a standard Fourier inversion formula. □

We stress that continuous-time formula (2.58) rests on nontrivial potential-analytic results, whereas discrete-time formula (3.152) is rather simple.

Corollary 3.4.1 *Let the assumptions and notation of Lemma 3.4.1 be in force. Denote by $(N_\rho(t))_{t \geq 0}$ a Poisson process on $[0, \infty)$ of intensity $\rho > 0$, which is independent of Y, and put $\widetilde{Y}(t) = Y(N_\rho(t))$ for $t \geq 0$. For $x^* \in G$, put $\widetilde{\sigma} := \widetilde{\sigma}_{x^*} := \inf\{t \geq 0 : \widetilde{Y}(t) = x^*\}$. Then*

$$\mathbb{E}^x e^{-\lambda \widetilde{\sigma}} = \frac{\hat{u}_\lambda(x, x^*)}{\hat{u}_\lambda(x^*, x^*)} = \frac{u_s(x, x^*)}{u_s(x^*, x^*)}, \quad x \in G, \; \lambda > 0, \tag{3.155}$$

3.4 Limit Theorems for Heavy-Tailed Random Walks with Membrane at 0

where $s = \rho/(\lambda + \rho)$ and $\hat{u}_\lambda(x, x^*) = \int_0^\infty e^{-\lambda t} \mathbb{P}\{\widetilde{Y}(t) = x^* \mid \widetilde{Y}(0) = x\} dt$. In particular, if $x^* = 0$, $Y(k) = S_\tau(k)$ for $k \in \mathbb{N}_0$ and $(Y(k))_{k \geq 0}$ lives on the lattice $G = b\mathbb{Z}$ for some $b > 0$, then

$$\mathbb{E}^x e^{-\lambda \widetilde{\sigma}} = \frac{\hat{u}_\lambda(-x)}{\hat{u}_\lambda(0)} = \frac{u_s(-x)}{u_s(0)}, \quad x \in b\mathbb{Z}, \quad \lambda > 0, \tag{3.156}$$

where $s = \rho/(\lambda + \rho)$ and

$$\hat{u}_\lambda(x) = \hat{u}_\lambda(x, 0) = \int_0^\infty e^{-\lambda t} \mathbb{P}\{\widetilde{Y}(t) = 0 \mid \widetilde{Y}(0) = x\} dt$$

$$= \int_0^\infty e^{-\lambda t} \mathbb{P}\{S_\tau(N_\rho(t)) = -x\} dt.$$

Proof The first equality in (3.155) follows from (4.18) and the argument used for the proof of (3.152). To prove the second equality in (3.155) we shall derive a formula relating the resolvent $R_\lambda^{\widetilde{Y}}$ of $\widetilde{Y}$ to the resolvent R_s of Y. By a repeated application of Fubini's theorem

$$R_\lambda^{\widetilde{Y}} f(x) = \mathbb{E}^x \int_0^\infty e^{-\lambda t} f(\widetilde{Y}(t)) dt = \int_0^\infty e^{-\lambda t} \sum_{k \geq 0} e^{-\rho t} \frac{(\rho t)^k}{k!} \mathbb{E}^x f(Y(k)) dt$$

$$= \sum_{k \geq 0} \frac{\rho^k}{k!} \mathbb{E}^x f(Y(k)) \int_0^\infty t^k e^{-(\lambda+\rho)t} dt = \frac{1}{\lambda + \rho} \sum_{k \geq 0} \left(\frac{\rho}{\lambda + \rho}\right)^k \mathbb{E}^x f(Y(k))$$

$$= \frac{1}{\lambda + \rho} R_s f(x),$$

where $s = \rho/(\lambda + \rho)$. Putting $f(x) = \mathbb{1}_{\{x^*\}}(x)$ for $x \in G$, we infer

$$\hat{u}_\lambda(x, x^*) = (\lambda + \rho)^{-1} u_s(x, x^*)$$

for $x \in G$ and the same s as before, thereby justifying the second equality in (3.155). Formula (3.156) is a specialization of (3.155). $\square$

Proof of (3.150) We shall prove (3.150) in an equivalent form:

$$\limsup_{v \to \infty} \sup_{x \in \mathbb{R}} \left| \mathbb{E}^{\lfloor xa(v) \rfloor} e^{-(\lambda/v)\sigma(S_\xi \circ N)} - \mathbb{E}^{\lfloor xa(v) \rfloor / a(v)} e^{-\lambda \sigma(U_\alpha)} \right| = 0. \tag{3.157}$$

With $s = v/(v + \lambda)$ and $\psi(\theta) := \mathbb{E}e^{i\theta \xi}$ for $\theta \in \mathbb{R}$, we use the following representation: for $x \in \mathbb{R}$,

$$\mathbb{E}^{\lfloor xa(v) \rfloor} e^{-(\lambda/v)\sigma(S_\xi \circ N)} = u_s(-\lfloor xa(v) \rfloor)/u_s(0)$$

$$= \int_{-\pi}^{\pi} \frac{e^{i\theta \lfloor xa(v) \rfloor}}{1 - s\psi(\theta)} d\theta \bigg/ \int_{-\pi}^{\pi} \frac{d\theta}{1 - s\psi(\theta)}$$

$$= \int_{-\pi}^{\pi} \frac{e^{i\theta \lfloor xa(v) \rfloor}}{\lambda + v(1 - \psi(\theta))} d\theta \bigg/ \int_{-\pi}^{\pi} \frac{d\theta}{\lambda + v(1 - \psi(\theta))}$$

$$= \int_{-\pi a(v)}^{\pi a(v)} \frac{e^{i\theta \lfloor xa(v) \rfloor / a(v)}}{\lambda + v(1 - \psi(\theta/a(v)))} d\theta \bigg/ \int_{-\pi a(v)}^{\pi a(v)} \frac{d\theta}{\lambda + v(1 - \psi(\theta/a(v)))}, \quad (3.158)$$

where the first equality is a specialization of (3.156) for $Y = S_\xi$, $\rho = 1$ and $b = 1$, and the second equality follows from (3.154) with $\tau = \xi$. Further, by (2.58) and (2.59),

$$\mathbb{E}^{\lfloor xa(v) \rfloor / a(v)} e^{-\lambda \sigma(U_\alpha)} = \int_0^\infty \frac{\cos(\theta \lfloor xa(v) \rfloor / a(v))}{\lambda + \theta^\alpha} d\theta \bigg/ \int_0^\infty \frac{d\theta}{\lambda + \theta^\alpha}$$

$$= \int_{\mathbb{R}} \frac{e^{i\theta \lfloor xa(v) \rfloor / a(v)}}{\lambda + |\theta|^\alpha} d\theta \bigg/ \int_{\mathbb{R}} \frac{d\theta}{\lambda + |\theta|^\alpha}.$$

Summarizing, (3.157) is a consequence of

$$\lim_{v \to \infty} \sup_{x \in \mathbb{Z}} \left| \int_{-\pi a(v)}^{\pi a(v)} \frac{e^{i\theta x \lfloor xa(v) \rfloor / a(v)}}{\lambda + v(1 - \psi(\theta/a(v)))} d\theta - \int_{\mathbb{R}} \frac{e^{i\theta \lfloor xa(v) \rfloor / a(v)}}{\lambda + |\theta|^\alpha} d\theta \right| = 0. \quad (3.159)$$

To prove (3.159), write, for any $x \in \mathbb{Z}$, any $A > 1$, some $\varepsilon \in (0, \pi)$ to be specified later and large enough v,

$$\left| \int_{-\pi a(v)}^{\pi a(v)} \frac{e^{i\theta \lfloor xa(v) \rfloor / a(v)}}{\lambda + v(1 - \psi(\theta/a(v)))} d\theta - \int_{\mathbb{R}} \frac{e^{i\theta \lfloor xa(v) \rfloor / a(v)}}{\lambda + |\theta|^\alpha} d\theta \right|$$

$$\leq \int_{-A}^{A} \left| \frac{1}{\lambda + v(1 - \psi(\theta/a(v)))} - \frac{1}{\lambda + |\theta|^\alpha} \right| d\theta + \int_{A \leq |\theta| \leq \varepsilon a(v)} \frac{d\theta}{|\lambda + v(1 - \psi(\theta/a(v)))|}$$

$$+ \int_{|\theta| > A} \frac{d\theta}{\lambda + |\theta|^\alpha} + \int_{\varepsilon a(v) \leq |\theta| \leq \pi a(v)} \frac{d\theta}{|\lambda + v(1 - \psi(\theta/a(v)))|}$$

$$=: I(v, A) + J(v, A) + K(A) + M(v).$$

A specialization of (3.139) to a one-dimensional convergence entails

$$\lim_{v \to \infty} v(1 - \psi(\theta/a(v))) = |\theta|^\alpha \quad (3.160)$$

3.4 Limit Theorems for Heavy-Tailed Random Walks with Membrane at 0

locally uniformly in θ, whence $\lim_{v\to\infty} I(v, A) = 0$.

Relation (3.160) entails

$$\lim_{v\to\infty} \frac{|1 - \psi(\theta/a(v))|}{|1 - \psi(1/a(v))|} = |\theta|^\alpha,$$

which shows that the functions $\theta \mapsto |1 - \psi(\theta)|$, $\theta > 0$ and $\theta \mapsto |1 - \psi(-\theta)|$, $\theta > 0$ are regularly varying at 0 of index α. By an analogue of Potter's bound (Theorem 1.5.6 in [13]), given $c \in (0, 1)$ and $\delta \in (0, \alpha - 1)$ there exists $\varepsilon > 0$ such that

$$v|1 - \psi(\theta/a(v))| \geq c(|\theta|^{\alpha+\delta} \wedge |\theta|^{\alpha-\delta}) \tag{3.161}$$

for all $\theta \in [-\varepsilon a(v), \varepsilon a(v)]$ and large v. Hence,

$$J(v, A) \leq \int_{|\theta|\geq A} \frac{d\theta}{c|\theta|^{\alpha-\delta}} \to 0, \quad A \to \infty.$$

Also, trivially,

$$\lim_{A\to\infty} K(A) = 0.$$

As a preparation for the subsequent presentation, we show that, for $\lambda, v > 0$ and $\vartheta \in \mathbb{R}$,

$$|\lambda + v(1 - \psi(\vartheta))| \geq v|1 - \psi(\vartheta)|. \tag{3.162}$$

Observing that $\operatorname{Re}\psi$ is a characteristic function of a proper probability distribution, we infer $v\operatorname{Re}(1 - \psi(\vartheta)) \geq 0$. Use now the inequality $|a+z| \geq |z|$ which holds true for $a \geq 0$ and complex z with $\operatorname{Re} z \geq 0$, for $a = \lambda$ and $z = v(1 - \psi(\vartheta))$.

Since the distribution of ξ is 1-arithmetic by assumption we conclude that $\psi(\theta) = 1$ for $\theta \in \mathbb{R}$ if, and only if, $\theta = 2\pi n$, $n \in \mathbb{Z}$. In particular, $\min_{\varepsilon \leq |\theta| \leq \pi} |1 - \psi(\theta)| > 0$. Thus, using (3.162) we conclude that

$$M(v) \leq \frac{a(v)}{v} \frac{2(\pi - \varepsilon)}{\min_{\varepsilon\leq|\theta|\leq\pi} |1 - \psi(\theta)|} \to 0, \quad v \to \infty$$

because a is regularly varying at ∞ of index $1/\alpha < 1$.

Combining fragments together we arrive at (3.159), which completes the proof of (3.150).

Proof of (3.147) It follows from (4.29) with $\kappa = \eta/a(v)$ that, for $\lambda > 0$,

$$\lambda R_\lambda^{\widetilde{\mathfrak{X}}_v} f(0) = \frac{f(0)/v + \mathbb{E}\big[(V_\lambda^{\widetilde{\mathfrak{X}}_v} f)(\eta/a(v))\big]}{1/v + \mathbb{E}\big[(V_\lambda^{\widetilde{\mathfrak{X}}_v} 1)(\eta/a(v))\big]} = \frac{f(0)/v + \mathbb{E}\big[(V_\lambda^{\widetilde{S}_v} f)(\eta/a(v))\big]}{1/v + \mathbb{E}\big[(V_\lambda^{\widetilde{S}_v} 1)(\eta/a(v))\big]}, \tag{3.163}$$

where the last equality is secured by (3.148). Comparing a specialization of formula (4.18) for $U_{\alpha,\beta}$ and (2.56) we infer, for $\lambda > 0$,

$$\lambda R_\lambda^{U_{\alpha,\beta}} f(0) = \frac{\int_\mathbb{R} V_\lambda^{U_\alpha} f(x) \eta^*(\mathrm{d}x)}{\int_\mathbb{R} V_\lambda^{U_\alpha} 1(x) \eta^*(\mathrm{d}x)},$$

where η^* is a measure defined in (2.57) with nonnegative $c_\pm$ satisfying (3.140) (so that necessarily $c_+ + c_- = 1$). Hence, (3.147) is equivalent to

$$\lim_{v \to \infty} \frac{\mathbb{E}\big[(V_\lambda^{\widetilde{S}_v} f)(\eta/a(v))\big]}{\mathbb{E}\big[(V_\lambda^{\widetilde{S}_v} 1)(\eta/a(v))\big]} = \frac{\int_\mathbb{R} V_\lambda^{U_\alpha} f(x) \eta^*(\mathrm{d}x)}{\int_\mathbb{R} V_\lambda^{U_\alpha} 1(x) \eta^*(\mathrm{d}x)} \tag{3.164}$$

for $f \in C_0(\mathbb{R})$.

Our proof of (3.164) is based on Lemma 2.3.3(b) with $\zeta = \eta$ and an auxiliary fact to be discussed next.

Lemma 3.4.2 *Given $\delta \in (0, (\alpha - 1) \wedge (2 - \alpha))$ there exist positive constants v_0 and c_2 which do not depend on x and v such that*

$$\mathbb{E}^{\lfloor xa(v) \rfloor}\Big(1 - e^{-(\lambda/v)\sigma(S_\xi \circ N)}\Big) \leq c_2\Big(|x| + \frac{1}{a(v)}\Big)^{\alpha - 1 - \delta}, \quad x \in \mathbb{R}, \quad v \geq v_0.$$

Proof Since the left-hand side is bounded from above by 1 it suffices to prove the inequality for x satisfying $|\lfloor xa(v) \rfloor| \leq a(v)$.

Let $\varepsilon > 0$ be the same as in (3.161). Using (3.158) and changing the variable we obtain, for $x \in \mathbb{R}$,

$$\mathbb{E}^{\lfloor xa(v) \rfloor}\Big(1 - e^{-(\lambda/v)\sigma(S_\xi \circ N)}\Big)$$

$$= \frac{\pi}{\varepsilon}\bigg|\int_{-\varepsilon a(v)}^{\varepsilon a(v)} \frac{1 - e^{i\pi\theta \lfloor xa(v)\rfloor/(\varepsilon a(v))}}{\lambda + v(1 - \psi(\pi\theta/(\varepsilon a(v))))} \mathrm{d}\theta\bigg| \bigg/ \bigg|\int_{-\pi a(v)}^{\pi a(v)} \frac{\mathrm{d}\theta}{\lambda + v(1 - \psi(\theta/a(v)))}\bigg|.$$

3.4 Limit Theorems for Heavy-Tailed Random Walks with Membrane at 0

In view of (3.159) the denominator on the right-hand side converges to $\int_{\mathbb{R}}(\lambda+|\theta|^\alpha)^{-1}d\theta$ as $v\to\infty$. Invoking (3.161) in combination with

$$\lim_{v\to\infty}\frac{|1-\psi(\pi\theta/(\varepsilon a(v)))|}{|1-\psi(\theta/a(v))|}=\left(\frac{\pi}{\varepsilon}\right)^\alpha$$

we arrive at a counterpart of (3.161): given $\delta\in(0,(\alpha-1)\wedge(2-\alpha))$ there exists $v_0>0$ such that

$$v|1-\psi(\pi\theta/(\varepsilon a(v)))|\geq c_1(|\theta|^{\alpha+\delta}\wedge|\theta|^{\alpha-\delta})$$

for all $\theta\in[-\varepsilon a(v),\varepsilon a(v)]$, $v\geq v_0$ and some finite positive constant c_1. Hence,

$$\left|\int_{-\varepsilon a(v)}^{\varepsilon a(v)}\frac{1-e^{i\pi\theta\lfloor xa(v)\rfloor/(\varepsilon a(v))}}{\lambda+v(1-\psi(\pi\theta/(\varepsilon a(v))))}d\theta\right|\leq\frac{2}{c_1}\int_{\mathbb{R}}\frac{|\sin(\pi\theta\lfloor xa(v)\rfloor/(2\varepsilon a(v)))|}{|\theta|^{\alpha+\delta}\wedge|\theta|^{\alpha-\delta}}d\theta$$

$$\leq\frac{2}{c_1}\left(\frac{\pi|\lfloor xa(v)\rfloor|}{2\varepsilon a(v)}\int_{-1}^{1}|\theta|^{1-\alpha-\delta}d\theta+\left(\frac{|\lfloor xa(v)\rfloor|}{a(v)}\right)^{\alpha-1-\delta}\int_{\mathbb{R}}\frac{|\sin(\pi\theta/(2\varepsilon))|}{|\theta|^{\alpha-\delta}}d\theta\right)$$

$$\leq c_2\left(\frac{|\lfloor xa(v)\rfloor|}{a(v)}\right)^{\alpha-1-\delta}\leq c_2\left(|x|+\frac{1}{a(v)}\right)^{\alpha-1-\delta}$$

for x satisfying $|\lfloor xa(v)\rfloor|\leq a(v)$. Here, we have used (3.162) for the first inequality. Our choice of δ ensures that all the integrals in the last centered formula converge. $\square$

Proof of (3.164) For fixed $\lambda>0$ and bounded continuous $f:\mathbb{R}\to\mathbb{R}$, put

$$\hat{V}^{\widetilde{S}_v}_\lambda f(x):=\int_0^\infty e^{-\lambda t}\mathbb{E}\big(f(S_\xi(N(vt))/a(v))|S_\xi(N(vt))=\lfloor xa(v)\rfloor\big)\mathbb{1}_{\{\sigma(\widetilde{S}_v)>t\}}dt$$

$$=v^{-1}\int_0^\infty e^{-(\lambda/v)t}\mathbb{E}\big(f(S_\xi(N(t))/a(v))|S_\xi(N(t))=\lfloor xa(v)\rfloor\big)\mathbb{1}_{\{\sigma(S_\xi\circ N)>t\}}dt \quad (3.165)$$

for $x\in\mathbb{R}$. Similarly, we define

$$\hat{V}^{\widetilde{S}_v}_\lambda 1(x):=\lambda^{-1}\mathbb{E}^{\lfloor xa(v)\rfloor/a(v)}(1-e^{-\lambda\sigma(\widetilde{S}_v)})=\lambda^{-1}\mathbb{E}^{\lfloor xa(v)\rfloor}(1-e^{-(\lambda/v)\sigma(S_\xi\circ N)}) \quad (3.166)$$

for $x\in\mathbb{R}$. The functions $\hat{V}^{\widetilde{S}_v}_\lambda f$ and $\hat{V}^{\widetilde{S}_v}_\lambda 1$ are piecewise constant interpolations of $V^{\widetilde{S}_v}_\lambda f$ and $V^{\widetilde{S}_v}_\lambda 1$, respectively, satisfying $\hat{V}^{\widetilde{S}_v}_\lambda f(x)=V^{\widetilde{S}_v}_\lambda f(x)$ and $\hat{V}^{\widetilde{S}_v}_\lambda 1(x)=V^{\widetilde{S}_v}_\lambda 1(x)$ for each $x\in(a(v))^{-1}\mathbb{Z}$.

Without loss of generality, we can assume that a is strictly increasing and continuous, so that the inverse function a^{-1} exists. Then v^{-1} and λ/v on the right-hand side of (3.165) are equal to $1/((a^{-1} \circ a)(v))$ and $\lambda/((a^{-1} \circ a)(v))$, respectively. Let $\lambda > 0$ be fixed and $f \in C_0(\mathbb{R})$ which particularly implies that f is uniformly continuous on $\mathbb{R}$. We intend to apply Lemma 2.3.3(b) with $\zeta = \eta$, $u = a(v)$, $g^{(1)} = V_\lambda^{U_\alpha} f$, $g^{(2)} = V_\lambda^{U_\alpha} 1$, $g_u^{(1)} = \hat{V}_\lambda^{\widetilde{S}_v} f$ and $g_u^{(2)} = \hat{V}_\lambda^{\widetilde{S}_v} 1$. The assumed monotonicity and continuity of a ensure that the so-defined $g_{a(v)}^{(1)}$ and $g_{a(v)}^{(2)}$ are functions of $a(v)$ alone.

Now we check that the so-defined functions satisfy the assumptions of part (b) of Lemma 2.3.3. (Uniform) Continuity of $g^{(2)}$ follows from

$$V_\lambda^{U_\alpha} 1(x) = \lambda^{-1} \mathbb{E}^x (1 - e^{-\lambda \sigma(U_\alpha)}), \quad x \in \mathbb{R} \tag{3.167}$$

in combination with (2.58) and (2.59). (Uniform) Continuity of $g^{(1)}$ is secured by formula (2.74), uniform continuity of $g^{(2)}$, and uniform continuity of f in combination with the inequality $\omega_{R_\lambda^{U_\alpha} f}(\delta) \leq \lambda^{-1} \omega_f(\delta)$ for $\delta > 0$. Here, as before, ω_h denotes the modulus of continuity of h. The uniform convergence

$$\lim_{u \to \infty} \sup_{x \in \mathbb{R}} |g_u^{(1)}(x) - g^{(1)}(x)| = \lim_{v \to \infty} \sup_{x \in \mathbb{R}} |\hat{V}_\lambda^{\widetilde{S}_v} f(x) - V_\lambda^{U_\alpha} f(x)| = 0$$

is guaranteed by (3.149) and uniform continuity of $V_\lambda^{U_\alpha} f = g^{(1)}$. Analogously, the relation

$$\lim_{u \to \infty} \sup_{x \in \mathbb{R}} |g_u^{(2)}(x) - g^{(2)}(x)| = \lim_{v \to \infty} \sup_{x \in \mathbb{R}} |\hat{V}_\lambda^{\widetilde{S}_v} 1(x) - V_\lambda^{U_\alpha} 1(x)| = 0$$

follows from (3.167), (3.150), and uniform continuity of $V_\lambda^{U_\alpha} 1 = g^{(2)}$. Uniform boundedness of $(g_u^{(2)})_{u > 0}$ follows from representation (3.166) and entails uniform boundedness of $(g_u^{(1)})_{u > 0}$ via

$$|\hat{V}_\lambda^{\widetilde{S}_v} f(x)| \leq \|f\| \hat{V}_\lambda^{\widetilde{S}_v} 1(x), \quad x \in \mathbb{R} \tag{3.168}$$

for $f \in C_0(\mathbb{R})$. We apply Lemma 3.4.2 with $\delta \in (0, (\alpha - 1 - \beta) \wedge (2 - \alpha))$. While the functions $g_u^{(2)}$ satisfy (2.67) with $\gamma = \alpha - 1 - \beta - \delta$, $c = \lambda^{-1} c_2$ (λ is from (3.166), c_2 is from Lemma 3.4.2), and $r = 1$, the functions $g_u^{(1)}$ do so (with a different c) as a consequence of (3.168).

Thus, all the conditions of part (b) of Lemma 2.3.3 are satisfied, and an application of that result yields (3.164) and thereupon (3.147). The proof of Theorem 3.4.1 (a) is complete. □

3.4.3 Proof of Theorem 3.4.1(b)

We shall work with a particular realization of the Markov chain $X = (X(n))_{n \geq 0}$, still denoted by X and defined by

$$X(n+1) = \begin{cases} X(n) + \xi_{n+1-T(n)}, & \text{if } X(n) \neq 0, \\ X(n) + \eta_{T(n)}, & \text{if } X(n) = 0 \end{cases} \quad (3.169)$$

for $n \in \mathbb{N}_0$, where $T(n) := \sum_{k=0}^{n} \mathbb{1}_{\{X(k)=0\}}$ and $\eta_1, \eta_2, \ldots$ are independent copies of η, which are also independent of $\xi_1, \xi_2, \ldots$. We claim that the so-defined X can equivalently be represented as follows:

$$X(n) = X(0) + S_\xi(n - T(n-1)) + S_\eta(T(n-1)), \quad n \in \mathbb{N}, \quad (3.170)$$

where $S_\xi(0) = S_\eta(0) = 0$ a.s.

To check this, write

$$X(n+1) - X(n) = S_\xi(n+1-T(n)) - S_\xi(n-T(n-1)) + S_\eta(T(n)) - S_\eta(T(n-1)).$$

Observe now that $X(n) \neq 0$ if, and only if, $T(n-1) = T(n)$ and that on this event

$$X(n+1) - X(n) = S_\xi(n+1-T(n)) - S_\xi(n-T(n)) = \xi_{n+1-T(n)},$$

which is in line with (3.169). On the other hand, $X(n) = 0$ if, and only if, $T(n-1) = T(n) - 1$ and on this event

$$X(n+1) - X(n) = S_\eta(T(n)) - S_\eta(T(n) - 1) = \eta_{T(n)},$$

which is again in agreement with (3.169).

Put $T(-1) := 0$. Using (3.170), with $X_v(n)$ replacing $X(n)$, $X_v(0)$ replacing $X(0)$, and $T_v(n)$ replacing $T(n)$, where T_v is a counterpart of T which corresponds to X_v, we conclude that relation (3.141) holds if we can show that

$$\left(\frac{S_\xi(\lfloor vt \rfloor) - T_v(\lfloor vt \rfloor - 1))}{a(v)} \right)_{t \geq 0} \implies U_\alpha, \quad v \to \infty \quad (3.171)$$

on $D([0, \infty))$ and, for all $t_0 > 0$,

$$\frac{\sup_{t \in [0, t_0]} |S_\eta(T_v(\lfloor vt \rfloor - 1))|}{a(v)} \xrightarrow{\mathbb{P}} 0, \quad v \to \infty. \quad (3.172)$$

Assume that we can prove that, for all $\delta > 0$ and all $t \geq 0$,

$$\frac{T_v(\lfloor vt \rfloor)}{v^{1-1/\alpha+\delta}} \xrightarrow{\mathbb{P}} 0, \quad v \to \infty, \tag{3.173}$$

which particularly implies that, for all $t \geq 0$, $v^{-1}T_v(\lfloor vt \rfloor) \xrightarrow{\mathbb{P}} 0$ as $v \to \infty$. By Lemma 3.1.1, this in combination with (3.139) entails (3.171).

To prove (3.172), write, for any $\gamma > 0$,

$$\frac{\sup_{t\in[0,t_0]} |S_\eta(T_v(\lfloor vt \rfloor - 1))|}{a(v)} \leq \frac{S_{|\eta|}(T_v(\lfloor vt_0 \rfloor))}{a(v)}$$

$$\leq \frac{S_{|\eta|}(T_v(\lfloor vt_0 \rfloor))}{a(v)} \mathbb{1}_{\{T_v(\lfloor vt_0 \rfloor) > \gamma v^{1-1/\alpha+\delta}\}} + \frac{S_{|\eta|}(\lfloor \gamma v^{1-1/\alpha+\delta} \rfloor)}{a(v)}.$$

In view of (3.173), the first term on the right-hand side converges to 0 in probability, as $v \to \infty$. To analyze the second term, recall that the function a is regularly varying at ∞ of index $1/\alpha$.

If $\mathbb{E}|\eta| < \infty$, then choosing any $\delta \in (0, 2/\alpha - 1)$ and invoking the weak law of large numbers for $S_{|\eta|}$ we infer

$$\frac{S_{|\eta|}(\lfloor \gamma v^{1-1/\alpha+\delta} \rfloor)}{a(v)} \xrightarrow{\mathbb{P}} 0, \quad v \to \infty. \tag{3.174}$$

If the distribution of η belongs to the domain of attraction of a β-stable distribution with $\beta \in (\alpha - 1, 1)$, then so does the distribution of $|\eta|$, and according to Theorem 4.4.1, $S_{|\eta|}(v)/c(v)$ converges in distribution to a positive β-stable random variable. For any $\delta \in (0, \alpha^{-1}(\beta - (\alpha - 1)))$ (such a choice is possible because $\beta > \alpha - 1$) $\lim_{v\to\infty} c(\lfloor \gamma v^{1-1/\alpha+\delta} \rfloor)/a(v) = 0$ and thereupon (3.174) holds true.

It remains to prove (3.173). Observe that, for $v, x > 0$ and $n \in \mathbb{N}_0$,

$$\mathbb{P}\{T_v(n) > x | X_v(0) \neq 0\} \leq \mathbb{P}\{T_v(n) > x | X_v(0) = 0\}.$$

In view of this, we assume in what follows that $X_v(0) = 0$ a.s. and write $T(n)$ for $T_v(n)$. Relation (3.173) holds if we can show that

$$\frac{T(n)}{n^{1-1/\alpha+\delta}} \xrightarrow{\mathbb{P}} 0, \quad n \to \infty. \tag{3.175}$$

For $k \in \mathbb{Z}$, put $\tau_k := \inf\{j \geq 0 : S_\xi(j) + k = 0\}$. Let $(\tau_k^{(1)})_{k\in\mathbb{Z}}$, $(\tau_k^{(2)})_{k\in\mathbb{Z}}, \ldots$ be independent copies of $(\tau_k)_{k\in\mathbb{Z}}$, which are also independent of $\eta_1, \eta_2, \ldots$ The random

3.4 Limit Theorems for Heavy-Tailed Random Walks with Membrane at 0

variable $T(n)$ has the same distribution as

$$T'(n) := 1 + \sum_{k \geq 1} \mathbb{1}_{\{(1+\tau^{(1)}_{-\eta_1})+\ldots+(1+\tau^{(k)}_{-\eta_k}) \leq n\}}$$

for each fixed $n \in \mathbb{N}_0$. Fix any $n_0 \in \mathbb{Z}\setminus\{0\}$ satisfying

$$p_0 := \mathbb{P}\{X(1) = n_0 \mid X(0) = 0\} = \mathbb{P}\{\eta = n_0\} > 0.$$

Put $\theta_0 := 0$ and $\theta_{i+1} := \inf\{j > \theta_i : \eta_j = n_0\}$ for $i \in \mathbb{N}_0$. The random variables θ_1, $\theta_2 - \theta_1, \ldots$ are independent and have a geometric distribution with success probability p_0, that is, $\mathbb{P}\{\theta_1 = k\} = (1-p_0)^{k-1} p_0$ for $k \in \mathbb{N}$. Also, $\theta_1, \theta_2, \ldots$ are independent of $(\tau^{(1)}_k)_{k \in \mathbb{Z}}, (\tau^{(2)}_k)_{k \in \mathbb{Z}}, \ldots$ Write

$$T'(n) - 1 = \sum_{i \geq 0} \sum_{k=\theta_i+1}^{\theta_{i+1}} \mathbb{1}_{\{(1+\tau^{(1)}_{-\eta_1})+\ldots+(1+\tau^{(k)}_{-\eta_k}) \leq n\}}$$

$$\leq \theta_1 + \sum_{i \geq 1}(\theta_{i+1} - \theta_i)\mathbb{1}_{\{(1+\tau^{(\theta_1)}_{-\eta_{\theta_1}})+\ldots+(1+\tau^{(\theta_i)}_{-\eta_{\theta_i}}) \leq n\}}$$

$$= \theta_1 + \sum_{i \geq 1}(\theta_{i+1} - \theta_i)\mathbb{1}_{\{(1+\tau^{(\theta_1)}_{-n_0})+\ldots+(1+\tau^{(\theta_i)}_{-n_0}) \leq n\}} \quad \text{a.s.}$$

The latter random variable has the same distribution as

$$\theta_1 + \sum_{i \geq 1}(\theta_{i+1} - \theta_i)\mathbb{1}_{\{(1+\tau^{(1)}_{-n_0})+\ldots+(1+\tau^{(i)}_{-n_0}) \leq n\}} = \sum_{k=1}^{T^*(n)}(\theta_k - \theta_{k-1}),$$

where $T^*(n) := 1 + \sum_{i \geq 1} \mathbb{1}_{\{(1+\tau^{(1)}_{-n_0})+\ldots+(1+\tau^{(i)}_{-n_0}) \leq n\}}$, for $n \in \mathbb{N}_0$, is independent of η_1, $\eta_2, \ldots$ Summarizing, to prove (3.175) it is enough to show that, for all $\delta > 0$,

$$n^{-(1-1/\alpha+\delta)} \sum_{k=1}^{T^*(n)}(\theta_k - \theta_{k-1}) \xrightarrow{\mathbb{P}} 0, \quad n \to \infty.$$

By the weak law of large numbers for the random walk $(\theta_i)_{i \geq 1}$, the latter holds provided that

$$\frac{T^*(n)}{n^{1-1/\alpha+\delta}} \xrightarrow{\mathbb{P}} 0, \quad n \to \infty.$$

According to Lemma 2.1 in [9], $\mathbb{P}\{\tau_0 > n\} \sim n^{-(1-1/\alpha)} L_1(n)$ as $n \to \infty$ for some L_1 slowly varying at ∞. By Theorem 1 on p. 378 in [153],

$$\lim_{n \to \infty} \frac{\mathbb{P}\{\tau_{-n_0} > n\}}{\mathbb{P}\{\tau_0 > n\}} = g(n_0) \in [0, \infty), \qquad (3.176)$$

where g is the potential kernel of S_ξ. By Theorem 2 on p. 361 of the same reference, $g(k) > 0$ for all $k \in \mathbb{Z}\setminus\{0\}$ and particularly $g(n_0) > 0$. This follows from the fact that ξ has a distribution with unbounded support, so that S_ξ cannot be a left- or right-continuous random walk.

The sequence $(T^*(n))_{n \geq 0}$ is the first-passage time (generalized inverse) sequence for $S_{1+\tau_{-n_0}}$. In view of (3.176) and $g(n_0) > 0$, the distribution tail of $1 + \tau_{-n_0}$ is regularly varying at ∞ of index $-(1 - 1/\alpha) \in (-1, 0)$. Then $\mathbb{P}\{\tau_{-n_0} > n\} T^*(n)$ converges in distribution as $n \to \infty$ to a random variable having a Mittag-Leffler distribution (the distribution of an inverse $(1 - 1/\alpha)$-subordinator evaluated at time 1); see, for instance, Theorem 7 in [50]. Since, for all $\delta > 0$ and any L^* slowly varying at ∞, $\lim_{n \to \infty} n^\delta L^*(n) = \infty$, relation (3.175) follows.

The proof of Theorem 3.4.1(b) is complete.

3.5 Multidimensional Random Walks with Membranes

Let $d \in \mathbb{N}$ and A be a given subset of $\mathbb{Z}^d$. We investigate X a $\mathbb{Z}^d$-valued random walk with membrane A. Its transition probabilities are as given by formula (3.1), with $\mathbb{Z}^d$ replacing $\mathbb{Z}$.

We start by discussing a trivial situation in which the asymptotic behavior of X is driven by that of S_ξ, just because the set A is hit by X finitely often.

Proposition 3.5.1 *Let $d \in \mathbb{N}$. Assume that, for a sequence of positive numbers $(a_n)_{n \geq 1}$ and a stochastic process $\mathcal{S}$,*

$$\left(\frac{S_\xi(\lfloor nt \rfloor)}{a_n}\right)_{t \geq 0} \Longrightarrow (\mathcal{S}(t))_{t \geq 0}, \quad n \to \infty \qquad (3.177)$$

in the J_1-topology on $D([0, \infty), \mathbb{R}^d)$, and that the perturbed random walk X visits A finitely often with probability one. Then

$$\left(\frac{X(\lfloor nt \rfloor)}{a_n}\right)_{t \geq 0} \Longrightarrow (\mathcal{S}(t))_{t \geq 0}, \quad n \to \infty$$

in the J_1-topology on $D([0, \infty), \mathbb{R}^d)$.

3.5 Multidimensional Random Walks with Membranes

The proof is similar to that of Theorem 3.1.1, hence omitted.

Following p. 287 in [83] or p. 20 in [153], we call the random walk S_ξ *aperiodic*, if no proper subgroup of $\mathbb{Z}^d$ contains all $x \in \mathbb{Z}^d$ satisfying $\mathbb{P}\{\xi = x\} > 0$. By Theorem 1 on p. 67 in [153], the walk S_ξ is aperiodic if, and only if, the characteristic function ψ defined by $\psi(z) := \mathbb{E}e^{i\langle z,\xi\rangle}$ for $z \in \mathbb{R}^d$ possesses the following property: $\psi(z) = 1$ if, and only if, each coordinate of z is an integer multiple of 2π. The class of aperiodic distributions on $\mathbb{Z}$ coincides with the class of 1-*arithmetic distributions*, that is, the distributions which are concentrated on the integers and not concentrated on any sparser centered lattice.

To exclude trivial cases we shall assume throughout this section that the following condition holds.

Condition $\mathcal{B}$: the random walk S_ξ is aperiodic, and, as far as the states of the Markov chain X are concerned, all points (states) of $\mathbb{Z}^d \setminus A$ are accessible from any point of A.

Under Condition $\mathcal{B}$, if the random walk S_ξ is transient, then so is the Markov chain X. Here are three examples, in which, under Condition $\mathcal{B}$, Theorem 3.5.1 applies if $A \subset \mathbb{Z}^d$ is a finite set.

Example 3.5.1 Let $d \geq 3$. Under the first part of Condition $\mathcal{B}$, S_ξ is a genuinely d-dimensional standard random walk. Hence, it is transient by Theorem 1 on p. 83 in [153].

As a consequence of Example 3.5.1, Theorem 3.5.1 applies whenever the set A is contained in a hyperplane $\{(0,0,0)\} \times \mathbb{Z}^{d-3}$, that is, its co-dimension is greater than or equal to 3.

Example 3.5.2 Let $d = 1$ and the distribution of ξ be 1-arithmetic. Assume that the function $t \mapsto \mathbb{P}\{|\xi| > t\}$ is regularly varying at ∞ of index $-\alpha$ for some $\alpha \in (0,1)$ and that

$$\mathbb{P}\{\xi > t\} \sim c_+ \mathbb{P}\{|\xi| > t\} \quad \text{and} \quad \mathbb{P}\{-\xi > t\} \sim c_- \mathbb{P}\{|\xi| > t\}, \quad t \to \infty$$

for some nonnegative c_+ and c_- summing up to 1. Then (3.177) holds with $a_n = b(n)$ for b regularly varying at ∞ of index $1/\alpha$ and $\mathcal{S} = \mathcal{S}_\alpha$ an α-stable Lévy process. Arguing as in the proof of Theorem 3.1.1 we conclude that $\sum_{n\geq 0} \mathbb{P}\{S_\xi(n) = 0\} < \infty$, thereby proving transience of S_ξ.

Example 3.5.3 Let $d = 2$ and

$$\mathbb{P}\{\xi = x\} = \frac{c}{1+|x|^{2+\alpha}}, \quad x \in \mathbb{Z}^2$$

for some $\alpha > 0$ and a constant $c := (\sum_{x \in \mathbb{Z}^2}(1+|x|^{2+\alpha})^{-1})^{-1}$. According to Section B.1 in [30], the random walk S_ξ is transient if, and only if, $\alpha \in (0,2)$. In the present situation condition (3.177) holds with $a_n = n^{1/\alpha}$ and $\mathcal{S}$ being a rotation invariant two-dimensional

α-stable Lévy process (its characteristic function is given by $z \mapsto e^{-c_\alpha |z|^\alpha}$ for $z \in \mathbb{R}^2$ and appropriate positive constant c_α).

Here are sufficient conditions of recurrence for one- and two-dimensional standard random walks.

Example 3.5.4 Let $d = 1$ and $\mathbb{E}\xi = 0$. Then S_ξ is recurrent; see Proposition 8 on p. 23 in [153].

Example 3.5.5 Let $d = 2$, $\mathbb{E}\xi = 0$, and $\mathbb{E}|\xi|^2 < \infty$. Then S_ξ is recurrent; see Theorem 1 on p. 83 in [153].

Our standing assumptions for the remainder of this section are $\mathbb{E}\xi = 0$ and $\mathbb{E}|\xi|^2 < \infty$. We shall present results about distributional convergence of X under Donsker's scaling in the following settings:

(a) $d = 2$ and $A \subset \mathbb{Z}^d$ is a finite set.
(b) A is a hyperplane $H = \{0\} \times \mathbb{Z}^{d-1}$.

Our results in the case (a) reveal that the scaling limit is a two-dimensional Brownian motion whenever the distribution tails of the perturbations are not too heavy. This means that the perturbations have no effect. This is not that surprising because, with probability one, a two-dimensional Brownian motion does not hit any fixed deterministic point. Thus, even though according to Example 3.5.5, S_ξ visits A infinitely often with probability one, the number of visits of X to A up to time n is of order $O(\log n)$; see Proposition 3.5.2. Theorem 3.5.2 exhibits an example of perturbations whose distribution tails are so heavy that Donsker's scaling limit of X does not exist. Now we explain why we deem the result quite unexpected. Observe that an effect of any fixed perturbation vanishes upon scaling by $\sqrt{n}$ and that a large jump from the membrane leads to a large return time to the membrane. Thus, extremely large jumps from the membrane should decrease the number of returns, hence an effect of all perturbations accumulated up to time n.

In case (b) we will assume that perturbations have a periodic structure. Note that the behavior of the first coordinate of the walk looks like one-dimensional perturbed random walk and it is natural to expect that the Donsker scaling limit of the first coordinate will be a skew Brownian motion. The other coordinates will be a multidimensional Brownian motion with drift $cL(t)$, where $c \in H$ is some vector and L is a local time of the limit process on H, which is a local time of the first coordinate at 0.

3.5.1 Two-Dimensional Random Walks with Finite Membranes

In this section we investigate X a two-dimensional random walk with membrane A which contains finitely many points with integer coordinates. We first treat the situation in which Donsker's scaling limit of X coincides with that of S_ξ and, as such, is not affected by the presence of perturbations.

Theorem 3.5.1 *Assume that $d = 2$, $\mathbb{E}\xi = 0$, $\mathbb{E}|\xi|^2 < \infty$, Condition $\mathcal{B}$ holds and*

$$\max_{x \in A} \mathbb{P}\{|\eta^{(x)}| > t\} = o\Big(\frac{1}{\log t}\Big), \quad t \to \infty. \tag{3.178}$$

Then

$$\Big(\frac{X(\lfloor nt \rfloor)}{\sqrt{n}}\Big)_{t \geq 0} \implies \big(W_\Gamma(t)\big)_{t \geq 0}, \quad n \to \infty$$

in the J_1-topology on $D([0, \infty), \mathbb{R}^2)$. Here, $W_\Gamma := (W_\Gamma(t))$ is a two-dimensional Wiener process with the characteristic function

$$\mathbb{E}e^{i\langle z, W_\Gamma(t)\rangle} = e^{-\frac{\langle \Gamma z, z\rangle t}{2}}, \quad z \in \mathbb{R}^2$$

and Γ is a covariance matrix of ξ.

Remark 3.5.1 It follows from Condition $\mathcal{B}$ that the matrix Γ is not degenerate.

Our next theorem is concerned with the situation in which the distribution tail of perturbations is extremely heavy. As a consequence, the distributions of Donsker's scaling of X are not tight. Heuristically, this is caused by the fact that, with a probability bounded away from 0, there is a huge jump from A which occurs early enough.

Theorem 3.5.2 *Let X be a simple random walk in $\mathbb{Z}^2$ with the membrane $A = \{0\}$ and with $X(0) = 0$ a.s. and probability $1/4$ of moving from any point of $\mathbb{Z}^2 \setminus \{0\}$ to the closest neighbor in $\mathbb{Z}^2$. Assume that, for a constant $a > 0$,*

$$\mathbb{P}\{|X(1)| > t\} \sim \frac{a}{\log \log t}, \quad t \to \infty. \tag{3.179}$$

Then

$$\liminf_{n \to \infty} \mathbb{P}\{\max_{1 \leq k \leq n} |X(k)| > n\} > 0. \tag{3.180}$$

In particular, the sequence of distributions of $(n^{-1}X(\lfloor nt \rfloor))_{t\geq 0}$ is not tight in the Skorokhod J_1-topology on $D([0,\infty), \mathbb{R}^2)$.

Remark 3.5.2 Skorokhod in [148] introduced four topologies on $D([0,\infty), \mathbb{R}^d)$: the J_1, J_2, M_1, and M_2-topologies. Relation (3.180) implies that the sequence of processes $(X(\lfloor nt \rfloor)/\sqrt{n})_{t\geq 0}$ admits no weak limit as $n \to \infty$ in any of Skorokhod's topologies.

Proof of Theorem 3.5.1

In what follows we assume, without loss of generality, that X is constructed via a two-dimensional version of formula (3.4). Our proof of Theorem 3.5.1 is essentially based on several auxiliary results given next. Recall the notation $T(n) = \sum_{k=0}^{n-1} \mathbb{1}_{\{X(k)\in A\}}$ for $n \in \mathbb{N}$, which appeared in (3.4).

Proposition 3.5.2 *Under the assumptions of Theorem 3.5.1, the sequence of random variables $(T(n)/\log n)_{n\geq 2}$ is bounded in probability.*

For a nonempty set $B \subseteq \mathbb{Z}^2$, put $\tau_B := \inf\{k \geq 1 : S_\xi(k) \in B\}$ with the usual convention that the infimum of the empty set is equal to $+\infty$. We shall write τ_0 for $\tau_{\{0\}}$. A proof of Proposition 3.5.2, in turn, relies on the following fact.

Lemma 3.5.1 *Under the assumptions of Theorem 3.5.1,*

$$\mathbb{P}\{\tau_0 > n\} \sim \frac{c}{2\pi\sqrt{\det \Gamma}\,\log n}, \quad n \to \infty,$$

where c is the greatest common divisor of the set $\{n \in \mathbb{N} : \mathbb{P}\{S_\xi(n) = 0\} > 0\}$.

Proof Recall that an aperiodic standard random walk S_ξ with $c = 1$ is called *strongly aperiodic*. Observe that the random walk $(S_\xi(cn))_{n\geq 0}$, with jumps having the distribution of $S_\xi(c)$, is strongly aperiodic. By a local limit theorem for strongly aperiodic standard random walks (for instance, Proposition 9 on p. 75 in [153]),

$$\mathbb{P}\{S_\xi(cn) = 0\} \sim \frac{c}{2\pi\sqrt{\det \Gamma}} \frac{1}{n}, \quad n \to \infty. \tag{3.181}$$

The remainder of the proof is analogous to the proof given in Example 1 on p. 167 in [153] for simple random walks in the plane. We provide a sketch for completeness.

In view of (3.181),

$$\sum_{k=0}^{cn} \mathbb{P}\{S_\xi(ck) = 0\} \sim \frac{c}{2\pi\sqrt{\det \Gamma}} \log n, \quad n \to \infty.$$

3.5 Multidimensional Random Walks with Membranes

For $n \in \mathbb{N}_0$, put $U_n := \mathbb{P}\{S_\xi(cn) = 0\}$ and $R_n := \mathbb{P}\{\tau_0 > n\}$. Using that $\sum_{k=0}^{cn} U_k R_{cn-k} = 1$ for $n \in \mathbb{N}_0$, we infer, for any nonnegative integer $\ell \leq n$,

$$R_{cn-c\ell}(U_0 + \ldots + U_{c\ell}) + U_{c\ell+1} + \ldots + U_{cn} \geq 1.$$

Choosing $\ell := \ell(n) = n - \lfloor n/\log n \rfloor$ we obtain

$$U_0 + \ldots + U_{c\ell} \sim \frac{c}{2\pi\sqrt{\det\Gamma}} \log \ell \sim \frac{c}{2\pi\sqrt{\det\Gamma}} \log(n - \ell), \quad n \to \infty$$

and

$$U_{c\ell+1} + \ldots + U_{cn} = O(1/\log n) \to 0, \quad n \to \infty.$$

Hence, given $\varepsilon \in (0, 1)$,

$$\frac{c}{2\pi\sqrt{\det\Gamma}} \log(cn - c\ell) R_{cn-c\ell} \geq 1 - \varepsilon$$

for large enough n and thereupon

$$\liminf_{n\to\infty} (\log n) R_{cn} \geq \frac{2\pi\sqrt{\det\Gamma}}{c}.$$

On the other hand,

$$\limsup_{n\to\infty} (\log n) R_{cn} \leq \frac{2\pi\sqrt{\det\Gamma}}{c}$$

follows from

$$1 = \sum_{k=0}^{cn} U_k R_{cn-k} \geq R_{cn}(U_0 + \ldots + U_{cn}).$$

Thus,

$$R_{cn} \sim \frac{c}{2\pi\sqrt{\det\Gamma}\log n}, \quad n \to \infty.$$

Monotonicity of $(R_k)_{k\geq 0}$ enables us to replace cn with n, thereby securing the claim. □

We are ready to prove Proposition 3.5.2.

Proof of Proposition 3.5.2 According to Example 3.5.5, the random walk S_ξ is recurrent. This taken together with the second part of Condition $\mathcal{B}$ ensures that the random walk X with membrane A is also recurrent. Fix any $v \notin A$. We first prove the claim for an auxiliary Markov chain $\widetilde{X} := (\widetilde{X}(n))_{n \geq 0}$ with $\widetilde{X}(0)$ having the same distribution as $X(0)$. The transition probabilities of $\widetilde{X}$ are given by

$$\mathbb{P}\{\widetilde{X}(1) = y \mid \widetilde{X}(0) = x\} = \begin{cases} \mathbb{P}\{X(1) = y|X(0) = x\} = \mathbb{P}\{\xi = y - x\}, & \text{if } x \notin A, \\ \mathbb{1}_{\{y=v\}}, & \text{if } x \in A. \end{cases} \quad (3.182)$$

Thus, jumps of $\widetilde{X}$ from the points outside A have the same distribution as jumps of X, that is, the distribution of ξ, and $\widetilde{X}$ jumps to the fixed state v upon hitting A.

Put $\tau_A^{\widetilde{X}} := \inf\{k \geq 1 : \widetilde{X}(k) \in A\}$ and note that $\tau_A^{\widetilde{X}} < \infty$ a.s. because the Markov chain $\widetilde{X}$ is recurrent. By the Kesten-Spitzer ratio theorem (Theorem 4a in [83]), for any $y \in \mathbb{Z}^2$,

$$\lim_{n \to \infty} \frac{\mathbb{P}\{\tau_{A-y} > n\}}{\mathbb{P}\{\tau_0 > n\}} = g_A(y), \quad (3.183)$$

where $g_A(y) := \lim_{z \to \infty} \sum_{n \geq 0} \mathbb{P}\{S_\xi(n) = z, \tau_A > n | S_\xi(0) = y\} \in [0, \infty)$. In view of this we infer

$$\lim_{n \to \infty} \frac{\mathbb{P}\{\tau_A^{\widetilde{X}} > n \mid \widetilde{X}(0) = y\}}{\mathbb{P}\{\tau_0 > n\}} = g_A(y), \quad y \notin A.$$

This in combination with Lemma 3.5.2 yields

$$\mathbb{P}\{\tau_A^{\widetilde{X}} > n \mid \widetilde{X}(0) = y\} \sim \frac{cg_A(y)}{2\pi\sqrt{\det\Gamma}\log n}, \quad n \to \infty. \quad (3.184)$$

Let ρ be a random variable with distribution being the conditional distribution of $(\tau_A^{\widetilde{X}} + 1)$ given $\widetilde{X}(0) = v$. Put $\widetilde{N}(t) := \sum_{k \geq 1} \mathbb{1}_{\{S_\rho(k) \leq t\}}$ for $t \geq 0$. The process $(\widetilde{N}(t))_{t \geq 0}$ is a renewal process that corresponds to the one-dimensional standard random walk S_ρ with jumps having a slowly varying distribution tail. As a consequence of the functional weak convergence proved in Theorem 2.1 of [76], $\widetilde{N}(n)/\log n$ converges in distribution as $n \to \infty$ to an exponentially distributed random variable Exp_v with mean $2\pi\sqrt{\det\Gamma}/(cg_A(v))$ if $g_A(v) > 0$. For $n \in \mathbb{N}_0$, put $\widetilde{T}(n) := \sum_{k=0}^{n} \mathbb{1}_{\{\widetilde{X}(k) \in A\}}$, so that $\widetilde{T}(n)$ is the number of visits of $\widetilde{X}$ to A up to and including time n. If $\widetilde{X}(0) \in A$, then $\widetilde{T}(n)$ has the same distribution as $1 + \widetilde{N}(n)$. If $\widetilde{X}(0) = y \notin A$, then $\widetilde{T}(n)$ has the same distribution as $(1 + \widetilde{N}(n - \kappa))\mathbb{1}_{\{\kappa \leq n\}}$, where κ is a random variable with distribution being the conditional distribution of $\tau_A^{\widetilde{X}}$ given $\widetilde{X}(0) = y$. It is assumed that κ is independent of S_ρ. We infer that

3.5 Multidimensional Random Walks with Membranes

in both cases $\widetilde{T}(n)/\log n$ converges in distribution to Exp_v as $n \to \infty$. The proof of the claim for $\widetilde{X}$ is complete if $g_A(v) > 0$.

We stress that there is no guarantee that the inequality $g_A(v) > 0$ holds true for all $v \notin A$. For instance, if $A = \{(0, 1), (0, -1), (1, 0), (-1, 0)\}$, $v = (0, 0)$, and $|\xi| = 1$ a.s., then $\tau_A = 1$ given $S_\xi(0) = (0, 0)$, that is, $g_A(v) = 0$. Nevertheless, we shall show that *there exists $v \notin A$ such that $g_A(v) > 0$.*

Let $m \in \mathbb{N}$ be the minimal number such that $A \subseteq [-m, m]^2 \cap \mathbb{Z}^2$. For the proofs of both Theorem 3.5.1 and Proposition 3.5.2, we put $\eta^{(x)} := \xi$ for $x \in ([-m, m]^2 \cap \mathbb{Z}^2) \setminus A$. This enables us to work with the membrane $A := [-m, m]^2 \cap \mathbb{Z}^2$ rather than original A. The advantage of this choice is justified by the result given next.

Lemma 3.5.2 *If*

$$A = [-m, m]^2 \cap \mathbb{Z}^2, \tag{3.185}$$

then $g_A(v) > 0$ for any $v \notin A$.

Proof Let $x, y \notin A$. In view of (3.185), there exist $k \in \mathbb{N}$ and a path of length k from x to y that do not visit A and has a positive probability. Hence, for some $c > 0$,

$$\mathbb{P}\{\tau_{A-x} > n + k\} \geq c\mathbb{P}\{\tau_{A-y} > n\}$$

and consequently, for any $r > 0$, there exists $k = k_r$ satisfying

$$c_r := \inf_{|x-y|\leq r,\, x,y \notin A} \frac{\mathbb{P}\{\tau_{A-x} > n + k_r\}}{\mathbb{P}\{\tau_{A-y} > n\}} > 0. \tag{3.186}$$

By Lemma 3.5.1, $\lim_{n\to\infty} \frac{\mathbb{P}\{\tau_0 > n+k\}}{\mathbb{P}\{\tau_0 > n\}} = 1$. In combination with (3.183) and (3.186) this entails that either $g_A(x) > 0$ for all $x \notin A$ or $g_A(x) = 0$ for all $x \notin A$. We shall prove that the first alternative prevails.

According to formula (1.34) in [83],

$$\sum_{z \in A} g_A(z) = 1.$$

Hence,

$$1 = \lim_{n\to\infty} \frac{\sum_{z\in A} \sum_{y\in\mathbb{Z}} \mathbb{P}\{\xi = y\}\mathbb{P}\{\sigma_{A-(z+y)} > n - 1\}}{\mathbb{P}\{\tau_0 > n\}}$$

$$= \lim_{n\to\infty} \frac{\sum_{z\in A} \sum_{y\in\mathbb{Z},\, y+z\notin A} \mathbb{P}\{\xi = y\}\mathbb{P}\{\tau_{A-(z+y)} > n - 1\}}{\mathbb{P}\{\tau_0 > n\}},$$

where $\sigma_B := \inf\{k \geq 0 : S_\xi(k) \in B\}$ for a nonempty set $B \subseteq \mathbb{Z}^2$. If the variable ξ takes finitely many values $y_1, \ldots, y_\ell$, say, then the last formula ensures that there exist $y \in \{y_1, \ldots, y_\ell\}$ and $z \in A$ satisfying $y + z \notin A$ and $\lim_{n \to \infty} \frac{\mathbb{P}\{\tau_{A-(z+y)} > n-1\}}{\mathbb{P}\{\tau_0 > n\}} > 0$. This proves $g_A(z + y) > 0$ and thereupon $g_A(x) > 0$ for all $x \notin A$.

Assume now that the variable ξ takes infinitely many values. To complete the proof of the lemma, it suffices to show that, for some $x \notin A$, $k \in \mathbb{N}$ and any $z \in A$,

$$\liminf_{n \to \infty} \frac{\mathbb{P}\{\tau_{A-x} > n + k\}}{\mathbb{P}\{\tau_{A-z} > n\}} > 0. \tag{3.187}$$

For brevity, we shall only prove (3.187) for $z = 0$. For $y = (y_1, y_2) \in \mathbb{Z}^2$, put $|y| := \max(|y_1|, |y_2|)$. For any $n \geq 2$ and any fixed $x \in \mathbb{Z}^2$ with $|x| \geq 1$,

$$\mathbb{P}\{\tau_A > n\} = \sum_{|y| > m} \mathbb{P}\{\xi = y\}\mathbb{P}\{\tau_{A-y} > n - 1\} =$$

$$\sum_{m < |y| \leq m+|x|} \mathbb{P}\{\xi = y\}\mathbb{P}\{\tau_{A-y} > n - 1\} + \sum_{|y| > m+|x|} \mathbb{P}\{\xi = y\}\mathbb{P}\{\tau_{A-y} > n - 1\} \leq$$

$$\sum_{m < |y| \leq m+|x|} \mathbb{P}\{\tau_{A-y} > n - 1\} + c_{|x|}^{-1} \sum_{|y| > m+|x|} \mathbb{P}\{\xi = y\}\mathbb{P}\{\tau_{A-(y+x)} > n - 1 + k_{|x|}\}$$

having utilized (3.186) for the last inequality. Here, $k_{|x|}$ and $c_{|x|}$ are as in (3.186). Another application of (3.186) to the first summand on the right-hand side of the last inequality yields

$$\mathbb{P}\{\tau_A > n\} \leq c_{m+2|x|}^{-1} \sum_{m < |y| \leq m+|x|} \mathbb{P}\{\tau_{A-x} > n - 1 + k_{m+2|x|}\}$$

$$+ c_{|x|}^{-1} \sum_{|y| > m+|x|} \mathbb{P}\{\xi = y\}\mathbb{P}\{\tau_{A-(y+x)} > n - 1 + k_{|x|}\}$$

$$\leq C_1\Big(\mathbb{P}\{\tau_{A-x} > n - 1 + k_{m+2|x|}\} + \sum_{y+x \notin A} \mathbb{P}\{\xi = y\}\mathbb{P}\{\tau_{A-(y+x)} > n - 1 + k_{|x|}\}\Big)$$

$$= C_1\big(\mathbb{P}\{\tau_{A-x} > n - 1 + k_{m+2|x|}\} + \mathbb{P}\{\tau_{A-x} > n + k_{|x|}\}\big)$$

$$\leq C_2 \mathbb{P}\{\tau_{A-x} > n - 1 + K\},$$

where C_1, C_2 and K are some positive constants. This proves (3.187). Hence, $g_A(x) > 0$ for any $x \notin A$.

The proof of Lemma 3.5.2 is complete. □

3.5 Multidimensional Random Walks with Membranes

Now we continue the proof of Proposition 3.5.2. Condition $\mathcal{B}$ implies that, for some $u \in A$ and $v \notin A$,

$$\mathbb{P}\Big\{\bigcup_{k\geq 1}\{X(1) \notin A, \ldots, X(k-1) \notin A, X(k) = u \mid X(0) = v\}\Big\} > 0.$$

Invoking Condition $\mathcal{B}$ once again we conclude that there exist $l \in \mathbb{N}$ and distinct elements $v_1, \ldots, v_{l-1} \in A$ other than u and v (the collection is empty if $l = 1$) such that

$$\mathbb{P}\{X(1) = v_1, \ldots, X(l-1) = v_{l-1}, X(l) = v \mid X(0) = u\} > 0.$$

We claim that, for all $c > 0$ and all $n \in \mathbb{N}$,

$$\mathbb{P}\Big\{\sum_{k=0}^{n} \mathbb{1}_{\{X(k)=u, X(k+1)=v_1, \ldots, X(k+l-1)=v_{l-1}, X(k+l)=v\}} > c\Big\}$$

$$\leq \mathbb{P}\Big\{\sum_{k=0}^{n} \mathbb{1}_{\{\widetilde{X}(k) \in A\}} > c\Big\}. \quad (3.188)$$

To prove (3.188), put

$$\widetilde{\sigma}_0 := 0, \quad \sigma_1 := \inf\{i \geq 0 : X(i) \in A\},$$

$$\widetilde{\sigma}_k := \inf\{i > \sigma_k : X(i) = v\}, \quad \sigma_{k+1} := \inf\{i > \widetilde{\sigma}_k : X(i) \in A\}, \quad k \in \mathbb{N},$$

$$\lambda(n) := \sum_{k \geq 0} \sum_{i=0}^{n} \mathbb{1}_{\{\widetilde{\sigma}_k \leq i \leq \sigma_{k+1}\}}, \quad n \in \mathbb{N}_0$$

and

$$\lambda^{\leftarrow}(n) := \inf\{k \geq 0 : \lambda(k) > n\}, \quad n \in \mathbb{N}_0.$$

Since the sequence $(X(\lambda^{\leftarrow}(n)))_{n\geq 0}$ has the same distribution as $(\widetilde{X}(n))_{n\geq 0}$ and

$$\sum_{k=0}^{n} \mathbb{1}_{\{X(k)=u, X(k+1)=v_1, \ldots, X(k+l-1)=v_{l-1}, X(k+l)=v\}} \leq \sum_{k=0}^{n} \mathbb{1}_{\{X(\lambda^{\leftarrow}(k)) \in A\}} \quad \text{a.s.},$$

(3.188) follows. Using now (3.188) together with the already proved claim for $\widetilde{X}$ we conclude that the sequence

$$\Big(\frac{\sum_{k=1}^{n} \mathbb{1}_{\{X(k)=u, X(k+1)=v_1, \ldots, X(k+l-1)=v_{l-1}, X(k+l)=v\}}}{\log n}\Big)_{n \geq 2} \quad (3.189)$$

is bounded in probability. Put $\zeta_0 := 0$, $\zeta_{k+1} := \inf\{i > \zeta_k : X(i) \in A\}$ for $k \in \mathbb{N}$ and $N(n) := \sum_{k=1}^n \mathbb{1}_{\{X(k) \in A\}}$ for $n \in \mathbb{N}$. Since X is recurrent, the random variables ζ_1, $\zeta_2, \ldots$ are a.s. finite, and $\lim_{n \to \infty} N(n) = +\infty$ a.s. The sequence $(X(\zeta_k))_{k \geq 1}$ is a Markov chain taking values in A. Condition $\mathcal{B}$ ensures that $(X(\zeta_k))_{k \geq 1}$ admits a unique stationary distribution $(\pi_x)_{x \in A}$ with $\pi_u > 0$. Hence, by the strong law of large numbers for Markov chains,

$$\lim_{n \to \infty} \frac{\sum_{k=1}^n \mathbb{1}_{\{X(\zeta_k)=u\}}}{n} = \pi_u \quad \text{a.s.}$$

and thereupon

$$\frac{\sum_{j=1}^n \mathbb{1}_{\{X(j)=u\}}}{N(n)} = \frac{\sum_{k=1}^{N(n)} \mathbb{1}_{\{X(\zeta_k)=u\}}}{N(n)} \to \pi_u, \quad n \to \infty \quad \text{a.s.}$$

This entails

$$\lim_{n \to \infty} \frac{\sum_{k=1}^n \mathbb{1}_{\{X(k)=u, X(k+1)=v_1, \ldots, X(k+l-1)=v_{l-1}, X(k+l)=v\}}}{N(n)}$$

$$= \pi_u \, \mathbb{P}\{X(1) = v_1, \ldots, X(l-1) = v_{l-1}, X(l) = v \mid X(0) = u\} > 0 \quad \text{a.s.}$$

This in combination with (3.189) completes the proof of Proposition 3.5.2. □

Proof of Theorem 3.5.1 It follows from (3.4) that, for all $t_0 > 0$ and $n \in \mathbb{N}$,

$$\sup_{t \in [0, t_0]} \left| \frac{X(\lfloor nt \rfloor)}{\sqrt{n}} - \frac{S_\xi(\lfloor nt \rfloor - T(\lfloor nt \rfloor))}{\sqrt{n}} \right| \leq \frac{X(0)}{\sqrt{n}} + \sum_{x \in A} \sum_{k=1}^{T^{(x)}(nt_0)} \frac{|\eta_k^{(x)}|}{\sqrt{n}} \quad \text{a.s.}$$

By a functional limit theorem for multidimensional standard random walks,

$$\left(\frac{S_\xi(\lfloor nt \rfloor)}{\sqrt{n}} \right)_{t \geq 0} \Longrightarrow \left(W_\Gamma(t) \right)_{t \geq 0}, \quad n \to \infty$$

in the J_1-topology on $D([0, \infty), \mathbb{R}^2)$. Proposition 3.5.2 entails

$$\frac{T(n)}{n} \overset{\mathbb{P}}{\to} 0, \quad n \to \infty.$$

Hence, by a straightforward multidimensional generalization of Lemma 3.1.1,

$$\left(\frac{S_\xi(\lfloor nt \rfloor - T(\lfloor nt \rfloor))}{\sqrt{n}} \right)_{t \geq 0} \Longrightarrow \left(W_\Gamma(t) \right)_{t \geq 0}, \quad n \to \infty$$

3.5 Multidimensional Random Walks with Membranes

in the J_1-topology on $D([0, \infty), \mathbb{R}^2)$. It remains to prove that

$$\sum_{x \in A} \sum_{k=1}^{T^{(x)}(nt_0)} \frac{|\eta_k^{(x)}|}{\sqrt{n}} \xrightarrow{\mathbb{P}} 0, \quad n \to \infty.$$

According to Proposition 3.5.2 and using the fact that the set A contains finitely many points with integer coordinates, it is enough to show that, for any $b > 0$ and all $x \in A$,

$$\sum_{k=1}^{\lfloor b \log n \rfloor} \frac{|\eta_k^{(x)}|}{\sqrt{n}} \xrightarrow{\mathbb{P}} 0, \quad n \to \infty.$$

Observe that $\sum_{k=1}^{\lfloor b \log n \rfloor} |\eta_k^{(x)}| \leq \lfloor b \log n \rfloor \max_{1 \leq k \leq \lfloor b \log n \rfloor} |\eta_k^{(x)}|$ a.s. and, for all $\varepsilon > 0$ and all $x \in A$,

$$\mathbb{P}\left\{ \frac{\log n}{\sqrt{n}} \max_{1 \leq k \leq \lfloor b \log n \rfloor} |\eta_k^{(x)}| \leq \varepsilon \right\} = \left(\mathbb{P}\left\{ |\eta^{(x)}| \leq \frac{\varepsilon \sqrt{n}}{\log n} \right\} \right)^{\lfloor b \log n \rfloor}$$

$$\stackrel{(3.178)}{=} \left(1 - o\left(1/\log\left(\frac{\varepsilon \sqrt{n}}{\log n} \right) \right) \right)^{\lfloor b \log n \rfloor} = \left(1 + o(1/\log n) \right)^{\lfloor b \log n \rfloor} \to 1$$

as $n \to \infty$. The proof of Theorem 3.5.1 is complete. □

Proof of Theorem 3.5.2

Put $l_1(t) := \log t$ and $l_2(t) := \log \log t$. Let $\eta_1, \eta_2, \ldots$ be independent copies of a random variable η with distribution $\mathbb{P}\{\eta = x\} = \mathbb{P}\{X(1) = x\}$ for $x \in \mathbb{Z}^2$. For each fixed $x \in \mathbb{Z}^2 \setminus \{0\}$, denote by $\tau(x)$ the first hitting time of 0 by a simple symmetric random walk in $\mathbb{Z}^2$ which starts at x and is independent of η. Let $\tau_1(x), \tau_2(x), \ldots$ be independent copies of $\tau(x)$, which are also independent of $\eta_1, \eta_2, \ldots$.

To prove (3.180) it is sufficient to check that

$$\liminf_{n \to \infty} \mathbb{P}\left\{ \bigcup_{k=1}^{l_2(n)} \left\{ \max_{1 \leq j \leq k-1} |\eta_j| \leq e^{\sqrt{l_1(n)}}, |\eta_k| > 2n, \sum_{j=1}^{k-1} \tau_j(\eta_j) \leq n \right\} \right\} > 0. \quad (3.190)$$

Here and hereafter, to simplify notation we do not write the integer parts in summation or union ranges. Observe that the last probability is equal to the probability of the event "there exists $k \leq l_2(n)$ such that the kth jump from 0 occurs before time n and its magnitude is larger than $2n$, whereas all the previous jumps do not exceed $e^{\sqrt{l_1(n)}}$".

We bound the aforementioned probability from below by

$$\mathbb{P}\bigg\{\bigcup_{k=1}^{l_2(n)}\Big\{\max_{1\leq j\leq k-1}|\eta_j|\leq e^{\sqrt{l_1(n)}},\ |\eta_k|>2n\Big\}\bigg\}$$
$$-\mathbb{P}\bigg\{\bigcup_{k=1}^{l_2(n)}\Big\{\max_{1\leq j\leq k-1}|\eta_j|\leq e^{\sqrt{l_1(n)}},\ \sum_{j=1}^{k-1}\tau_j(\eta_j)>n\Big\}\bigg\}. \quad (3.191)$$

The first term is equal to

$$\mathbb{P}\bigg\{\bigcup_{k=1}^{l_2(n)}\Big\{\max_{1\leq j\leq k-1}|\eta_j|\leq e^{\sqrt{l_1(n)}},\ |\eta_k|>2n\Big\}\bigg\}$$
$$=\sum_{k=1}^{l_2(n)}\mathbb{P}\Big\{\max_{1\leq j\leq k-1}|\eta_j|\leq e^{\sqrt{l_1(n)}},\ |\eta_k|>2n\Big\}$$
$$=\sum_{k=1}^{l_2(n)}\Big(\mathbb{P}\big\{|\eta|\leq e^{\sqrt{l_1(n)}}\big\}\Big)^{k-1}\mathbb{P}\{|\eta|>2n\}$$
$$=\frac{\Big(1-\big(\mathbb{P}\{|\eta|\leq e^{\sqrt{l_1(n)}}\}\big)^{\lfloor l_2(n)\rfloor}\Big)\mathbb{P}\{|\eta|>2n\}}{\mathbb{P}\{|\eta|>e^{\sqrt{l_1(n)}}\}}.$$

We have used the fact that the events

$$\Big\{\max_{1\leq j\leq k-1}|\eta_j|\leq e^{\sqrt{l_1(n)}},\ |\eta_k|>2n\Big\}$$

are disjoint for different k. This entails that the probability of the corresponding union is equal to the sum of probabilities.

We estimate the second term of (3.191) with the help of Boole's inequality:

$$\mathbb{P}\bigg\{\bigcup_{k=1}^{l_2(n)}\Big\{\max_{1\leq j\leq k-1}|\eta_j|\leq e^{\sqrt{l_1(n)}},\ \sum_{j=1}^{k-1}\tau_j(\eta_j)>n\Big\}\bigg\}$$
$$\leq\sum_{k=1}^{l_2(n)}\mathbb{P}\Big\{\max_{1\leq j\leq k-1}|\eta_j|\leq e^{\sqrt{l_1(n)}},\ \sum_{j=1}^{k-1}\tau_j(\eta_j)>n\Big\}.$$

3.5 Multidimensional Random Walks with Membranes

Combining fragments together we infer

$$\mathbb{P}\Big\{ \bigcup_{k=1}^{l_2(n)} \Big\{ \max_{1\le j\le k-1} |\eta_j| \le e^{\sqrt{l_1(n)}},\ |\eta_k| > n,\ \sum_{j=1}^{k-1} \tau_j(\eta_j) \le n \Big\}\Big\}$$

$$\ge \frac{\big(1 - (\mathbb{P}\{|\eta| \le e^{\sqrt{l_1(n)}}\})^{\lfloor l_2(n) \rfloor}\big)\mathbb{P}\{|\eta| > n\}}{\mathbb{P}\{|\eta| > e^{\sqrt{l_1(n)}}\}}$$

$$- \sum_{k=1}^{l_2(n)} \mathbb{P}\Big\{ \max_{1\le j\le k-1} |\eta_j| \le e^{\sqrt{l_1(n)}},\ \sum_{j=1}^{k-1} \tau_j(\eta_j) > n \Big\}.$$

It follows from

$$\mathbb{P}\{|\eta| > e^{\sqrt{l_1(n)}}\} \sim \frac{a}{l_2(e^{\sqrt{l_1(n)}})} = \frac{a}{l_1(\sqrt{l_1(n)})} = \frac{2a}{l_2(n)}, \quad n \to \infty$$

that the first summand on the right-hand side of the last inequality converges to $(1 - e^{-2a})/2$ as $n \to \infty$. We intend to prove that the second summand vanishes. As a preparation, we write

$$\sum_{k=1}^{l_2(n)} \mathbb{P}\Big\{ \max_{1\le j\le k-1} |\eta_j| \le e^{\sqrt{l_1(n)}},\ \sum_{j=1}^{k-1} \tau_j(\eta_j) > n \Big\}$$

$$\le \sum_{k=1}^{l_2(n)} \mathbb{P}\Big\{ \bigcup_{j=1}^{l_2(n)} \{|\eta_j| \le e^{\sqrt{l_1(n)}},\ \tau_j(\eta_j) > \sqrt{n}\}\Big\}$$

$$\le (l_2(n))^2 \mathbb{P}\{\eta \le e^{\sqrt{l_1(n)}},\ \tau(\eta) > \sqrt{n}\}.$$

According to formulas (2.16) and (2.17) in [48], for any $x \in \mathbb{Z}^2$ satisfying $|x| \le a_n = O(n^{1/3})$ there exists a constant $B > 0$ such that

$$\mathbb{P}\{\tau(x) > n\} \le Bl_1(a_n)/l_1(n).$$

This entails

$$(l_2(n))^2 \mathbb{P}\{|\eta| \le e^{\sqrt{l_1(n)}},\ \tau(\eta) > \sqrt{n}\}$$

$$= (l_2(n))^2 \sum_{x \in \mathbb{Z}^2 : |x| \le e^{\sqrt{l_1(n)}}} \mathbb{P}\{\eta = x\}\mathbb{P}\{\tau(x) > \sqrt{n}\}$$

$$\le (l_2(n))^2 \frac{Bl_1(e^{\sqrt{l_1(n)}})}{l_1(\sqrt{n})} = \frac{2B(l_2(n))^2}{\sqrt{l_1(n)}} \to 0, \quad n \to \infty,$$

thereby completing the proof of (3.190). The last claim of the theorem concerning tightness (as well as the claim made in Remark 3.5.2) follows from the fact that the supremum functional is continuous in all the four Skorokhod topologies.

3.5.2 Multidimensional Random Walks with Periodic Membranes Concentrated on Hyperplanes

Our approach is based on techniques developed in Sect. 3.3. As in the one-dimensional case, we prefer to formulate the model for Walsh-type processes and then derive a result for a multidimensional random walk perturbed on a hyperplane as a particular case. We shall omit the proofs because they are almost identical with the proofs in Sect. 3.3.

We start with the model description. Similarly to Sect. 3.3 we introduce the notation

$$\mathcal{Z}^m := \mathbb{Z} \times \{1, \ldots, m\} \quad \text{and} \quad \mathcal{N}^m := \mathbb{N} \times \{1, \ldots, m\}.$$

A Markov chain

$$(Z(k))_{k \geq 0} = ((R(k), l(k), Y(k)))_{k \geq 0}$$

on[22] $\mathcal{Z}^m \times \mathbb{Z}^d$ is defined according to the rules formulated in Sect. 3.3.1. More precisely,

(**A**$_0$) For $1 \leq j \leq m$, integer-valued random variables $\xi^{(j)}$ and $\mathbb{Z}^d$-valued random vectors $\eta^{(j)}$ have zero means and finite second moments. Put $v_j^2 := \operatorname{Var} \xi^{(j)} > 0$ for $1 \leq j \leq m$.

(**A**$_1$) For all $x \in \mathbb{N}, \widetilde{x} \in \mathbb{Z}, i \in \{1, \ldots, m\}$ and $y, \widetilde{y} \in \mathbb{Z}^d$,

$$\mathbb{P}\{(R(1), l(1), Y(1)) = (x + \widetilde{x}, i, y + \widetilde{y}) \,|\, (R(0), l(0), Y(0)) = (x, i, y)\}$$
$$= \mathbb{P}\{(\xi^{(i)}, \eta^{(i)}) = (\widetilde{x}, \widetilde{y})\}.$$

(**A**$_2$) For all integer $x \leq 0$, $\mathbb{P}\{R(1) \in \mathbb{N} \,|\, R(0) = x\} = 1$.

(**A**$_3$) There exists $C > 0$ such that, for all $(x, i, y) \in \mathcal{Z}^m \times \mathbb{Z}^d$ with $x \leq 0$,

$$\mathbb{E}(R(1) \,|\, (R(0), l(0), Y(0)) = (x, i, y)) \leq C(1 + |x|);$$
$$\mathbb{E}(|Y(1) - Y(0)| \,|\, (R(0), l(0), Y(0)) = (x, i, y)) \leq C(1 + |x|).$$

(**A**$_4$) The states $\mathcal{N}^m \times \mathbb{Z}^d$ of the Markov chain $(Z(k))_{k \geq 0}$ are connected.

---

[22] For each $k \in \mathbb{N}_0$, the vector $(R(k), l(k))$ takes values in $\mathcal{Z}^m$.

3.5 Multidimensional Random Walks with Membranes

To formulate our last assumption we need more notation. For $k_1, \ldots, k_d \in \mathbb{N}$ fixed, let U be the box in $\mathbb{Z}^d$ defined by

$$U := ([0, k_1) \times \cdots \times [0, k_d)) \cap \mathbb{Z}^d.$$

For $y = (y_1, \ldots, y_d) \in \mathbb{Z}^d$, let $\hat{y} \in U$ with

$$\hat{y}_i := y_i \,(\bmod\, k_i), \quad i = 1, \ldots, d$$

and then put $\hat{y} := (\hat{y}_1, \ldots, \hat{y}_d)$. Below we also write $\hat{V}$ for a $\mathbb{Z}^d$-valued random vector V meaning that $\hat{V} \in U$ a.s. and $\hat{V}_i := V_i \,(\bmod\, k_i)$ for $i = 1, \ldots, d$.

Assumption $\mathbf{A}_5$ given next means that jumps from the set $(\mathcal{Z}^m \times \mathbb{Z}^d) \setminus (\mathcal{N}^m \times \mathbb{Z}^d)$ into $(\mathcal{N}^m \times \mathbb{Z}^d)$ exhibit certain periodicity.

($\mathbf{A}_5$) (periodicity of entrances into $\mathcal{N}^m \times \mathbb{Z}^d$). There exist $k_1, \ldots, k_d \in \mathbb{N}$ such that for all integer $x \leq 0$, $i, j \in \{1, \ldots, m\}$ and $y, \widetilde{y} \in \mathbb{Z}^d$,

$$\mathbb{P}\{(R(1), l(1), Y(1)) = (\widetilde{x}, j, y + \widetilde{y}) \mid (R(0), l(0), Y(0)) = (x, i, y)\}$$
$$= \mathbb{P}\{(R(1), l(1), Y(1)) = (\widetilde{x}, j, \hat{y} + \widetilde{y}), \mid (R(0), l(0), Y(0)) = (x, i, \hat{y})\}.$$

Denote by σ_n the time of the nth exit of R from $\mathbb{N}$, that is,

$$\sigma = \sigma_1 := \inf\{k \geq 0 : R(k) \leq 0\}, \quad \sigma_{n+1} = \inf\{k > \sigma_n : R(k) \leq 0\}.$$

Put $\tau_n := \sigma_n + 1$ for $n \in \mathbb{N}$. It follows from $\mathbf{A}_2$ that τ_n is the time of the nth entrance of R into $\mathbb{N}$. For $n \in \mathbb{N}$, put

$$Z_{\text{exit}}(n) := (R(\sigma_n), l(\sigma_n), \hat{Y}(\sigma_n)) \quad \text{and} \quad Z_{\text{entr}}(n) := (R(\tau_n), l(\tau_n), \hat{Y}(\tau_n)).$$

The random sequences $Z_{\text{exit}} := (Z_{\text{exit}}(n))_{n \geq 1}$ and $Z_{\text{entr}} := (Z_{\text{entr}}(n))_{n \geq 1}$ are Markov chains defined by the exits from $\mathcal{N}^m \times \mathbb{Z}^d$ and entrances to $\mathcal{N}^m \times \mathbb{Z}^d$ with the phase spaces $(\mathcal{Z}^m \setminus \mathcal{N}^m) \times U$ and $\mathcal{N}^m \times U$, respectively.

The following result can be proved similarly to Lemmas 3.3.1 and 3.3.2.

Lemma 3.5.3 *The Markov chains Z_{exit} and Z_{entr} are ergodic. Furthermore, their stationary distributions π_{exit} and π_{entr} have finite means.*

For notational simplicity we shall write, similarly to (3.81), the expectation with respect to the stationary distribution π_{exit} of a real-valued measurable function f defined on the phase space of the process Z_{exit} as

$$\mathbb{E}^{\pi_{\text{exit}}} f(R(0), l(0), \hat{Y}(0)) := \sum_{i \in \mathbb{Z}} \sum_{j=1}^{m} \sum_{k \in U} f(i, j, k) \pi_{\text{exit}}(\{i, j, k\}).$$

The same convention applies to the expectation with respect to π_{entr}.

For each $n \in \mathbb{N}$, let $((R^{(n)}(k), l^{(n)}(k), Y^{(n)}(k)))_{k \geq 0}$ be a Markov chain having the same transition probabilities as $((R(k), l(k), Y(k)))_{k \geq 0}$ but possibly satisfying a different initial condition. For $t \geq 0$, put

$$\mathcal{X}^{(n)}(t) := \left(\frac{R^{(n)}(\lfloor nt \rfloor)}{\sqrt{n}} \mathbb{1}_{\{l^{(n)}(\lfloor nt \rfloor)=1\}}, \ldots, \frac{R^{(n)}(\lfloor nt \rfloor)}{\sqrt{n}} \mathbb{1}_{\{l^{(n)}(\lfloor nt \rfloor)=m\}} \right),$$

$$\mathcal{L}^{(n)}(t) := n^{-1/2} \sum_{\tau_k \leq \lfloor nt \rfloor} \frac{R^{(n)}(\tau_k) - R^{(n)}(\sigma_{k+1})}{v_{l_{(n)}(\tau_k)}}, \quad \mathcal{Y}^{(n)}(t) := \frac{Y^{(n)}(\lfloor nt \rfloor)}{\sqrt{n}}.$$

Define $\psi : \mathbb{R}^m \to \mathbb{R}^m$ by $\psi(x_1, \ldots, x_m) := (x_1/v_1, \ldots, x_m/v_m)$ for $(x_1, \ldots, x_m) \in \mathbb{R}^m$. For the vector $\eta^{(j)}$, let $\Sigma_{\eta^{(j)}}$ denote its covariance matrix. For the vectors $\xi^{(j)}$ and $\eta^{(j)}$, let $\Sigma_{\xi^{(j)}, \eta^{(j)}}$ denote their cross-covariance matrix. Recall that $\langle Z, Z \rangle$ and $\langle Z_1, Z_2 \rangle$ denote the predictable quadratic variation of Z and the predictable quadratic covariation of Z_1 and Z_2, respectively, provided these are well-defined.

The following generalization of Theorem 3.3.2 is the main result of this section.

Theorem 3.5.3 *Suppose* $\mathbf{A}_0$–$\mathbf{A}_5$ *and that*

$$(\psi(\mathcal{X}^{(n)}(0)), \mathcal{Y}^{(n)}(0)) \xrightarrow{\mathbb{P}} (x, y), \quad n \to \infty$$

for some $(x, y) \in ([0, \infty))^m \times \mathbb{R}^d$. *Then*

$$(\psi(\mathcal{X}^{(n)}), \mathcal{L}^{(n)}, \mathcal{Y}^{(n)}) \implies (\mathcal{X}^{(\infty)}, \mathcal{L}^{(\infty)}, \mathcal{Y}^{(\infty)}), \quad n \to \infty$$

in the product topology on $(D([0, \infty), \mathbb{R}^m))^2 \times D([0, \infty), \mathbb{R}^d)$, *where the limit process is a continuous semimartingale. More precisely,*

- $(\mathcal{X}^{(\infty)}(t))_{t \geq 0} = (W_{\mathbf{p}}(t))_{t \geq 0}$ *is a Walsh Brownian motion starting at x with parameter* $\mathbf{p} = (p_1, \ldots, p_m)$ *given by*

3.5 Multidimensional Random Walks with Membranes

$$p_k = \frac{\mathbb{E}^{\pi_{\text{entr}}} R(0) v_k^{-1} \mathbb{1}_{\{l(0)=k\}} - \mathbb{E}^{\pi_{\text{exit}}} R(0) v_k^{-1} \mathbb{1}_{\{l(0)=k\}}}{\mathbb{E}^{\pi_{\text{entr}}} R(0) v_{l(0)}^{-1} - \mathbb{E}^{\pi_{\text{exit}}} R(0) v_{l(0)}^{-1}}$$

$$= \frac{\mathbb{E}^{\pi_{\text{entr}}} (R(0) - R(\sigma)) v_k^{-1} \mathbb{1}_{\{l(0)=k\}}}{\mathbb{E}^{\pi_{\text{entr}}} (R(0) - R(\sigma)) v_{l(0)}^{-1}}$$

$$= \frac{\mathbb{E}^{\pi_{\text{exit}}} (R(1) v_{l(1)}^{-1} \mathbb{1}_{\{l(1)=k\}} - R(0) v_{l(0)}^{-1} \mathbb{1}_{\{l(0)=k\}})}{\mathbb{E}^{\pi_{\text{exit}}} (R(1) v_{l(1)}^{-1} - R(0) v_{l(0)}^{-1})}, \quad k = 1, \ldots, m.$$

- $(\mathcal{L}^{(\infty)}(t))_{t\geq 0} = (L_0^{W_{\mathbf{p}}}(t))_{t\geq 0}$ is the local time of $(W_{\mathbf{p}}(s))_{s\geq 0}$ at 0.
- The process $(\mathcal{Y}^{(\infty)}(t))_{t\geq 0}$ satisfies

$$\mathcal{Y}^{(\infty)}(t) = B(y,t) + c L_0^{W_{\mathbf{p}}}(t), \quad t \geq 0, \tag{3.192}$$

where

$$c := \frac{\mathbb{E}^{\pi_{\text{exit}}}(Y(1) - Y(0))}{\mathbb{E}^{\pi_{\text{exit}}}(R(\tau_1) v_{l(\tau_1)}^{-1} - R(0) v_{l(0)}^{-1})} = \frac{\mathbb{E}^{\pi_{\text{exit}}}(Y(1) - Y(0))}{\mathbb{E}^{\pi_{\text{entr}}}(R(0) - R(\sigma)) v_{l(0)}^{-1}} \tag{3.193}$$

and $B := (B(y,t))_{t\geq 0}$ is a continuous martingale satisfying $B(y,0) = y$ and

$$\langle B, B \rangle(t) = \sum_{j=1}^{m} \Sigma_{\eta^{(j)}} \int_0^t \mathbb{1}_{\{\mathcal{X}_j^{(\infty)}(s) > 0\}} ds, \quad t \geq 0,$$

$$\langle \mathcal{X}^{(\infty)}, B \rangle(t) = \sum_{j=1}^{m} v_j^{-1} \Sigma_{\xi^{(j)}, \eta^{(j)}} \int_0^t \mathbb{1}_{\{\mathcal{X}_j^{(\infty)}(s) > 0\}} ds, \quad t \geq 0. \tag{3.194}$$

Example 3.5.6 Let $m = 2$ and $\mathbb{P}\{(\xi^{(j)}, \eta^{(j)}) = e\} = \frac{1}{2(1+d)}$ for $j = 1, 2$ and any $e \in \mathbb{Z}^{1+d}$ with $|e| = 1$, that is, the underlying standard random walks jump to neighboring elements of the lattice with equal probabilities. In particular, this entails $\mathbb{P}\{R(\sigma_n) = 0\} = 1$ for all $n \in \mathbb{N}$. One can check that, for $j = 1, 2$, $v_j^2 = 1/(1+d)$, the matrix $(1+d)\Sigma_{\eta^{(j)}}$ is the $d \times d$ identity matrix and $\Sigma_{\xi^{(j)}, \eta^{(j)}}$ is the $1 \times d$ zero matrix.

We already know that Assumption $\mathbf{A}_0$ holds true in the present setting and let so do assumptions $\mathbf{A}_1$–$\mathbf{A}_5$. Then, by Theorem 3.5.3, the process $\mathcal{X}^{(\infty)} := (\mathcal{X}_1^{(\infty)}, \mathcal{X}_2^{(\infty)})$ is a Walsh Brownian motion with parameter (p_1, p_2), where

$$p_i = \mathbb{E}^{\pi_{\text{entr}}}(R(0) \mathbb{1}_{\{l(0)=i\}}) / \mathbb{E}^{\pi_{\text{entr}}} R(0), \quad i = 1, 2.$$

Hence, Theorem 2.2.4 implies that $\mathcal{X}_1^{(\infty)} - \mathcal{X}_2^{(\infty)} =: W_\gamma^{\text{skew}}$ is a skew Brownian motion with permeability parameter $\gamma := p_1 - p_2$. It follows from Theorem 3.5.3 in combination with the last observation that the scaling limit of $\sqrt{1+d}\,(\mathcal{X}_1^{(n)} - \mathcal{X}_2^{(n)}, \mathcal{Y}^{(n)})$ as $n \to \infty$ is $(W_\gamma^{\text{skew}}, W_Y + cL_0^{W_\gamma^{\text{skew}}})$, where W_γ^{skew} is a skew Brownian motion with

$$\gamma = \mathbb{E}^{\pi_{\text{entr}}}(R(0)(\mathbb{1}_{\{l(0)=1\}} - \mathbb{1}_{\{l(0)=2\}}))/\mathbb{E}^{\pi_{\text{entr}}} R(0) \in [-1, 1],$$

W_Y is a d-dimensional standard Brownian motion that is independent of W_γ^{skew} and

$$c = \mathbb{E}^{\pi_{\text{exit}}}(Y(1) - Y(0))/\mathbb{E}^{\pi_{\text{entr}}} R(0).$$

Getting back from the limit result to the original random walks on Euclidean space, we note that the sequence $\left((R(k)(\mathbb{1}_{\{l(k)=1\}} - \mathbb{1}_{\{l(k)=2\}}), Y(k))\right)_{k\geq 0}$ can be identified with a simple random walk on $\mathbb{Z}^{1+d}$ which is perturbed on a "two-sided" periodic membrane located on $H := \{0\} \times \mathbb{Z}^d$. Here, "two-sided membrane" means that the distribution of the next jump from the membrane depends, among others, on the half-space from which the walk enters H.

Now we adapt the set of Assumptions **B** of Theorem 3.3.3 to the multidimensional case $\mathbb{Z}^{1+d}$ in the situation when the membrane is located on a strip $\{-m, -m+1, \ldots, m\} \times \mathbb{Z}^d$ and has a periodic structure. We denote by $((X(k), Y(k)))_{k\geq 0}$ the corresponding Markov chain taking values in $\mathbb{Z} \times \mathbb{Z}^d$.

Now we are ready to state adapted assumptions. Actually, assumption **B**$_0$ does not have a counterpart among the assumptions of Sect. 3.3.1. It rules out the possibility that the sequence $((X(k), Y(k)))_{k\geq 0}$ jumps over the membrane, thereby simplifying significantly an analysis of the embedded exit and entrance Markov chains.

(**B**$_0$) The variables $\xi_\pm^X$ take integer values and satisfy $\xi_+^X \geq -2m - 1$ and $\xi_-^X \leq 2m + 1$ a.s.

(**B**$_1$) $\mathbb{E}\xi_\pm^X = \mathbb{E}\xi_\pm^Y = 0$, $v_\pm := \text{Var}(\xi_\pm^X) \in (0, \infty)$, $\mathbb{E}(\xi_\pm^Y)^2 < \infty$ and, for all $x, y \in \mathbb{Z}$ and all $\widetilde{y} \in \mathbb{Z}^d$,

$$\mathbb{P}\left\{(X(1), Y(1)) = (x + \widetilde{x}, y + \widetilde{y}) \,\big|\, (X(0), Y(0)) = (x, y)\right\}$$
$$= \mathbb{P}\{\xi_+^X = \widetilde{x}, \xi_+^Y = \widetilde{y},\}, \quad \widetilde{x} > m, \quad \widetilde{x} \in \mathbb{Z},$$
$$\mathbb{P}\left\{(X(1), Y(1)) = (x + \widetilde{x}, y + \widetilde{y}) \,\big|\, (X(0), Y(0)) = (x, y)\right\}$$
$$= \mathbb{P}\{\xi_-^X = \widetilde{x}, \xi_-^Y = \widetilde{y},\}, \quad \widetilde{x} < -m, \quad \widetilde{x} \in \mathbb{Z}.$$

3.5 Multidimensional Random Walks with Membranes

(**B$_2$**) For all $(x, y) \in \{-m, \ldots, m\} \times \mathbb{Z}^d$,

$$\mathbb{P}\{|X(k)| > m \text{ for some } k \in \mathbb{N} | (X(0), Y(0)) = (x, y)\} = 1.$$

(**B$_3$**) For all $(x, y) \in \{-m, \ldots, m\} \times \mathbb{Z}^d$ and $(\widetilde{x}, \widetilde{y}) \in \mathbb{Z}^{1+d}$,

$$\mathbb{P}\{(X(1), Y(1)) = (x + \widetilde{x}, y + \widetilde{y}) | (X(0), Y(0)) = (x, y)\} =$$
$$\mathbb{P}\{\eta^{(x)} = \widetilde{x}, \eta^{(\hat{y})} = \widetilde{y}\}$$

and $\mathbb{E}(|\eta^{(x)}| + |\eta^{(\hat{y})}|) < \infty$.

(**B$_4$**) The states $\mathbb{Z}^{1+d} \setminus (\{-m, \ldots, m\} \times \mathbb{Z}^d)$ of the Markov chain $(X(n), Y(n))_{n \geq 0}$ are connected.

(**B$_5$**) There exist $k_1, \ldots, k_d \in \mathbb{N}$ such that for all integer $x \in \{-m, \ldots, m\}, \widetilde{x} \in \mathbb{Z}$ and $y, \widetilde{y} \in \mathbb{Z}^d$,

$$\mathbb{P}\{(X(1), Y(1)) = (\widetilde{x}, y + \widetilde{y}) | (X(0), Y(0)) = (x, y)\}$$
$$= \mathbb{P}\{(X(1), Y(1)) = (\widetilde{x}, \hat{y} + \widetilde{y}), | (X(0), Y(0)) = (x, \hat{y})\}.$$

We recall that the notation $\hat{y}$ was introduced in the paragraph following the line in which assumption **A$_4$** was formulated.

As in Sect. 3.3.1, we introduce the embedded Markov chains

$$(X_{\text{exit}}(\iota), Y_{\text{exit}}(\iota))_{\iota \geq 1} = (X(\widetilde{\sigma}_\iota), \hat{Y}(\widetilde{\sigma}_\iota))_{\iota \geq 1},$$
$$(X_{\text{entr}}(\iota), Y_{\text{entr}}(\iota))_{\iota \geq 0} = (X(\widetilde{\tau}_\iota), \hat{Y}(\widetilde{\tau}_\iota))_{\iota \geq 0}.$$

Along the lines of the proofs of Lemmas 3.3.1 and 3.3.2 it can be shown that these Markov chains admit unique stationary distributions π_{exit} and π_{entr}, respectively, which are integrable.

For each $n \in \mathbb{N}$, let $(X^{(n)}(k), Y^{(n)}(k))_{k \geq 0}$ be a Markov chain having the same transition probabilities as $(X(k), Y(k))_{k \geq 0}$ but possibly satisfying a different initial condition. For $t \geq 0$, put

$$\mathcal{X}^{(n)}(t) := \frac{X^{(n)}(\lfloor nt \rfloor)}{\sqrt{n}},$$

$$\mathcal{L}^{(n)}(t) := n^{-1/2} \sum_{\tau_k^{(n)} \leq \lfloor nt \rfloor} \frac{|X^{(n)}(\tau_k^{(n)}) - X^{(n)}(\sigma_{k+1}^{(n)})|}{v_{l^{(n)}(\tau_k^{(n)})}},$$

$$\mathcal{Y}^{(n)}(t) := \frac{Y^{(n)}(\lfloor nt \rfloor)}{\sqrt{n}}.$$

Theorem 3.5.4 *Suppose* $\mathbf{B_0}$–$\mathbf{B_5}$ *and that*

$$\left(\varphi(\mathcal{X}^{(n)}(0)), \mathcal{Y}^{(n)}(0)\right) \xrightarrow{\mathbb{P}} (x, y), \quad n \to \infty$$

for some $(x, y) \in \mathbb{R}^{1+d}$, *where* φ *is as defined in* (3.71). *Then*

$$\left(\varphi(\mathcal{X}^{(n)}(t)), \mathcal{L}^{(n)}(t), \mathcal{Y}^{(n)}(t)\right)_{t \geq 0} \Longrightarrow \left(\mathcal{X}^{(\infty)}(t), \mathcal{L}^{(\infty)}(t), \mathcal{Y}^{(\infty)}(t)\right)_{t \geq 0}, \quad n \to \infty$$

in the product topology on $(D([0, \infty), \mathbb{R}))^2 \times D([0, \infty), \mathbb{R}^d)$, *where the limit process is a continuous semimartingale. More precisely,*

- $(\mathcal{X}^{(\infty)}(t))_{t \geq 0} = (W_\gamma^{\mathrm{skew}}(t))_{t \geq 0}$ *is a skew Brownian motion starting at* x *with permeability parameter* γ *as given in* (3.91).
- $(\mathcal{L}^{(\infty)}(t))_{t \geq 0} = (L_0^{W_\gamma^{\mathrm{skew}}}(t))_{t \geq 0}$ *is the local time of* $(W_\gamma^{\mathrm{skew}}(s))_{s \geq 0}$ *at* 0.
- *The process* $(\mathcal{Y}^{(\infty)}(t))_{t \geq 0}$ *satisfies*

$$\mathcal{Y}^{(\infty)}(t) = B(y, t) + c\mathcal{L}^{(\infty)}(t), \quad t \geq 0,$$

where[23]

$$c := \frac{\mathbb{E}^{\pi_{\mathrm{exit}}}(Y(1) - Y(0))}{\mathbb{E}^{\pi_{\mathrm{exit}}}|X(\tau_1)v_{l(\tau_1)}^{-1} - X(0)v_{l(0)}^{-1}|} = \frac{\mathbb{E}^{\pi_{\mathrm{exit}}}(Y(1) - Y(0))}{\mathbb{E}^{\pi_{\mathrm{entr}}}|X(0) - X(\sigma_1)|v_{l(0)}^{-1}}$$

and $B = (B(y, t))_{t \geq 0}$ *is a continuous martingale satisfying* $B(y, 0) = y$ *and*

$$\langle B, B \rangle(t) = \int_0^t \left(\Sigma_{\xi_+}^Y \mathbb{1}_{\{\mathcal{X}^{(\infty)}(s) > 0\}} + \Sigma_{\xi_-}^Y \mathbb{1}_{\{\mathcal{X}^{(\infty)}(s) < 0\}}\right) ds, \quad t \geq 0,$$

$$\langle \mathcal{X}^{(\infty)}, B \rangle(t) = \int_0^t \left(v_+^{-1} \Sigma_{\xi_+, \xi_+}^{X,Y} \mathbb{1}_{\{\mathcal{X}^{(\infty)}(s) > 0\}} + v_-^{-1} \Sigma_{\xi_-, \xi_-}^{X,Y} \mathbb{1}_{\{\mathcal{X}^{(\infty)}(s) < 0\}}\right) ds$$

for $t \geq 0$.

Bibliographic Comments

Recurrence/transience of what we called in Sect. 3.1.2 an oscillating random walk was discussed in [25]. Actually, such a Markov chain is closely related to the oscillating random

---

[23] Observe that $\mathcal{Y}^{(\infty)}$ and c are given by formulas similar to (3.192) and (3.193).

3.5 Multidimensional Random Walks with Membranes

walk as defined by Kemperman in [82], yet slightly different. For each $\alpha \in [0, 1]$, the original oscillating random walk $(X_n^{(\alpha)})_{n \geq 0}$ is given by

$$X^{(\alpha)}(n+1) := \begin{cases} X^{(\alpha)}(n) + \xi_{n+1}, & \text{if } X^{(\alpha)}(n) \geq 1, \\ B_{n+1}^{(\alpha)} \xi_{n+1} + (1 - B_{n+1}^{(\alpha)}) \eta_{n+1}, & \text{if } X^{(\alpha)}(n) = 0, \\ X^{(\alpha)}(n) + \eta_{n+1}, & \text{if } X^{(\alpha)}(n) \leq 1 \end{cases}$$

for $n \in \mathbb{N}_0$, where $X_0^{(\alpha)} \in \mathbb{Z}$ is deterministic and $(\xi_k)_{k \geq 1}$ and $(\eta_k)_{k \geq 1}$ are independent sequences of independent identically distributed integer-valued random variables, both independent of $(B_k^{(\alpha)})_{k \geq 1}$, a sequence of independent Bernoulli random variables with $\mathbb{P}\{B_k^{(\alpha)} = 1\} = \alpha$. Properties of the oscillating random walks and related models, in particular, recurrence/transience, were investigated in [44, 82, 109, 142, 166]. Under the assumptions that the variables ξ_1 and η_1 have zero means and finite power moments of order $3 + \delta$ and that their distributions are non-lattice, a functional limit theorem for $(X^{(\alpha)}(n))$, properly scaled, was proved in [167]. The weak limit is a skew Brownian motion with explicitly given permeability parameter.

Closing the discussion of oscillating random walks we mention Corollary 8.4 in [66], which is a functional limit theorem for $(X^*(\lfloor nt \rfloor)/\sqrt{n})_{t \geq 0}$ as $n \to \infty$ with the weak limit being an *oscillating Brownian motion*. The sequence $(X^*(k))_{k \geq 0}$ differs from our oscillating random walk $(X_k)_{k \geq 0}$ as defined in Sect. 3.1.2 in that, for $k \in \mathbb{N}_0$, $X^*(k+1) := X^*(k) + \theta_{k+1}$ whenever $X^*(k) = 0$. Here, $\theta, \theta_1, \ldots$ are independent identically distributed real-valued random variables which are independent of $(\xi_k)_{k \geq 1}$ and $(\eta_k)_{k \geq 1}$; $\mathbb{E}\xi = \mathbb{E}\eta = \mathbb{E}\theta = 0$, $\text{Var}\, \xi < \infty$, $\text{Var}\, \eta < \infty$ and $\text{Var}\, \theta \in (0, \infty)$. Other results of this flavor can be found in [81].

Theorem 3.2.1 given in Sect. 3.2.1 treats several natural models of random walks with reflection, with perturbations of finite mean, and proves that their Donsker's scaling limits are reflected Brownian motions. We stress that only for Lindley's model does a limit theorem of this flavor follow directly from continuity of the Skorokhod reflection map and the continuous mapping theorem; see Theorem 3.1.2 and the discussion preceding it. Our argument in Sect. 3.2.1 is reasonably simple and should be useful from the methodological viewpoint. We prove that the corresponding random sequences are weakly relatively compact and that the limit process is a solution to the Skorokhod reflection problem, with the driving noise being a Brownian motion. The distributional convergence is then secured by uniqueness of a solution to the Skorokhod reflection problem. An alternative proof of Theorem 3.2.1 for *linearly interpolated* MODEL OSCRW can be found in [126]. By using an approach different from ours a version of Theorem 3.2.1 for MODEL REFLRW was proved in [117] under more restrictive assumptions that the distribution of ξ is non-lattice and $\mathbb{E}(\xi_-)^3 < \infty$.

Most of the results of Sect. 3.2.2 are based on the recent article [132], which gives a general unified approach towards investigating both deterministic and random sequences with different reflection rules. According to Theorem 2 in [63], a weak limit of Moran's model of a dam is a sticky reflected Brownian motion. The second part of Example 3.2.2 recovers this fact and discusses further limit results for Moran's model.

Under the assumption $\eta > 0$ a.s. the result of Theorem 3.2.3 presented in Sect. 3.2.3 was obtained in[24] [71] for MODELS LINDREFILL, OSCRW, and REGEN with a different proof. We refer to [71] for a discussion of various properties of the limit process $W_x^{\text{refl}, \beta}$ (including formula (3.45)). The limit process of Theorem 3.2.3 appeared in Theorem 1.1 of [128] as Donsker's scaling limit for a random walk with the membrane {0} (see Definition 3.1.1). In [128], random variables ξ and η are integer-valued and ξ is bounded from below by -1. The cited article invokes a representation arising in a generalized Skorokhod reflection problem. That approach fails in the setting of Theorem 3.2.3 where ξ and η are real-valued, for no reduction to the generalized Skorokhod reflection problem seems to be possible. Theorem 3.2.4 from Sect. 3.2.4 was obtained in [126] with a proof which is similar to that given here.

Main results of Sect. 3.3 which are based on the paper [120] cover almost all possible scenarios in the situation when the jumps from a membrane have a finite mean. The approach taken here which is based on a martingale characterization of a scaling limit was worked out in [69].

Diffusion approximations by scaling limits of Markov chains is a classical topic of the theory of stochastic processes; see, for instance, [57] for the case of diffusions with smooth parameters. An approximation of a general one-dimensional diffusion by birth-and-death continuous-time Markov chains was investigated by Stone [154] with the help of analysis of a scale measure and a speed measure. We stress that (not birth-and-death) Markov chain approximations of diffusions with membranes, say a skew or a Walsh Brownian motion, are much harder to deal with. Earlier results in particular situations were obtained by different methods:

- Classical probability methods (weak convergence of finite-dimensional distributions and weak relative compactness) [21, 65, 117, 167].
- Methods based on martingale problems [54, 55, 120].
- Methods which exploit constructing processes via excursions, their concatenations, transformations, etc. [96, 110, 127, 128, 131, 178, 179].
- A resolvent approach [17, 20, 41, 91].
- A semigroup approach and a convergence of generators technique [114].

Each method has its advantages. For instance, methods based on operating with excursions provide transparent probabilistic insight into converging and limit processes and their

---

[24] We use the opportunity to point out that the assumptions of Theorems 1.1 and 1.2 in [71], should read $\tilde{S}_v(0)/(\sigma v^{1/2}) \xrightarrow{\mathbb{P}} x$, $\hat{S}_v(0)/(\sigma v^{1/2}) \xrightarrow{\mathbb{P}} x$, and $\grave{S}_v(0)/(\sigma v^{1/2}) \xrightarrow{\mathbb{P}} x$ rather than $\tilde{S}_v(0)/v^{1/2} \xrightarrow{\mathbb{P}} x$, $\hat{S}_v(0)/v^{1/2} \xrightarrow{\mathbb{P}} x$, and $\grave{S}_v(0)/v^{1/2} \xrightarrow{\mathbb{P}} x$.

3.5 Multidimensional Random Walks with Membranes

interplay. The results in [129] derived with the help of such methods allow one to prove limit theorems for singular diffusions, where the limit objects are skew Bessel processes [130] or a snapping out Brownian motion [103]. Many models of anomalous diffusions are obtained by subordinating some random walks or Markov chains; see, for instance, [85, 105–107]. Functional limit theorems for such models are often derived by the methods which are systematically discussed in the present book: first limit theorems for subordinated and subordinating processes are proved, and then the Skorokhod representation theorem and results on continuity of the composition are applied.

The exit and entrance Markov chains play an important role in Sect. 3.3. These are essentially used for calculating the parameters of limit skew or Walsh Brownian motions. Also, these helped us to rediscover some results on distributional convergence of scaled additive functionals (the number of crossings the membrane, etc.) to local times at 0 of the limit processes. For a detailed investigation of the exit and entrance chains which correspond to general Markov chains, we refer to [111].

Section 3.4 is based on the recent article [41].

To the best of our knowledge, multidimensional locally perturbed random walks have never been investigated in the same generality as the one-dimensional ones. The papers [119, 158] were concerned with membranes containing finitely many points with integer coordinates. It was shown that if the jumps from the membrane admit a moment of some positive order, then the scaling limit of a random walk with membrane is a Brownian motion, that is, the perturbations do not affect the limit. Our findings presented in Sect. 3.5.1 which is based on the article [42] can be thought of as generalizations of the latter results. Theorem 3.5.2 showing that if the perturbations are too heavy, then there is no limit of two-dimensional perturbed random walk appeared in [42] as well. We believe that our argument is more probabilistic than the proofs given in [119, 158].

Under the additional assumption that the walk S_ξ is strongly aperiodic, the result of our Lemma 3.5.1 can be found in Lemma 2.1 of [23]. Although it is likely the claim is known in full generality, we have been unable to locate its complete form in the literature.

The results given in Sect. 3.5.2 are generalizations of those in [22]. After the paper [22] has been published, a novel approach discussed in details in Sect. 3.3 was worked out. Thus, the presentation in Sect. 3.5.2 follows the aforementioned approach.

We refer to Chapter 14 in [175] for a survey on the limit behavior of scaled random walks in an orthant and applications of the corresponding results to queuing theory. We stress that all the proofs given there are based on the continuity of a multidimensional Skorokhod reflection map. As far as we know, [4] is the only work, which succeeded in proving a functional limit theorem for integer-valued random walks in a two-dimensional quarter-plane, in the situation where the Skorokhod reflection map is undefined. It is assumed in [4] that the random walk can only jump to a neighboring state. Recurrence/transience integer-valued random walks in a quarter-plane were investigated in [108] in more general situations, where transition probabilities of the walk were perturbed in a regular way in fixed strips around the axes. Our methods can be used to determine Donsker's scaling limit for perturbed random walks from [108] if they are transient or Donsker's scaling limit for the walks stopped at zero. The general case is still an open problem.

Auxiliary Results

> *Begin at the beginning, the King said, very gravely, and go on till you come to the end: then stop.*
>
> Lewis Carroll, Alice's Adventures in Wonderland (1865)

Abstract The chapter presents a collection of essential facts and definitions frequently used throughout the text. We recall the definitions of the Skorokhod J_1-topology on the space of càdlàg functions, provide a brief overview of stable distributions and their domains of attraction, and discuss generalized inverse functions and their properties. Additionally, we review fundamental results on weak convergence of probability measures, including the continuous mapping theorem, the Skorokhod representation theorem, Donsker's invariance principle, and convergence of Markov processes. Towards the end of the chapter, we delve into Itô's excursion theory, a more advanced topic which is instrumental in our construction of a skew stable Lévy process. Typically, Itô's excursion theory is not covered in standard textbooks.

4.1 Probability Measures and Weak Convergence

Let (E, ρ) be a separable metric space and $\mathcal{E}$ its Borel sigma-algebra generated by the open subsets of (E, ρ). A sequence of probability measures $(\mathbb{P}_n)_{n \geq 1}$ on the measurable space $(E, \mathcal{E})$ converges *weakly* to a probability measure $\mathbb{P}_0$, if

$$\lim_{n \to \infty} \int_E f(x) \mathbb{P}_n(\mathrm{d}x) = \int_E f(x) \mathbb{P}_0(\mathrm{d}x) \tag{4.1}$$

for every $f \in C_b(E)$.

Let $(\Omega, \mathcal{F}, \mathbb{P})$ be a probability space and $(\theta_n)_{n \geq 0}$ a sequence of E-valued random elements defined on $(\Omega, \mathcal{F}, \mathbb{P})$ such that, for each $n \in \mathbb{N}_0$, θ_n has distribution $\mathbb{P}_n$. This means that $\theta_n : \Omega \to E$ is $\mathcal{F}/\mathcal{E}$-measurable and

$$\mathbb{P}\{\theta_n \in A\} = \mathbb{P}_n(A), \quad A \in \mathcal{E}, \quad n \in \mathbb{N}_0.$$

If (4.1) holds, then the random elements θ_n are said to converge in distribution to θ_0. This is denoted by

$$\theta_n \Longrightarrow \theta_0, \quad n \to \infty.$$

We proceed with a lemma that will prove useful on many occasions. Its proof that we omit follows from the definition of convergence in distribution and the corresponding results on convergence of sequences of real numbers from analysis.

Lemma 4.1.1

(a) Assume that a family of random elements $(\theta_v)_{v>0}$ is indexed by a continuous parameter v. Then

$$\theta_v \Longrightarrow \theta, \quad v \to \infty$$

if, and only if, for every sequence $(v_n)_{n \geq 1}$ satisfying $\lim_{n \to \infty} v_n = \infty$,

$$\theta_{v_n} \Longrightarrow \theta, \quad n \to \infty.$$

(b) The convergence in distribution

$$\theta_n \Longrightarrow \theta, \quad n \to \infty$$

holds if, and only if, for every subsequence $(\theta_{n_k})_{k \geq 1}$, there exists a further subsubsequence $(\theta_{n_{k_l}})_{l \geq 1}$ such that

$$\theta_{n_{k_l}} \Longrightarrow \theta, \quad l \to \infty.$$

A family $\mathcal{I}$ of probability measures on $(E, \mathcal{E})$ is called *relatively compact* if every sequence $(\mathbb{P}_n)_{n \geq 1} \subseteq \mathcal{I}$ contains a subsequence $(\mathbb{P}_{n_k})_{k \geq 1} \subseteq \mathcal{I}$ that converges to a probability measure Q on $(E, \mathcal{E})$. The measure Q need not be in $\mathcal{I}$.

The first part of Lemma 4.1.1 shows that when dealing with convergence in distribution one can work, without loss of generality, with sequences rather than families indexed by a continuous parameter. The second part will mainly be used in the following setting.

4.1 Probability Measures and Weak Convergence

Assume that the sequence of distributions of $(\theta_n)_{n\geq 1}$ is relatively compact, and we want to prove that $(\theta_n)_{n\geq 1}$ converges in distribution to a limit θ. Take any subsequence $(\theta_{n_k})_{k\geq 1}$ and select a convergent subsequence $(\theta_{n_{k_l}})_{l\geq 1}$, which exists thanks to relative compactness. Then we need to show that θ is the distributional limit of $(\theta_{n_{k_l}})_{l\geq 1}$. The fact that the sequence $(\theta_{n_{k_l}})_{l\geq 1}$ is convergent in distribution rather than just relatively compact often leads to technical simplifications. For instance, it allows an application of the Skorokhod representation theorem (Theorem 4.1.2).

A family of probability measures $\mathcal{I}$ is called *tight* if, for each $\varepsilon > 0$, there exists a compact set $K \subset E$ such that

$$\sup_{\mathbb{P}\in\mathcal{I}} \mathbb{P}\{E \setminus K\} \leq \varepsilon.$$

Theorem 4.1.1 (The Prokhorov Theorem) *If $\mathcal{I}$ is tight, then $\mathcal{I}$ is relatively compact. If $\mathcal{I}$ is relatively compact and E is complete and separable, then $\mathcal{I}$ is tight.*

The following two lemmas reveal additional properties of convergence in distribution in the case when the limit is nonrandom.

Lemma 4.1.2 *Let $(\theta_n)_{n\geq 1}$ be a sequence of random elements with values in a metric space E and $c \in E$ a nonrandom constant. Then convergence in distribution $\theta_n \Longrightarrow c$ as $n \to \infty$ is equivalent to convergence in probability $\theta_n \xrightarrow{\mathbb{P}} c$ as $n \to \infty$.*

Lemma 4.1.3 *Let $(\theta_n)_{n\geq 0}$ and $(\rho_n)_{n\geq 1}$ be sequences of random elements with values in separable metric spaces E and F, respectively, and $c \in F$ a nonrandom element. Assume that $\theta_n \Longrightarrow \theta_0$ and $\rho_n \Longrightarrow c$ as $n \to \infty$. Then there is also joint convergence in distribution of pairs on $E \times F$*

$$(\theta_n, \rho_n) \Longrightarrow (\theta_0, c), \quad n \to \infty.$$

The following fundamental theorem allows us to treat convergence in distribution as an a.s. convergence. Its proofs can be found in [151] or Theorem 3.30 in [77]. Recall that a sequence of E-valued random elements $(\theta_n)_{n\geq 1}$ converges $\mathbb{P}$-a.s. to θ_0, if

$$\mathbb{P}\{\omega : \lim_{n\to\infty} \theta_n(\omega) = \theta_0(\omega)\} = 1.$$

To simplify the notation we often suppress the dependence on the underlying probability measure $\mathbb{P}$ and simply write a.s. instead of $\mathbb{P}$-a.s.

Theorem 4.1.2 (The Skorokhod Representation Theorem) *Let $(\theta_n)_{n \geq 0}$ be a sequence of random elements taking values in a separable metric space E and*

$$\theta_n \Longrightarrow \theta_0, \quad n \to \infty.$$

Then there exists a probability space and a sequence $(\tilde{\theta}_n)_{n \geq 0}$ defined on this probability space such that $\tilde{\theta}_n \stackrel{d}{=} \theta_n$ for $n \in \mathbb{N}_0$ and

$$\lim_{n \to \infty} \tilde{\theta}_n = \tilde{\theta}_0 \quad \text{a.s.}$$

The Skorokhod representation theorem has various important corollaries, the next one being of utter importance.

Corollary 4.1.1 (Continuous Mapping Theorem) *Let $(\theta_n)_{n \geq 0}$ be a sequence of random elements taking values in a separable metric space E and satisfying*

$$\theta_n \Longrightarrow \theta_0, \quad n \to \infty,$$

and f a measurable map from E to a metric space F. Assume that f is continuous on a measurable set U such that $\mathbb{P}\{\theta_0 \in U\} = 1$. Then

$$f(\theta_n) \Longrightarrow f(\theta_0), \quad n \to \infty.$$

The proof immediately follows from Theorem 4.1.2 and the fact that the sequence of versions $(\tilde{\theta}_n)_{n \geq 0}$ satisfies

$$\lim_{n \to \infty} f(\tilde{\theta}_n) = f(\tilde{\theta}_0) \quad \text{a.s.}$$

A similar argument may be used to prove the following result which is a generalization of Slutsky's lemma.

Corollary 4.1.2 *Let $(\theta_n, \rho_n)_{n \geq 0}$ be a sequence of pairs of random elements taking values in a separable metric space E and satisfying*

$$(\theta_n, \rho_n) \Longrightarrow (\theta_0, c), \quad n \to \infty,$$

where c is a nonrandom constant. Also, let f be a measurable map from E to a metric space F. Assume that a measurable set U is such that $\mathbb{P}\{\theta_0 \in U\} = 1$ and, for each $x \in U$ and any sequence $(x_n, y_n) \to (x, c)$ as $n \to \infty$,

$$\lim_{n \to \infty} f(x_n, y_n) = f(x, c),$$

that is, f is continuous at (x, c) whenever $x \in U$. Then

$$f(\theta_n, \rho_n) \Longrightarrow f(\theta_0, c), \quad n \to \infty.$$

4.2 The Space of Continuous Functions $C([0, \infty), \mathbb{R}^d)$

4.2.1 Weak Convergence and the Locally Uniform Topology

Fix $T > 0$ and let $C([0, T], \mathbb{R}^d)$ be the space of continuous $\mathbb{R}^d$-valued functions defined on $[0, T]$ and endowed with the uniform topology generated by the metric

$$\rho_T(f, g) = \sup_{t \in [0, T]} \|f(t) - g(t)\|, \quad f, g \in C([0, T], \mathbb{R}^d).$$

Let $(X_n)_{n \geq 1}$ be a sequence of random elements taking values in $C([0, T], \mathbb{R}^d)$. We say that the sequence $(X_n)_{n \geq 1}$ converges weakly in the sense of finite-dimensional distributions to a random element $X_0 \in C([0, T], \mathbb{R}^d)$ if, for any $m \in \mathbb{N}$ and any $0 \leq t_1 < t_2 < \cdots < t_m \leq T$,

$$(X_n(t_1), \ldots, X_n(t_m)) \Longrightarrow (X_0(t_1), \ldots, X_0(t_m)), \quad n \to \infty. \qquad (4.2)$$

For $n \in \mathbb{N}$, let P_n denote the distribution of X_n. If (4.2) holds and the sequence $(P_n)_{n \geq 1}$ is tight, then the Prokhorov theorem implies that

$$X_n \Longrightarrow X_0, \quad n \to \infty$$

on $C([0, T], \mathbb{R}^d)$.

The case of an unbounded domain $[0, \infty)$ can be treated similarly. We endow the space $C([0, \infty), \mathbb{R}^d)$ of continuous $\mathbb{R}^d$-valued functions defined on $[0, \infty)$ with the topology of locally uniform convergence. This means that a sequence of functions $(f_n)_{n \geq 1}$, with $f_n \in C([0, \infty), \mathbb{R}^d)$ for $n \geq 1$, converges to $f_0 \in C([0, \infty), \mathbb{R}^d)$ if, and only if, for each fixed $T > 0$,

$$\lim_{n \to \infty} \sup_{t \in [0, T]} \|f_n(t) - f_0(t)\| = 0.$$

It is known that $C([0, \infty), \mathbb{R}^d)$ equipped with the topology of locally uniform convergence is metrizable as a complete separable metric space. One of the possible metrics is given by

$$\rho_\infty(f, g) = \int_0^\infty e^{-t} (\sup_{s \in [0, t]} \|f(s) - g(s)\| \wedge 1) dt.$$

A sequence $(X_n)_{n\geq 1}$ of random elements taking values in $C([0,\infty),\mathbb{R}^d)$ converges weakly to a random element X_0 if, and only if, for each fixed $T > 0$,

$$X_n\big|_{[0,T]} \Longrightarrow X_0\big|_{[0,T]}, \quad n \to \infty$$

on $C([0,T],\mathbb{R}^d)$. This fact follows, for instance, from the Skorokhod representation theorem.

4.2.2 The Wiener Measure and Donsker's Invariance Principle

We say that a random element $W = (W(t))_{t\geq 0}$ taking values in $C([0,\infty)) = C([0,\infty),\mathbb{R})$ is a *standard Brownian motion* if:

- For fixed $t > 0$, the random variable $W(t)$ has a normal distribution with mean zero and variance t, that is,

$$\mathbb{P}\{W(t) \in dx\} = \frac{1}{\sqrt{2\pi t}} e^{-x^2/(2t)} dx, \quad x \in \mathbb{R}.$$

- For any $m \in \mathbb{N}$ and any $0 \leq t_0 \leq t_1 \leq \cdots \leq t_m < \infty$, the random variables

$$W(t_1) - W(t_0), W(t_2) - W(t_1), \ldots, W(t_m) - W(t_{m-1})$$

are mutually independent.

The distribution of W is called the *Wiener measure*. The proof of existence and uniqueness of such a measure on $C([0,\infty))$ can be found in Section 8 of [12].

The importance of Brownian motion is revealed by Donsker's invariance principle.

Theorem 4.2.1 (Donsker's Theorem in $C([0,\infty))$) *Let $(\xi_n)_{n\geq 1}$ be a sequence of independent copies of a random variable ξ with mean zero and $\mathbb{E}\xi^2 = \sigma^2 \in (0,\infty)$. Put*

$$S_\xi(0) = 0, \quad S_\xi(n) := \xi_1 + \ldots + \xi_n, \quad n \in \mathbb{N}$$

and then

$$X_n(t) := \frac{1}{\sigma\sqrt{n}} S_\xi(\lfloor nt \rfloor) + (nt - \lfloor nt \rfloor)\frac{1}{\sigma\sqrt{n}} \xi_{\lfloor nt \rfloor + 1}, \quad t \geq 0,$$

4.3 The Skorokhod Space $D([0, \infty), \mathbb{R}^d)$

so that X_n is a continuous function which linearly interpolates the points

$$\left(\frac{i}{n}, \frac{S_\xi(i)}{\sigma\sqrt{n}}\right), \quad i \in \mathbb{N}_0.$$

Then

$$X_n \implies W, \quad n \to \infty$$

on $C([0, \infty))$.

4.3.1 The J_1-Topology

Fix $T > 0$ and denote by $D([0, T], \mathbb{R}^d)$ the space of right-continuous $\mathbb{R}^d$-valued functions with finite left limits at all points of $(0, T]$. This space is often called the space of càdlàg functions which is an acronym for the French "continue à droite, limite à gauche."

The space $D([0, T], \mathbb{R}^d)$ can be endowed with various topologies. However, throughout the book we only use the ubiquitous Skorokhod J_1-topology, whose construction we are now going to recall. The following equalities determine equivalent metrics on $D([0, T], \mathbb{R}^d)$:

$$d^T(f, g) := \inf_{\lambda \in \Lambda_T} \max\left(\sup_{t \in [0, T]} |f(\lambda(t)) - g(t)|, \sup_{t \in [0, T]} |\lambda(t) - t|\right),$$

$$d_0^T(f, g) := \inf_{\lambda \in \Lambda_T} \max\left(\sup_{t \in [0, T]} |f(\lambda(t)) - g(t)|, \sup_{t \neq s} \left|\log\frac{\lambda(t) - \lambda(s)}{t - s}\right|\right),$$

where Λ_T is the set of strictly increasing continuous mappings of $[0, T]$ onto itself

The following result is borrowed from Theorem 12.2 in [12].

Proposition 4.3.1 *The space $D([0, T], \mathbb{R}^d)$ endowed with the metric $d_0^T(f, g)$ is complete and separable.*

The topology generated by the metric d_0^T is called the *Skorokhod J_1-topology* on $D([0, T], \mathbb{R}^d)$. An important remark is in order here. If $d \geq 2$, then $D([0, T], \mathbb{R}^d)$ as a set coincides with the Cartesian product $(D([0, T], \mathbb{R}))^d$. However, the Skorokhod J_1-topology on $D([0, T], \mathbb{R}^d)$ is strictly stronger than the product topology on $(D([0, T], \mathbb{R}))^d$.

The passage from a compact domain $[0, T]$ to $[0, \infty)$ is essentially the same as for the space of continuous functions. In what follows, $D([0, \infty), \mathbb{R}^d)$ denotes the space of right-continuous $\mathbb{R}^d$-valued functions defined on $[0, \infty)$ and having left limits at all positive reals. However, one needs to be careful with continuity at T.

The next theorem is due to Lindvall; see Theorem 1(b) in [99].

Proposition 4.3.2 *For each $n \in \mathbb{N}_0$, let $f_n \in D([0, \infty), \mathbb{R}^d)$. There exists a complete and separable metric d^∞ on $D([0, \infty), \mathbb{R}^d)$ such that the following statements are equivalent:*

(I) $\lim_{n \to \infty} f_n = f_0$ *on* $(D([0, \infty), \mathbb{R}^d), d^\infty)$.
(II) There exist

$$\lambda_n \in \Lambda_{0,\infty} := \{\lambda : \lambda \text{ is a strictly increasing continuous function}$$
$$\text{defined on } [0, \infty) \text{ such that } \lambda(0) = 0, \lambda(+\infty) = +\infty\}$$

which satisfy, for each $T > 0$,

$$\lim_{n \to \infty} \max \left\{ \sup_{u \in [0, T]} |f_n(\lambda_n(u)) - f_0(u)|, \sup_{u \in [0, T]} |\lambda_n(u) - u| \right\} = 0.$$

(III) For each $T > 0$ which is a continuity point of f_0, $\lim_{n \to \infty} f_n\big|_{[0, T]} = f_0\big|_{[0, T]}$ on $(D([0, T], \mathbb{R}^d), d^T)$.

The topology generated by d^∞ is called the J_1-topology on $D([0, \infty), \mathbb{R}^d)$. We proceed by giving another characterization of the J_1-convergence, which can be found in Proposition 6.5 of [49].

Proposition 4.3.3 *For each $n \in \mathbb{N}_0$, let $f_n \in D([0, \infty), \mathbb{R}^d)$. Then $\lim_{n \to \infty} f_n = f_0$ in the J_1-topology on $D([0, \infty), \mathbb{R}^d)$ if, and only if, for each $u_0 \geq 0$ and any sequence $(u_n)_{n \geq 1}$ of nonnegative numbers satisfying $\lim_{n \to \infty} u_n = u_0$, the following conditions hold:*

(I) The set of limit points of $(f_n(u_n))_{n \geq 1}$ is a subset of $\{f_0(u_0), f_0(u_0-)\}$.
(II) If $\lim_{n \to \infty} f_n(u_n) = f_0(u_0)$, then $\lim_{n \to \infty} f_n(v_n) = f_0(u_0)$ for any sequence $(v_n)_{n \geq 1}$ satisfying $v_n \geq u_n$ for $n \in \mathbb{N}$ and $\lim_{n \to \infty} v_n = u_0$.
(III) If $\lim_{n \to \infty} f_n(u_n) = f_0(u_0-)$, then $\lim_{n \to \infty} f_n(v_n) = f_0(u_0-)$ for any sequence $(v_n)_{n \geq 1}$ satisfying $v_n \leq u_n$ for $n \in \mathbb{N}$ and $\lim_{n \to \infty} v_n = u_0$.

It can be checked that the locally uniform convergence on $D([0, \infty), \mathbb{R}^d)$ implies the J_1-convergence on $D([0, \infty), \mathbb{R}^d)$. Moreover, as the next proposition demonstrates, the two types of convergence are equivalent whenever the limit function is continuous.

4.3 The Skorokhod Space $D([0, \infty), \mathbb{R}^d)$

Proposition 4.3.4 *Assume that* $\lim_{n \to \infty} f_n = f_0$ *on* $D([0, \infty), \mathbb{R}^d)$, *where* f_0 *is a continuous function. Then, for all* $T > 0$,

$$\lim_{n \to \infty} \sup_{t \in [0, T]} |f_n(t) - f_0(t)| = 0.$$

4.3.2 Convergence of Probability Measures on $D([0, \infty), \mathbb{R}^d)$

Working with random elements of $D([0, \infty), \mathbb{R}^d)$ requires choosing an appropriate sigma-algebra on $D([0, \infty), \mathbb{R}^d)$. For our purposes the usual Borel sigma-algebra generated by open subsets of $D([0, \infty), \mathbb{R}^d)$ with respect to the J_1-topology suffices. We note that the space of continuous functions $C([0, \infty), \mathbb{R}^d)$ is a (Borel) measurable subset of $D([0, \infty), \mathbb{R}^d)$ and so are the sets of increasing, decreasing, and bounded càdlàg functions, as well as the other spaces which appear in the present book; see, for instance, p. 429 in [175].

Similarly to what has been stated for the space $C([0, \infty), \mathbb{R}^d)$, the Prokhorov theorem implies that a sequence of $D([0, \infty), \mathbb{R}^d)$-valued random elements $(X_n)_{n \geq 1}$ converges in distribution to $X_0 \in D([0, \infty), \mathbb{R}^d)$ if, and only if, the sequence of distributions of $(X_n)_{n \geq 1}$ is tight and (4.2) holds for all $t_1, \ldots, t_m$ in which X_0 is a.s. continuous.

A version of Donsker's invariance principle in the space $D([0, \infty), \mathbb{R})$ reads as follows:

Theorem 4.3.1 (Donsker's Theorem in $D([0, \infty), \mathbb{R})$) *Under the assumptions and notation of Theorem 4.2.1,*

$$\left(\frac{S_\xi(\lfloor nt \rfloor)}{\sigma \sqrt{n}} \right)_{t \geq 0} \Longrightarrow (W(t))_{t \geq 0}, \quad n \to \infty$$

in the J_1-topology on $D([0, \infty), \mathbb{R})$.

4.3.3 Continuity of Mappings on $D([0, \infty), \mathbb{R}^d)$

The next result follows from Proposition 4.3.3.

Proposition 4.3.5 (Continuity of Concatenations) *For each* $n \in \mathbb{N}_0$, *let* $f_n, g_n \in D([0, \infty), \mathbb{R}^d)$ *and* $t_n \geq 0$. *Assume that:*

- $\lim_{n \to \infty} t_n = t_0 > 0$.
- $\lim_{n \to \infty} f_n(\cdot \wedge t_n) = f_0(\cdot \wedge t_0)$ *in the J_1-topology on* $D([0, \infty), \mathbb{R}^d)$.
- $\lim_{n \to \infty} g_n(t_n + \cdot) = g_0(t_0 + \cdot)$ *in the J_1-topology on* $D([0, \infty), \mathbb{R}^d)$.

For $n \in \mathbb{N}_0$, $t \geq 0$, put $h_n(t) := f_n(t)\mathbb{1}_{\{t < t_n\}} + g_n(t)\mathbb{1}_{\{t \geq t_n\}}$. Then $\lim_{n \to \infty} h_n = h_0$ in the J_1-topology on $D([0, \infty), \mathbb{R}^d)$.

We frequently use in the text continuity of sums in $D([0, \infty), \mathbb{R}^d)$ as discussed in Theorem 4.1 of [174] and reproduced in the following proposition. For $f \in D([0, \infty), \mathbb{R}^d)$, denote by $\mathrm{Disc}(f) := \{a \in (0, \infty) : f(a-) \neq f(a)\}$ the set of discontinuities of f.

Proposition 4.3.6 (Continuity of Sums) *For $n \in \mathbb{N}_0$, let $f_n, g_n \in D([0, \infty), \mathbb{R}^d)$.*

- *If $\lim_{n \to \infty} f_n = f_0$ and $\lim_{n \to \infty} g_n = g_0$ in the J_1-topology on $D([0, \infty), \mathbb{R}^d)$, and $\mathrm{Disc}(f_0) \cap \mathrm{Disc}(g_0) = \varnothing$, then $\lim_{n \to \infty}(f_n + g_n) = f_0 + g_0$ in the J_1-topology on $D([0, \infty), \mathbb{R}^d)$.*
- *If $\lim_{n \to \infty}(f_n, g_n) = (f_0, g_0)$ in the J_1-topology on $D([0, \infty), \mathbb{R}^{2d})$, then $\lim_{n \to \infty}(f_n + g_n) = f_0 + g_0$ in the J_1-topology on $D([0, \infty), \mathbb{R}^d)$.*

The next four results are borrowed from Theorem 13.2.2 on p. 430 in [175], Proposition 2.3 in [155], and Lemmas 3.3 and 3.5 in [71], respectively.

Proposition 4.3.7 (Convergence of Compositions I) *For $n \in \mathbb{N}_0$, let $(f_n, g_n) \in D([0, \infty), \mathbb{R}^d) \times D([0, \infty), \mathbb{R})$. Assume that, for $n \in \mathbb{N}$, g_n are nonnegative and nondecreasing; either g_0 is continuous and strictly increasing, or f_0 is continuous; and $\lim_{n \to \infty}(f_n, g_n) = (f_0, g_0)$ in the product J_1-topology on $D([0, \infty), \mathbb{R}^d) \times D([0, \infty), \mathbb{R})$. Then $\lim_{n \to \infty} f_n \circ g_n = f_0 \circ g_0$ in the J_1-topology on $D([0, \infty), \mathbb{R}^d)$, where $\circ$ denotes composition.*

For a nondecreasing càdlàg unbounded function g we denote by $g^{\leftarrow}$ its generalized inverse defined by $g^{\leftarrow}(t) := \inf\{s \geq 0 : g(s) > t\}$; see Sect. 4.3.4 for some properties of generalized inverse functions.

Proposition 4.3.8 (Convergence of Compositions II) *For $n \in \mathbb{N}_0$, let $(f_n, g_n) \in D([0, \infty), \mathbb{R}^d) \times D([0, \infty), \mathbb{R})$. Assume that, for $n \in \mathbb{N}_0$, g_n are nonnegative, nondecreasing, and unbounded; g_0 is strictly increasing; and $\lim_{n \to \infty}(f_n, g_n) = (f_0, g_0)$ in the J_1-topology on $D([0, \infty), \mathbb{R}^d \times \mathbb{R})$. Then $\lim_{n \to \infty} f_n \circ g_n^{\leftarrow} = f_0 \circ g_0^{\leftarrow}$ in the J_1-topology on $D([0, \infty), \mathbb{R}^d)$.*

Proposition 4.3.9 (Convergence of Compositions III) *For $n \in \mathbb{N}_0$, let $(f_n, g_n) \in D([0, \infty), \mathbb{R}) \times D([0, \infty), \mathbb{R})$. Assume that, for $n \in \mathbb{N}$, g_n are nonnegative and nondecreasing; g_0 is continuous; if $g_0(t) \in \mathrm{Disc}(f_0)$ for some $t \geq 0$, then $g_0(u) \neq g_0(t)$ for all $u \neq t$; and $\lim_{n \to \infty}(f_n, g_n) = (f_0, g_0)$ in the product J_1-topology on $D([0, \infty), \mathbb{R}) \times D([0, \infty), \mathbb{R})$. Then $\lim_{n \to \infty} f_n \circ g_n = f_0 \circ g_0$ in the J_1-topology on $D([0, \infty), \mathbb{R})$.*

4.3 The Skorokhod Space $D([0, \infty), \mathbb{R}^d)$

Proposition 4.3.10 (Convergence of Compositions IV) *For $n \in \mathbb{N}_0$, let $(f_n, g_n) \in D([0, \infty), \mathbb{R}) \times D([0, \infty), \mathbb{R})$. Assume that, for $n \in \mathbb{N}$, g_n are nonnegative and nondecreasing, and that, for all $T > 0$,*

$$\lim_{n \to \infty} \sup_{t \in [0, T]} |g_n(t) - t| = 0, \tag{4.3}$$

and

$$\lim_{n \to \infty} f_n \circ g_n = f_0 \tag{4.4}$$

in the J_1-topology on $D([0, \infty), \mathbb{R})$. For $n \in \mathbb{N}$, denote by $(t_k^{(n)})_{k \geq 1}$ elements of the set $\mathrm{Disc}(g_n)$ and, for $k \in \mathbb{N}$, put $u_k^{(n)} := g_n(t_k^{(n)}-)$ and $v_k^{(n)} := g_n(t_k^{(n)})$. If, in addition to (4.3) and (4.4), for all $T > 0$,

$$\lim_{n \to \infty} \sup_{k \geq 1} \sup_{s \in [u_k^{(n)}, v_k^{(n)}] \cap [0, T]} |f_n(s) - f_n(u_k^{(n)}-)| = 0, \tag{4.5}$$

then

$$\lim_{n \to \infty} f_n = f_0$$

in the J_1-topology on $D([0, \infty), \mathbb{R})$.

Remark 4.3.1 Let $((f_n, g_n))_{n \geq 0}$ be a sequence of random processes, whose paths a.s. satisfy the assumptions of one of the four propositions above (under the assumptions of Proposition 4.3.9 we put $d = 1$). An appeal to Theorem 4.1.2 enables us to deduce the weak convergence $f_n \circ g_n \Longrightarrow f_0 \circ g_0$ or $f_n \circ g_n^{\leftarrow} \Longrightarrow f_0 \circ g_0^{\leftarrow}$ on $D([0, \infty), \mathbb{R}^d)$ as $n \to \infty$.

Theorem 4.1.2 is applicable in the settings of Propositions 4.3.7–4.3.10 because all the function spaces appearing in the text (the spaces of continuous functions, monotone functions, bounded càdlàg functions) are measurable subsets of $D([0, \infty), \mathbb{R}^d)$; see, for instance, p. 429 in [175].

Our last group of results in this section is concerned with continuity of suprema functionals. As in Theorem 2.1.2, for $f \in D([0, \infty), \mathbb{R})$, we denote by $m(f)$ the running supremum of the negative part of f:

$$(m(f))(t) = \sup_{0 \leq s \leq t} ((f(s))_-), \quad t \geq 0.$$

Also, let $m^*(f)$ denote the running supremum of f itself:

$$(m^*(f))(t) := \sup_{0 \leq s \leq t} f(s), \quad t \geq 0,$$

so that $m(f) = m^*(f_-)$. Proposition 4.3.11 is a consequence of Theorem 13.4.1 in [175] and the fact that $\lim_{n\to\infty} f_n = f_0$ in the J_1-topology on $D([0,\infty), \mathbb{R})$ implies that $\lim_{n\to\infty} (f_n)_- = (f_0)_-$ in the J_1-topology on $D([0,\infty), \mathbb{R})$.

Proposition 4.3.11 *For $n \in \mathbb{N}_0$, let $f_n \in D([0,\infty), \mathbb{R})$. Assume that $\lim_{n\to\infty} f_n = f_0$ in the J_1-topology on $D([0,\infty), \mathbb{R})$. Then*

$$\lim_{n\to\infty} m^*(f_n) = m^*(f_0) \quad \text{and} \quad \lim_{n\to\infty} m(f_n) = m(f_0)$$

in the J_1-topology on $D([0,\infty), \mathbb{R})$.

A proof of the first half of the next proposition (formula (4.6)) can be found in Example 12.1 in [12]. For the second half (formula (4.7)) see the bottom of p. 296 in [175].

Proposition 4.3.12 *For $n \in \mathbb{N}_0$, let $f_n \in D([0,\infty), \mathbb{R})$. Assume that $\lim_{n\to\infty} f_n = f_0$ in the J_1-topology on $D([0,\infty), \mathbb{R})$, and that f_0 is continuous at $T > 0$. Then*

$$\lim_{n\to\infty} \sup_{t \in [0,T]} |f_n(t) - f_n(t-)| = \sup_{t \in [0,T]} |f_0(t) - f_0(t-)| \qquad (4.6)$$

and

$$\lim_{n\to\infty} \inf_{t \in [0,T]} (f_n(t) - f_n(t-)) = \inf_{t \in [0,T]} (f_0(t) - f_0(t-)). \qquad (4.7)$$

In particular, if f_0 does not have negative jumps, then the limit of infima of the negative jumps of f_n is equal to 0.

4.3.4 Generalized Inverse Functions

We collect in this section various properties of the generalized inverse functions which are an important ingredient of the generalized Skorokhod reflection problem. Our presentation is based, for the most part, on Chapter 13.6 of [175] and [46].

Let F be a nondecreasing càdlàg unbounded function defined on $[0, \infty)$ with $F(0) \geq 0$. Recall that we have defined the generalized inverse function of F by

$$F^{\leftarrow}(t) = \inf\{s \geq 0 : F(s) > t\}, \quad t \geq 0.$$

4.3 The Skorokhod Space $D([0, \infty), \mathbb{R}^d)$

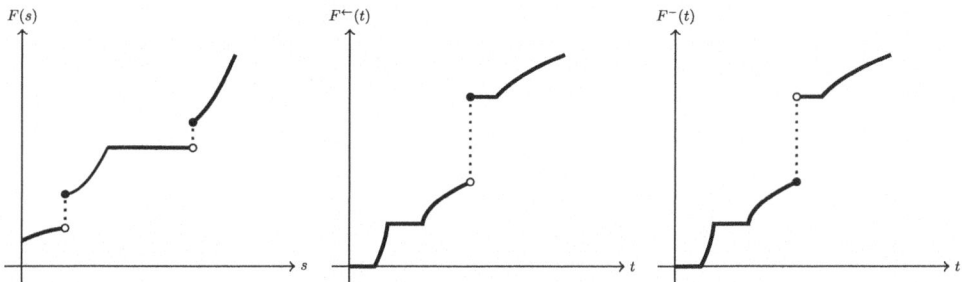

Fig. 4.1 A function F and its generalized inverses $F^{\leftarrow}$ and F^-

One can find in the literature an alternative definition with the non-strict inequality. Such an inverse will be denoted by F^-, that is,

$$F^-(t) := \inf\{s \geq 0 : F(s) \geq t\}, \quad t \geq 0. \tag{4.8}$$

Since F is assumed càdlàg, the generalized inverse $F^{\leftarrow}$ is also càdlàg, whereas F^- is left-continuous with finite limits from the right (càglàd); see Lemma 13.6.2 in [175]. The two functions $F^{\leftarrow}$ and F^- differ at most at countably many points and are related by

$$F^{\leftarrow}(t) = F^-(t+), \quad F^{\leftarrow}(t-) = F^-(t), \quad t \geq 0.$$

We stipulate hereafter that $F(0-) = 0$. To draw a graph of $F^{\leftarrow}$ we reflect the graph of F with respect to the diagonal of the first quadrant and take into account that jumps of F correspond to intervals of constancy of $F^{\leftarrow}$, whereas intervals of constancy of F correspond to jumps of $F^{\leftarrow}$; see Fig. 4.1 and Proposition 4.3.13 (V). From the definition it is clear that

$$0 \leq F^{\leftarrow}(F(t)) \leq t \quad \text{and} \quad t \leq F(F^{\leftarrow}(t)), \quad t \geq 0.$$

The values of the functions

$$[0, \infty) \ni t \mapsto F(F^{\leftarrow}(t)) - t \in [0, \infty) \quad \text{and} \quad [0, \infty) \ni t \mapsto t - F((F^{\leftarrow}(t))-) \in [0, \infty)$$

at a point $t_0 \geq 0$ are called the *overshoot* and the *undershoot* of F at t_0, respectively; see Fig. 4.2.

The following result is a combination of Lemma 13.6.4 in [175] and Proposition 1 in [46].

Proposition 4.3.13 *Assume that* $F : [0, \infty) \to [0, \infty)$ *is a nondecreasing, càdlàg, and unbounded function with* $F(0) \geq 0$. *Then, for* $t \geq 0$,

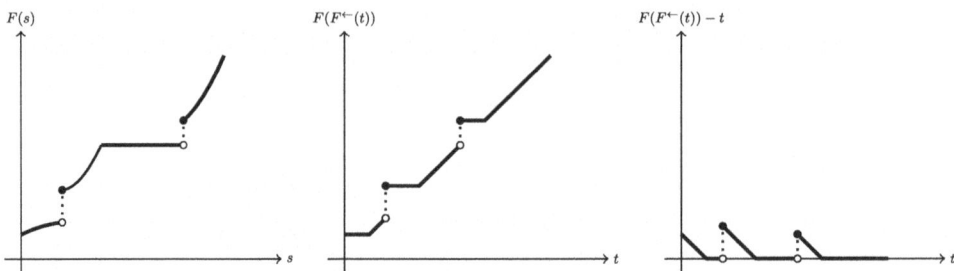

Fig. 4.2 A function F, the composition $t \mapsto F(F^{\leftarrow}(t))$, and the overshoot $t \mapsto F(F^{\leftarrow}(t)) - t$

(I) $F(s) \geq t$ if, and only if, $F^-(t) \leq s$.
(II) $0 \leq F^{\leftarrow}(F(t)) - t \leq F^{\leftarrow}(F(t)) - F^{\leftarrow}((F(t))-)$; in particular, $F^{\leftarrow}(F(t)) = t$ if, and only if, $F(t)$ is a point of continuity of $F^{\leftarrow}$.
(III) $0 \leq F(F^{\leftarrow}(t)) - t \leq F(F^{\leftarrow}(t)) - F((F^{\leftarrow}(t))-)$; in particular, $F(F^{\leftarrow}(t)) = t$ if, and only if, $F^{\leftarrow}(t)$ is a point of continuity of F.
(IV) $(F^-(t), F^-(t+)) \subseteq \{s \geq 0 : F(s) = t\} \subseteq [F^-(t), F^-(t+)]$.
(V) $(F^{\leftarrow}(t-), F^{\leftarrow}(t)) \subseteq \{s \geq 0 : F(s) = t\} \subseteq [F^{\leftarrow}(t-), F^{\leftarrow}(t)]$.
(VI) F is continuous if, and only if, either of the two equivalent conditions holds:
  (VI1) $F^{\leftarrow}$ is strictly increasing.
  (VI2) F^- is strictly increasing.
(VII) F is strictly increasing if, and only if, either of the two equivalent conditions holds:
  (VII1) $F^{\leftarrow}$ is continuous.
  (VII2) F^- is continuous.

The following result establishes continuity of the inverse mapping $F \mapsto F^{\leftarrow}$. Note that if the range is endowed with the J_1-topology, the inverse mapping is not continuous everywhere; see Example 13.6.1 in [175]. The proposition below is a combination of Corollary 13.6.3 and Corollary 13.6.4 in [175] in which we replaced weaker M_1- and M_2-topologies with the stronger J_1-topology.

Proposition 4.3.14 *For $n \in \mathbb{N}_0$, let $F_n \in D([0, \infty), \mathbb{R})$, $F_n(0) \geq 0$ and F_n is unbounded. Assume that $\lim_{n \to \infty} F_n = F_0$ in the J_1-topology on $D([0, \infty), \mathbb{R})$.*

(I) *If $F_0^{\leftarrow}(0) = 0$, then $\lim_{n \to \infty} F_n^{\leftarrow} = F_0^{\leftarrow}$ in the M_1-topology on $D([0, \infty), \mathbb{R})$.*
(II) *If F_0 is strictly increasing, then $\lim_{n \to \infty} F_n^{\leftarrow} = F_0^{\leftarrow}$ in the J_1-topology on $D([0, \infty), \mathbb{R})$. Moreover,[1] $\lim_{n \to \infty} F_n^{\leftarrow} = F_0^{\leftarrow}$ locally uniformly on $[0, \infty)$. Also, the last limit relation holds true, with F^- replacing $F^{\leftarrow}$.*

---

[1] This fragment is not stated in the aforementioned corollaries from [175]. It follows from the fact that $F_0^{\leftarrow}$ is continuous; see Proposition 4.3.13(VI).

4.4 Stable Distributions

A real-valued random variable X is said to have a *stable distribution* if, for any positive numbers a_1 and a_2, there exist a positive number c and a real number b such that

$$a_1 X_1 + a_2 X_2 \stackrel{\mathrm{d}}{=} cX + b, \tag{4.9}$$

where X_1 and X_2 are independent copies of X. It is known that the constants a_1, a_2 and c ought to satisfy

$$a_1^\gamma + a_2^\gamma = c^\gamma \tag{4.10}$$

for some $\gamma \in (0, 2]$. Moreover, (4.9) implies that there exists a sequence $(b_n)_{n \geq 1}$ such that

$$X_1 + X_2 + \cdots + X_n \stackrel{\mathrm{d}}{=} n^{1/\gamma} X + b_n, \quad n \in \mathbb{N},$$

where $(X_n)_{n \geq 1}$ is a sequence of independent copies of X. A stable distribution is called *strictly stable*, if (4.9) holds with $b = 0$.

In what follows we adhere to the terminology used in the book [145]; see also Chapter 4.5 in [175]. The family of stable distributions is parameterized by the four parameters $(\gamma, \mathfrak{s}, \beta, \mu)$. Here, $\gamma \in (0, 2]$ is as in (4.10), and it is called the *index of stability*; $\mathfrak{s} > 0$ is the *scale parameter*, $\beta \in [-1, 1]$ is the *skewness parameter*, and $\mu \in \mathbb{R}$ is the *shift parameter*. The characteristic function of a stable distribution $U_\gamma^{\mathfrak{s},\beta,\mu}$ with parameters $(\gamma, \mathfrak{s}, \beta, \mu)$ is given by

$$\int_\mathbb{R} \exp(izx) U_\gamma^{\mathfrak{s},\beta,\mu}(\mathrm{d}x)$$
$$= \begin{cases} \exp(-\mathfrak{s}^\gamma |z|^\gamma (1 - i\beta(\operatorname{sign} z)\tan(\pi\gamma/2)) + i\mu z), & \text{if } \gamma \in (0, 1) \cup (1, 2]; \\ \exp(-\mathfrak{s}|z|(1 + i\beta(2/\pi)(\operatorname{sign} z)\log|z|) + i\mu z), & \text{if } \gamma = 1 \end{cases}$$
(4.11)

for $z \in \mathbb{R}$. The distribution $U_\gamma^{\mathfrak{s},\beta,\mu}$ is strictly stable, if either $\gamma \neq 1$ and $\mu = 0$ or $\gamma = 1$ and $\beta = 0$. The distribution $U_\gamma^{\mathfrak{s},0,0}$ is called *symmetric stable*. The distribution $U_2^{\mathfrak{s},\beta,\mu}$ is normal with mean μ and variance $2\mathfrak{s}^2$, irrespective of β. The distribution $U_\gamma^{\mathfrak{s},1,\mu}$ is called *totally skewed to the right* or *spectrally positive*, and the distribution $U_\gamma^{\mathfrak{s},-1,\mu}$ is called *totally skewed to the left* or *spectrally negative*. The distributions $U_\gamma^{\mathfrak{s},1,0}$ and $U_\gamma^{\mathfrak{s},-1,0}$) with $\gamma \in (0, 1)$ are concentrated on the positive and the negative halflines, respectively, and $\int_{[0,\infty)} \exp(-vx) U_\gamma^{\mathfrak{s},1,0}(\mathrm{d}x) = \exp(-(\mathfrak{s}v)^\gamma / \cos((\pi\gamma)/2))$ for $v \geq 0$.

For $\gamma \in (0, 2)$, a version of the Lévy-Khintchine formula for the characteristic function of $U_\gamma^{\mathfrak{s},\beta,\mu}$ reads

$$\int_\mathbb{R} \exp(izx) U_\gamma^{\mathfrak{s},\beta,\mu}(dx) = \exp\left(i\mathfrak{m}z + \int_\mathbb{R} \left(e^{izx} - 1 - izx\mathbb{1}_{(-1,1)}(x)\right)\Pi(dx)\right),$$

where $\mathfrak{m} \in \mathbb{R}$ is an appropriate constant, and the Lévy measure Π is given by

$$\Pi(dx) := \begin{cases} c_+ x^{-1-\gamma} dx, & x > 0, \\ c_- |x|^{-1-\gamma} dx, & x < 0 \end{cases} \quad (4.12)$$

with $c_\pm$ being nonnegative constants which are determined by γ, $\mathfrak{s}$, and β and take different forms depending on whether $\gamma \in (0, 1)$, $\gamma = 1$, and $\gamma \in (1, 2)$. For instance, $c_\pm := \mathfrak{s}(1 \pm \beta)/\pi$ in the case $\gamma = 1$. Here is a concrete example. The distribution $U_1^{\pi/2,-1,0}$ is a spectrally negative stable distribution with the characteristic function $z \mapsto \exp(-(\pi/2)|z| + iz\log|z|)$ for $z \in \mathbb{R}$ and the Lévy measure Π defined by $\Pi(dx) = |x|^{-2} \mathbb{1}_{(-\infty,0)}(x) dx$.

The characteristic function of the symmetric stable (non-Gaussian) distribution $U_\gamma^{\mathfrak{s},0,0}$ takes the form $\int_\mathbb{R} \exp(ixz) U_\gamma^{\mathfrak{s},0,0}(dx) = \exp(-\mathfrak{s}^\gamma |z|^\gamma)$ for $z \in \mathbb{R}$ and corresponds to the choice $\mathfrak{m} = 0$ and

$$c_\pm = \begin{cases} \frac{\mathfrak{s}^\gamma \gamma(1-\gamma)}{2\Gamma(2-\gamma)\cos((\pi\gamma)/2)}, & \text{if } \gamma \in (0,1) \cup (1,2); \\ \frac{\mathfrak{s}}{\pi}, & \text{if } \gamma = 1, \end{cases}$$

in (4.12). The explicit form of the constants $c_\pm$ follows from equations (3.18), (3.19), and (5.12) in Chapter XVII.5 of [52]. In the case $\gamma \in (1, 2)$ it can also be found by direct calculations with the help of formula (2.61).

4.4.1 Domains of Attraction of Stable Distributions

The distribution of a real-valued random variable τ is said to belong to *the domain of attraction* of the stable distribution $U_\gamma^{\mathfrak{s},\beta,\mu}$, if there exist functions $c : (0, \infty) \to (0, \infty)$ and $b : (0, \infty) \to \mathbb{R}$ such that the variables $(S_\tau(n) - b(n))/c(n)$ converge in distribution as $n \to \infty$ to a random variable with the distribution $U_\gamma^{\mathfrak{s},\beta,\mu}$. Here, as before, $S_\tau(n) = \tau_1 + \ldots + \tau_n$ and $\tau_1, \tau_2, \ldots$ are independent copies of τ. By adjusting b and c, one may put, without loss of generality, $\mathfrak{s} = 1$ and $\mu = 0$.

The following classical theorem goes back to Gnedenko and Kolmogorov [58]. The present form is borrowed from Theorem 4.5.1 in [175].

4.4 Stable Distributions

Theorem 4.4.1 *The distribution of a random variable τ belongs to the domain of attraction of $U_\gamma^{1,\beta,0}$ for some $\gamma \in (0, 2)$ and some $\beta \in [-1, 1]$ if, and only if, the following two conditions hold true:*

(I) *The function $x \mapsto \mathbb{P}\{|\tau| > x\}$ is regularly varying at infinity of index $-\gamma$.*
(II)
$$\lim_{x \to +\infty} \frac{\mathbb{P}\{\tau > x\}}{\mathbb{P}\{|\tau| > x\}} = \frac{1+\beta}{2}.$$

The normalizing function c is any function satisfying

$$\lim_{x \to \infty} x \mathbb{P}\{|\tau| > c(x)\} = C_\gamma, \qquad (4.13)$$

where

$$C_\gamma := \begin{cases} \frac{1-\gamma}{\Gamma(2-\gamma)\cos(\pi\gamma/2)}, & \text{if } \gamma \neq 1, \\ 2/\pi, & \text{if } \gamma = 1 \end{cases}$$

and Γ is the gamma function. The centering function b can be chosen according to the rule

$$b(x) := \begin{cases} 0, & \text{if } \gamma \in (0, 1), \\ xc(x) \int_\mathbb{R} \sin(y/c(x)) \mathbb{P}\{\tau \in dy\}, & \text{if } \gamma = 1, \\ (\mathbb{E}\tau)x, & \text{if } \gamma \in (1, 2) \end{cases}$$

for $x > 0$.

Remark 4.4.1 It is known that functions c satisfying (4.13) do exist. If, for instance, $\mathbb{P}\{|\tau| > x\} \sim Ax^{-\gamma}$ as $x \to \infty$ for $\gamma \in (0, 2)$ and a constant $A \in (0, \infty)$, then one may take $c(x) = ((A/C_\gamma)x)^{1/\gamma}$. In general, c is a function which is regularly varying at ∞ of index $1/\gamma$. The distribution of a real-valued random variable τ is said to belong to the *normal domain of attraction* of a γ-stable distribution, if $c(x) = \text{const}\, x^{1/\gamma}$.

Without any extra assumptions, the one-dimensional convergence appearing in the definition of the domain of attraction can be strengthened to weak convergence in the Skorokhod space. By a *Lévy process* we understand a real-valued càdlàg process starting at 0 and having independent and stationary increments. A Lévy process is called *stable* if its one-dimensional distributions are stable.

Theorem 4.4.2 *Under the assumptions and notation of Theorem 4.4.1,*

$$\left(\frac{S_\tau(\lfloor vt \rfloor) - b(v)t}{c(v)}\right)_{t\geq 0} \Longrightarrow (\mathcal{S}_\gamma(t))_{t\geq 0}, \quad v \to \infty$$

in the J_1-topology on $D([0, \infty), \mathbb{R})$, where $\mathcal{S}_\gamma = (\mathcal{S}_\gamma(t))_{t\geq 0}$ is a stable Lévy process such that $\mathcal{S}_\gamma(1)$ has the distribution $U_\gamma^{1,\beta,0}$.

4.5 Convergence of Markov Processes

Let $Z = (Z(t))_{t\geq 0}$ be a time-homogeneous Markov process taking values in a locally compact separable metric space E with a family of transition probabilities

$$P(t, x, A) = \mathbb{P}\{Z(t) \in A \mid Z(0) = x\}$$

for $t \geq 0$, $x \in E$ and Borel sets $A \subseteq E$. Introduce the semigroup associated with Z via

$$P_t f(x) = \mathbb{E}^x f(Z(t)) = \int_E f(y) P(t, x, \mathrm{d}y), \quad t \geq 0$$

and the resolvent via

$$R_\lambda f(x) = \int_0^\infty \mathrm{e}^{-\lambda t} P_t f(x) \mathrm{d}t = \int_0^\infty \mathrm{e}^{-\lambda t} \mathbb{E}^x f(Z(t)) \mathrm{d}t, \quad \lambda > 0, \qquad (4.14)$$

where $f : E \to \mathbb{R}$ is a bounded measurable function.

Let $C_0(E)$ be the Banach space of continuous functions defined on E and vanishing at infinity, endowed with the supremum norm $\|f\| = \sup_{x \in E} |f(x)|$.

Definition 4.5.1 The Markov process Z is a Feller process if its semigroup $(P_t)_{t\geq 0}$ is Feller, that is, if $(P_t)_{t\geq 0}$ is strongly continuous on $C_0(E)$. The latter means that $\lim_{t \to 0+} \|P_t f - f\| = 0$ for each $f \in C_0(E)$.

Recall that a Markov process $(Z(t))_{t\geq 0}$ is a strong Markov process with respect to a filtration $(\mathcal{F}_t)_{t\geq 0}$, if for all a.s. finite $(\mathcal{F}_t)_{t\geq 0}$-stopping times τ, all measurable bounded functions f on E and all $t \geq 0$

$$\mathbb{E}\big(f(Z(\tau + t)) \mid \mathcal{F}_\tau\big) = \int_E f(y) P(t, Z(\tau), \mathrm{d}y).$$

It is known that each Feller process is also a strong Markov process with respect to the filtration $(\mathcal{F}_{t+})_{t\geq 0}$ defined by $\mathcal{F}_{t+} := \cap_{\varepsilon > 0} \mathcal{F}_{t+\varepsilon}$ for $t \geq 0$, and that there exists a version

4.5 Convergence of Markov Processes

of Z with càdlàg paths; see, for instance, Theorem 2.7 in Chapter 4 of [49]. In what follows we shall always assume that the paths of Z are càdlàg.

Now we state two general results on weak convergence of Feller processes. The first of these, Theorem 4.5.1, follows from Theorem 2.5 in Chapter 4 of [49] and Theorem 4.2 in Chapter 3 of [121]. It treats a simpler case in which the converging Feller processes take values in a common space.

Theorem 4.5.1 *Let $(Z_n)_{n\geq 0}$ be a sequence of Feller processes with paths in the space $D([0, \infty), \mathbb{R}^d)$. Denote the semigroup of Z_n by $(P_t^{(n)})_{t\geq 0}$ and the resolvent by $(R_\lambda^{(n)})_{\lambda>0}$. Assume that the sequence of random variables $(Z_n(0))_{n\geq 1}$ converges in distribution as $n \to \infty$ to $Z_0(0)$ and that one of the following equivalent conditions holds:*

(I) *For each $f \in C_0(\mathbb{R}^d)$ and each $t \geq 0$,*

$$\lim_{n\to\infty} \sup_{x\in\mathbb{R}^d} |P_t^{(n)} f(x) - P_t^{(0)} f(x)| = 0.$$

(II) *For each $f \in C_0(\mathbb{R}^d)$ and each $\lambda > 0$,*

$$\lim_{n\to\infty} \sup_{x\in\mathbb{R}^d} |R_\lambda^{(n)} f(x) - R_\lambda^{(0)} f(x)| = 0.$$

Then $Z_n \Longrightarrow Z_0$ as $n \to \infty$ in the J_1-topology on $D([0, \infty), \mathbb{R}^d)$.

The second result of this flavor, Theorem 4.5.2, is a modification of Theorem 4.5.1, in which the converging Feller processes are allowed to take values in different spaces. We apply Theorem 4.5.2 in the proofs of Theorems 2.2.3, 2.3.1(a), and 3.4.1(a) to particular cases, where the limit processes Z_0 take values in $\mathbb{R}$, whereas the converging Markov processes Z_n take values in $\mathbb{R}\setminus\{0\}$ in the first two theorems and in the grid $\mathbb{Z}/c_n$ for a divergent sequence $(c_n)_{n\geq 1}$ in the third theorem.

Theorem 4.5.2 *Let $(E_n)_{n\geq 1}$ be a sequence of Borel subsets of $\mathbb{R}^d$, Z_0 be a Feller process taking values in $\mathbb{R}^d$, and, for $n \in \mathbb{N}$, Z_n a Markov process with paths in $D([0, \infty), E_n)$. For $n \in \mathbb{N}_0$, denote the semigroup of Z_n by $(P_t^{(n)})_{t\geq 0}$ and the resolvent by $(R_\lambda^{(n)})_{\lambda>0}$. Assume that the sequence of random variables $(Z_n(0))_{n\geq 1}$ converges in distribution as $n \to \infty$ to $Z_0(0)$, and that one of the following equivalent conditions holds:*

(I) *For each $f \in C_0(\mathbb{R}^d)$ and each $t \geq 0$,*

$$\lim_{n\to\infty} \sup_{x\in E_n} |P_t^{(n)} f(x) - P_t^{(0)} f(x)| = 0. \tag{4.15}$$

(II) *For each $f \in C_0(\mathbb{R}^d)$ and each $\lambda > 0$,*

$$\lim_{n \to \infty} \sup_{x \in E_n} |R_\lambda^{(n)} f(x) - R_\lambda^{(0)} f(x)| = 0. \tag{4.16}$$

Then $Z_n \Longrightarrow Z_0$ as $n \to \infty$ in the J_1-topology on $D([0, \infty), \mathbb{R}^d)$.

If condition (I) prevails, the result follows from a specialization of Theorem 2.11 on p. 172 in [49]. If condition (II) holds, the argument follows the lines of the proof of Trotter's approximation theorem (Theorem 4.2 on p. 85 in [121]). We refer to the Appendix of the recent article [41] for more details.

Remark 4.5.1 As before, let E be a locally compact separable metric space. Theorems 4.5.1 and 4.5.2 continue to hold, with the space E replacing $\mathbb{R}^d$. The J_1-topology on $D([0, \infty), E)$ is defined analogously.

We have already encountered scenarios where a strong Markov process Z behaves like a standard process, say a Brownian motion, until hitting a fixed point x^* from which Z exhibits a "nonstandard" exit. Let

$$\sigma_{x^*}(Z) := \inf\{t \geq 0 : Z(t) = x^*\}$$

be the first hitting time of x^*, and

$$V_\lambda f(x) = V_\lambda^Z f(x) := \mathbb{E}^x \int_0^{\sigma_{x^*}(Z)} e^{-\lambda t} f(Z(t)) dt \tag{4.17}$$

the resolvent of the process killed at x^*.

The following statement provides a representation of the resolvent of Z (and particularly a limit process Z_0 in Theorems 4.5.1 and 4.5.2) in terms of the resolvent of the killed process and the value $R_\lambda f(x^*)$. This representation reduces verification of the uniform convergence of resolvents in Theorems 4.5.1 and 4.5.2 to the three simpler problems: (a) uniform convergence of the resolvents of killed processes, (b) convergence of the Laplace transforms of hitting times, and (c) convergence of values of the resolvents at the point x^*.

Lemma 4.5.1 *For any strong Markov process Z and any bounded measurable function $f : \mathbb{R}^d \to \mathbb{R}$*

$$R_\lambda f(x) = V_\lambda f(x) + \mathbb{E}^x e^{-\lambda \sigma_{x^*}(Z)} R_\lambda f(x^*), \quad x \in \mathbb{R}^d, \quad \lambda > 0. \tag{4.18}$$

Proof This is a consequence of the strong Markov property:

$$R_\lambda f(x) = \mathbb{E}^x \left(\int_0^{\sigma_{x^*}(Z)} + \int_{\sigma_{x^*}(Z)}^\infty \right) e^{-\lambda t} f(Z(t)) dt$$

$$= V_\lambda f(x) + \mathbb{E}^x e^{-\lambda \sigma_{x^*}(Z)} \int_0^\infty e^{-\lambda t} f(Z(t + \sigma_{x^*}(Z))) dt$$

$$= V_\lambda f(x) + \mathbb{E}^x e^{-\lambda \sigma_{x^*}(Z)} R_\lambda f(x^*).$$

□

The next statement follows from Proposition 4.3.5 and Theorem 4.1.2.

Proposition 4.5.1 *For $n \in \mathbb{N}_0$, let $Z_n = (Z_n(t))_{t \geq 0}$ be a strong Markov process with paths in $D([0, \infty), \mathbb{R}^d)$ and τ_n an a.s. finite stopping time with respect to the natural filtration of Z_n. Assume that*

$$(Z_n(\cdot \wedge \tau_n), \tau_n) \implies (Z_0(\cdot \wedge \tau_0), \tau_0), \quad n \to \infty$$

in the product topology on $D([0, \infty), \mathbb{R}^d) \times \mathbb{R}$ and

$$Z_n(\cdot + \tau_n) \implies Z_0(\cdot + \tau_0), \quad n \to \infty$$

in the J_1-topology on $D([0, \infty), \mathbb{R}^d)$. Then $Z_n \implies Z_0$ as $n \to \infty$ in the J_1-topology on $D([0, \infty), \mathbb{R}^d)$.

4.6 Itô's Excursion Theory

Let Z be a time-homogeneous strong Markov process adapted to a filtration $(\mathcal{F}_t)_{t \geq 0}$. Introduce the notation $\mathbb{P}^x\{\cdot\} := \mathbb{P}\{\cdot | Z(0) = x\}$ and assume that for any starting point $x \in \mathbb{R}$ the process Z hits a fixed point x^* with $\mathbb{P}^x$-probability one. In this section we present a construction of strong Markov processes that behave like Z until hitting x^*. Our basic object will be a process Z killed upon hitting x^*. Such a process is called the *minimal process*, and its semigroup denoted by $(P_t^0)_{t \geq 0}$ is called the *minimal semigroup*. Any strong Markov process that behaves like Z until hitting x^* will be called an *extension of the minimal process* and the corresponding semigroup will be called an *extension of the minimal semigroup*. If an extension is a Feller process, then we call it a *Feller extension*.

As warm-up examples, we point out several processes which are extensions of the minimal process pertaining to a standard Brownian motion W killed upon hitting $x^* = 0$:

(a) The Brownian motion W itself.
(b) A reflected Brownian motion W^{refl} (equivalently, the absolute value of the Brownian motion $|W|$).
(c) A solution to the generalized Skorokhod reflection problem $\mathcal{GSR}(W, F)$, where F is an increasing subordinator which is independent of W; see Sect. 2.1.2.
(d) A skew Brownian motion W_γ^{skew}, where $|\gamma| \leq 1$; see Sect. 2.2.1.

Another important example of an extension is called a *holding and jumping Brownian motion*. Let $x \in \mathbb{R}$, $(\kappa_j)_{j\geq 1}$ be independent copies of a random variable κ with $\mathbb{P}\{\kappa = 0\} = 0$ and $(\varepsilon_k)_{k\geq 0}$ be independent copies of a random variable with the exponential distribution of mean $1/\alpha$. Assume that the sequences $(\kappa_j)_{j\geq 1}$ and $(\varepsilon_k)_{k\geq 0}$ and a standard Brownian motion W are mutually independent. Construct a process $W_{\alpha,\kappa}$ as follows. Put

$$\sigma_0 := \inf\{t \geq 0 : x + W(t) = 0\}, \quad W_{\alpha,\kappa}(t) := x + W(t), \; t \in [0, \sigma_0],$$

$$\tau_n := \sigma_{n-1} + \varepsilon_{n-1}, \quad \sigma_n := \inf\{t \geq \tau_n : \kappa_n + W(t) - W(\tau_n) = 0\}, \quad n \in \mathbb{N},$$

and further

$$W_{\alpha,\kappa}(t) := \begin{cases} 0, & \text{if } t \in [\sigma_{n-1}, \tau_n), \\ \kappa_n + W(t) - W(\tau_n), & \text{if } t \in [\tau_n, \sigma_n) \end{cases} \text{ for } n \in \mathbb{N}.$$

Thus, for each $n \in \mathbb{N}_0$, the process $W_{\alpha,\kappa}$ hits 0 at the time σ_n, spends at 0 the random time ε_n, and then jumps to the position κ_{n+1} at the time τ_{n+1}. The increments of W and $W_{\alpha,\kappa}$ on the intervals $[\tau_n, \sigma_n)$ coincide. It follows from the memoryless property of the exponential distribution that $W_{\alpha,\kappa}$ is a Markov process.

We shall call *excursions* consecutive zero-free fragments of the path of $W_{\alpha,\kappa}$. In particular, for $n \in \mathbb{N}$, the nth excursion is

$$e_n(t) := W_{\alpha,\kappa}((\tau_n + t) \wedge \sigma_n), \quad t \geq 0.$$

We assume in what follows that $x = 0$, so that $W_{\alpha,\kappa}(0) = 0$ a.s.

The excursions $e_1, e_2, \ldots$ are independent and identically distributed stochastic processes. The process $W_{\alpha,\kappa}$ can be obtained by consequently gluing the excursions $(e_n)_{n\geq 1}$ with the "exponential" delays $(\varepsilon_n)_{n\geq 1}$ at 0. It turns out that each extension of W (or a general Markov process Z) can be obtained by gluing together some "excursions." Observe that this procedure is nontrivial even for the original Brownian motion W. Indeed, there are infinitely many excursions of W in any neighborhood of 0 with probability one. Hence, there is no natural enumeration of excursions. The excursion e_n for the holding and jumping Brownian motion can be identified with a Brownian motion starting at κ_n. Excursions of the original Brownian motion cannot be thought of as some stochastic

4.6 Itô's Excursion Theory

processes starting at 0. Actually, these will be defined on a measurable space with a sigma-finite measure satisfying some type of Markov property (see compatibility conditions stated below).

Now we review some basic facts of Itô's excursion theory for Markov processes. Let $Z = (Z(t))_{t \geq 0}$ be a real-valued Feller process equipped with an accompanying family of probability distributions $(\mathbb{P}^x)_{x \in \mathbb{R}}$ given by $\mathbb{P}^x := \mathbb{P}\{Z \in \cdot | Z(0) = x\}$ for $x \in \mathbb{R}$. Thus, Z is a Markov process whose transition semigroup $(P_t)_{t \geq 0}$ is strongly continuous on $C_0(\mathbb{R})$. We shall always assume that all Markov processes under consideration are standard; see the definition in Section 1.9 of [15].

Recall that $\sigma(Z)$ is the first hitting time of 0 by Z and introduce two processes associated with Z.

(i) Put $\bar{Z}(t) := Z(t \wedge \sigma(Z))$ for $t \geq 0$, that is, $(\bar{Z}(t))_{t \geq 0}$ is the process Z stopped at 0. Denote by $\bar{\mathbb{P}}^x$ for $x \in \mathbb{R}$ and $(\bar{P}_t)_{t \geq 0}$ its distribution and semigroup, respectively.

(ii) The process Z killed at 0 with the semigroup $(P_t^0)_{t \geq 0}$ and the transition probabilities

$$P^0(t, x, A) = \mathbb{P}^x\{Z(t) \in A, \, t < \sigma(Z)\} = \mathbb{P}\{Z(t) \in A, \, t < \sigma(Z) | Z(0) = x\}$$

for $t \geq 0$ and $x \in \mathbb{R}$.

According to the aforementioned terminology the killed process and its semigroup are *minimal process* and *minimal semigroup*, respectively. Here is an important remark. When constructing extensions the basic object that we start with is the killed process, and a natural problem is to describe *all* possible Feller processes Z with the minimal semigroup $(P_t^0)_{t \geq 0}$ or equivalently the stopped semigroup $(\bar{P}_t)_{t \geq 0}$. On the other hand, sometimes we use the term "extension of a Markov process" rather than "extension of a Markov process killed at 0," thereby slightly abusing the terminology.

We shall impose the following assumptions:

(**A**$_1$) $\bar{Z}$ is recurrent at 0, that is, $\bar{\mathbb{P}}^x\{\sigma(\bar{Z}) < \infty\} = 1$ for all $x \in \mathbb{R}$.

(**A**$_2$) With $\bar{\mathbb{E}}^x$ denoting the expectation with respect to the probability measure $\bar{\mathbb{P}}^x$, the function

$$x \mapsto \bar{\mathbb{E}}^x e^{-\lambda \sigma(\bar{Z})} \text{ is continuous for each } \lambda > 0 \text{ and } \lim_{|x| \to \infty} \bar{\mathbb{E}}^x e^{-\sigma(\bar{Z})} = 0. \quad (4.19)$$

We shall always assume that the semigroups $(P_t^0)_{t \geq 0}$ and $(\bar{P}_t)_{t \geq 0}$ are strongly continuous on $C_0(\mathbb{R})$.

Example 4.6.1 Assumptions **A**$_1$ and **A**$_2$ hold true for a symmetric α-stable Lévy process with $\alpha \in (1, 2]$; see, for instance, p. 63 in [10] or Example 43.42 in [146].

We are getting back to our practice of writing $D([0, \infty))$ as a shorthand for $D([0, \infty), \mathbb{R})$.

Definition 4.6.1 For a function $u \in D([0, \infty))$, put $\sigma(u) := \inf\{t > 0 : u(t) = 0\}$. A function $u \in D([0, \infty))$ satisfying $u(t) = 0$ for $t \geq \sigma(u)$ will be called an *excursion*, and $\sigma(u)$ will be called the *length of excursion u*.

Let $\hat{\mathbb{P}}$ be a sigma-finite measure on the Borel sigma-algebra of $D([0, \infty))$ supported by the set of excursions. For a Borel set $A \subseteq D([0, \infty))$ and a bounded measurable mapping $f : D([0, \infty)) \to \mathbb{R}$, put

$$\hat{\mathbb{P}}(f; A) := \int_A f(u)\hat{\mathbb{P}}(du), \quad \hat{\mathbb{P}}(f) := \int_{D([0,\infty))} f(u)\hat{\mathbb{P}}(du).$$

For a Borel set $B \subseteq D([0, \infty))$, we write $\hat{\mathbb{P}}\{B\}$ for $\hat{\mathbb{P}}(\mathbb{1}_B)$.

(A$_3$) $\hat{\mathbb{P}}(1 - e^{-\sigma}) := \int_{D([0,\infty))} (1 - e^{-\sigma(u)})\hat{\mathbb{P}}(du) \leq 1$ and $\hat{\mathbb{P}}\{|u(0)| > x\} < \infty$ for $x > 0$.

Let $N := \sum_k \delta_{(s_k, u_k)}$ be a Poisson random measure on $[0, \infty) \times D([0, \infty))$ with intensity measure $\text{LEB} \otimes \hat{\mathbb{P}}$. Here, $\delta_{(s,u)}$ denotes the Dirac measure concentrated at (s, u). Put $m := 1 - \hat{\mathbb{P}}(1 - e^{-\sigma})$ and

$$\tau(s) := ms + \sum_{s_k \leq s} \sigma(u_k) = ms + \int_{[0,s]} \int_{D([0,\infty))} \sigma(u) N(dz, du), \quad s \geq 0. \quad (4.20)$$

We call m the *delay coefficient*. The integral on the right-hand side of (4.20) is a.s. convergent because the first part of **A$_3$** in combination with the inequality $1 - e^{-x} \geq (x \wedge 1)/2$ for $x \geq 0$ entails

$$\int_{D([0,\infty))} (\sigma(u) \wedge 1)\hat{\mathbb{P}}(du) < \infty.$$

Furthermore, the process $(\tau(s))_{s \geq 0}$ is a subordinator.

Assume that at least one of the following conditions holds:

(A$_4$) $m > 0$ or the measure $\hat{\mathbb{P}}$ is infinite.

Condition **A$_4$** is equivalent to the fact that the subordinator $(\tau(s))_{s \geq 0}$ is strictly increasing.

Using the Poisson point process N we point out Itô's construction of a (not necessarily Markov) process

$$Z(t) := \begin{cases} u_k(t - \tau(s_k-)), & \text{if } t \in [\tau(s_k-), \tau(s_k)), \\ 0, & \text{if } t \notin \cup_k [\tau(s_k-), \tau(s_k)); \end{cases} \quad (4.21)$$

4.6 Itô's Excursion Theory

that is, the excursions u_k are merged, with possible delays at 0, in a way that an excursion u_j is used before u_k if, and only if, $s_j < s_k$. We state right away that, under additional natural assumptions given in Theorem 4.6.1 (I), this construction produces a Feller process.

Remark 4.6.1 The condition $m = 0$ is equivalent to the fact that the process Z defined in (4.21) has a zero sojourn at 0, that is,

$$\int_0^\infty \mathbb{1}_{\{Z(s)=0\}} ds = 0 \text{ a.s.}$$

This follows from the fact that Lebesgue measure of $\mathbb{R} \setminus \cup_k [\tau(s_k-), \tau(s_k))$ is zero if, and only if, $m = 0$; see Theorem 4 on p. 77 in [10].

Remark 4.6.2 The process Z given in (4.21) was constructed under the assumption $m + \hat{\mathbb{P}}(1 - e^{-\sigma}) = 1$. The latter is just a normalization condition which allows us to select m and $\hat{\mathbb{P}}$ in a canonical way. Indeed, assume that

$$m + \hat{\mathbb{P}}(1 - e^{-\sigma}) = a \in (0, \infty) \setminus \{1\}. \tag{4.22}$$

The process τ_a defined by $\tau_a(s) := \tau(a^{-1}s)$ for $s \geq 0$ is a subordinator with the delay coefficient $m_a := a^{-1}m$ and the Poisson random measure $N_a := \sum_k \delta_{(as_k, u_k)}$ with intensity measure $(a^{-1}\text{LEB}) \otimes \hat{\mathbb{P}} = \text{LEB} \otimes (a^{-1}\hat{\mathbb{P}}) =: \text{LEB} \otimes \hat{\mathbb{P}}_a$. Clearly, $m_a + \hat{\mathbb{P}}_a(1 - e^{-\sigma}) = 1$, and the processes obtained with the help of Itô's construction (formula (4.21)) using τ and τ_a coincide because $\tau_a(as_k) = \tau(s_k)$ and $\tau_a(as_k-) = \tau(s_k-)$ for all k.

Finally, we state the last assumption.

(**A$_5$**) The measure $\hat{\mathbb{P}}$ is *compatible with the minimal semigroup* $(P_t^0)_{t \geq 0}$. This means that, for all $s, t \geq 0$, all bounded measurable functions $g : \mathbb{R} \to \mathbb{R}$ with $g(0) = 0$ and all $\Lambda_s \in \sigma(u(r), r \in [0, s])$,

$$\hat{\mathbb{P}}(g(u(t+s)); \Lambda_s \cap \{\sigma(u) > s\}) = \hat{\mathbb{P}}\left((P_t^0 g)(u(s)); \Lambda_s \cap \{\sigma(u) > s\}\right); \tag{4.23}$$

and, for any Borel set A which does not contain 0,

$$\hat{\mathbb{P}}(g(u(t)); u(0) \in A) = \hat{\mathbb{P}}\left((P_t^0 g)(u(0)); u(0) \in A\right).$$

More details on the assumption **A$_5$** can be found in Sections V.1 and V.2(d) of [14].

Remark 4.6.3 Note that condition (4.23) is equivalent to

$$\hat{\mathbb{P}}(g(u(t+s)); \Lambda_s \cap \{\sigma(u) > s\}) = \hat{\mathbb{P}}\left((\bar{P}_t g)(u(s)); \Lambda_s \cap \{\sigma(u) > s\}\right)$$

for all $s, t \geq 0$, all bounded measurable functions $g : \mathbb{R} \to \mathbb{R}$, and all $\Lambda_s \in \sigma(u(r), r \in [0, s])$. Here, the only differences in comparison to (4.23) is that P_t^0 is replaced with $\bar{P}_t$ and that $g(0) = 0$ is not required.

In particular, this implies that for each fixed $s > 0$, the process $(u(s+t))_{t \geq 0}$ is Markov, where u is a random element of $D([0, \infty))$ picked according to a probability measure $\hat{\mathbb{P}}\{\cdot \cap \{\sigma(u) > s\}\}/\hat{\mathbb{P}}\{\sigma(u) > s\}$. The transition probabilities of $(u(s+t))_{t \geq 0}$ coincide with $(\bar{P}_t)_{t > 0}$, the transition probabilities of the stopped process.

The next result is, for the most part, Theorem 2.10 on p. 145 of [14].

Theorem 4.6.1 *Suppose* $\mathbf{A}_1$ *and* $\mathbf{A}_2$.

(I) *Let Z be as defined in (4.21), with a measure $\hat{\mathbb{P}}$ satisfying* $\mathbf{A}_3$–$\mathbf{A}_5$. *Then Z is a Feller extension of the minimal process with $Z(0) = 0$. Moreover, Z has a zero sojourn at 0 if, and only if, $m = 0$.*

(II) *Each Feller extension Z (starting at 0) of the minimal process can be obtained with the help of merging excursions under a suitable measure $\hat{\mathbb{P}}$ satisfying conditions* $\mathbf{A}_3$–$\mathbf{A}_5$. *Moreover, the measure $\hat{\mathbb{P}}$ is determined uniquely by the distribution of Z.*

In the following, we interchangeably call $\hat{\mathbb{P}}$ the *characteristic measure* (of the Poisson point measure N) or the *excursion measure*.

Remark 4.6.4 An extension Z of the minimal process with $Z(0) \neq 0$ can be constructed by merging two independent parts. One part is the fragment of the path before hitting 0. Plainly, it is determined uniquely by the minimal process. The other part is constructed with the help of Itô's construction.

Usually, checking condition $\mathbf{A}_5$ is the most difficult. Below we give a few examples of excursion measures that satisfy the compatibility conditions.

Example 4.6.2 The distribution $\bar{\mathbb{P}}^x$ of the process stopped at 0 satisfies $\mathbf{A}_5$, and so does the mixture

$$\bar{\mathbb{P}}^\theta := \int_{\mathbb{R} \setminus \{0\}} \bar{\mathbb{P}}^x \, \theta(\mathrm{d}x) \tag{4.24}$$

4.6 Itô's Excursion Theory

for any sigma-finite measure θ on $\mathbb{R} \setminus \{0\}$. If additionally

$$\int_{\mathbb{R}\setminus\{0\}} \mathbb{E}^x(1-e^{-\sigma})\theta(dx) \leq 1,$$

then $\mathbf{A}_3$ for $\bar{\mathbb{P}}^\theta$ also holds true. If θ is an infinite measure, then $\bar{\mathbb{P}}^\theta$ satisfies the second part of condition $\mathbf{A}_4$. If θ is a finite measure, then the first part of condition $\mathbf{A}_4$ holds for $\bar{\mathbb{P}}^\theta$ under the additional assumption

$$\int_{\mathbb{R}\setminus\{0\}} \mathbb{E}^x(1-e^{-\sigma})\theta(dx) < 1.$$

Example 4.6.3 Let $\alpha > 0$ be a fixed parameter, κ a random variable satisfying $\mathbb{P}\{\kappa = 0\} = 0$, and Z a Feller process which visits 0 with probability one for any starting point x. Consider a *holding and jumping process* $Z_{\alpha,\kappa}$ whose construction is similar to that of a holding and jumping Brownian motion. Starting at a point x the process $Z_{\alpha,\kappa}$ behaves like Z until first hitting 0. Then it stays at 0 for a random period of time having the exponential distribution of mean $1/\alpha$. Afterwards the process makes a jump with an amplitude having the distribution of κ, that we denote by P_κ, and then it behaves like Z until the next visit to 0. The evolution just described then iterates. It is assumed that the durations of stays at 0, the amplitudes of the jumps from 0, and the evolution following the jumps are independent of the previous history. It can be proved that $Z_{\alpha,\kappa}$ is a strong Markov process and that the corresponding excursion measure $\hat{\mathbb{P}}$ is given by $\hat{\mathbb{P}} = \bar{\mathbb{P}}^\theta$. Here, θ is a finite measure that is uniquely determined by

$$\theta(dx)/\theta(\mathbb{R}) = P_\kappa(dx) \tag{4.25}$$

and

$$\alpha = m^{-1}\theta(\mathbb{R}), \quad m = 1 - \int_\mathbb{R} \mathbb{E}^x(1-e^{-\sigma z})\theta(dx). \tag{4.26}$$

Conversely, it can be checked that if Z is a Feller process with the excursion measure $\hat{\mathbb{P}} = \bar{\mathbb{P}}^\theta$, where θ is a finite measure satisfying $\int_\mathbb{R} \mathbb{E}^x(1-e^{-\sigma z})\theta(dx) < 1$, then Z is a holding and jumping process. Actually, a more general fact holds true (see p. 145 in [14]) that $\hat{\mathbb{P}}$ is finite if, and only if, $\hat{\mathbb{P}} = \bar{\mathbb{P}}^\theta$, where θ is a finite measure.

Now we provide a formula for $R_\lambda^{Z_{\alpha,\kappa}}$ the resolvent of a holding and jumping process $Z_{\alpha,\kappa}$. For notational simplicity, put $\sigma := \sigma_{Z_{\alpha,\kappa}}$. Recall from (4.18) that for any bounded measurable functions $f : \mathbb{R} \to \mathbb{R}$

$$R_\lambda^{Z_{\alpha,\kappa}} f(x) = V_\lambda f(x) + \mathbb{E}^x e^{-\lambda\sigma} R_\lambda^{Z_{\alpha,\kappa}} f(0), \tag{4.27}$$

where V_λ is the resolvent of the killed process. Assuming that a formula for V_λ is known, we only calculate $R_\lambda^{Z_{\alpha,\kappa}} f(0)$.

Let τ be the time of the first jump of $Z_{\alpha,\kappa}$ from 0. Then, by the strong Markov property,

$$R_\lambda^{Z_{\alpha,\kappa}} f(0) = \mathbb{E}^0 \Big(\int_0^\tau + \int_\tau^\infty \Big) e^{-\lambda t} f(Z_{\alpha,\kappa}(t)) dt = \lambda^{-1} \mathbb{E}(1 - e^{-\lambda \tau}) f(0)$$

$$+ \mathbb{E} e^{-\lambda \tau} \mathbb{E}^\kappa \int_0^\infty e^{-\lambda t} f(Z_{\alpha,\kappa}(t)) dt = \frac{1}{\alpha + \lambda} f(0) + \frac{\alpha}{\alpha + \lambda} \langle P_\kappa, R_\lambda^{Z_{\alpha,\kappa}} f \rangle. \quad (4.28)$$

As before, for a measurable function $f : \mathbb{R} \to \mathbb{R}$ and a measure ν on $\mathbb{R}$ we write $\langle \nu, f \rangle$ for $\int_\mathbb{R} f(x) \nu(dx)$ provided that the integral is well-defined. Integrating both sides of (4.27) with respect to the distribution of κ yields

$$\langle P_\kappa, R_\lambda^{Z_{\alpha,\kappa}} f \rangle = \langle P_\kappa, V_\lambda f \rangle + \mathbb{E}^\kappa e^{-\lambda \sigma} R_\lambda^{Z_{\alpha,\kappa}} f(0).$$

Substituting this into (4.28) and then solving for $R_\lambda^{Z_{\alpha,\kappa}} f(0)$, we obtain

$$\lambda R_\lambda^{Z_{\alpha,\kappa}} f(0) = \frac{\alpha^{-1} f(0) + \langle P_\kappa, V_\lambda f \rangle}{\alpha^{-1} + \lambda^{-1} \mathbb{E}^\kappa (1 - e^{-\lambda \sigma})} = \frac{\alpha^{-1} f(0) + \langle P_\kappa, V_\lambda f \rangle}{\alpha^{-1} + \langle P_\kappa, V_\lambda 1 \rangle}. \quad (4.29)$$

Example 4.6.4 Let $Z_{\alpha,\kappa}$ be a holding and jumping process which corresponds to a finite measure θ and a constant $m \in [0, 1)$ as given in (4.25) and (4.26). Then the resolvent of $Z_{\alpha,\kappa}$ satisfies the equality

$$\lambda R_\lambda^{Z_{\alpha,\kappa}} f(0) = \frac{m f(0) + \langle \theta, V_\lambda f \rangle}{m + \langle \theta, V_\lambda 1 \rangle}. \quad (4.30)$$

Observe that formula (4.30) makes sense even for infinite (sigma-finite) measures θ on $\mathbb{R} \setminus \{0\}$ satisfying $\int_{\mathbb{R} \setminus \{0\}} \mathbb{E}^x (1 - e^{-\sigma z}) \theta(dx) < \infty$.

It follows from Theorems 2.6 and 2.8 and the proof of Theorem 2.10 in Chapter V of [14] that if θ is an infinite measure such that $m := 1 - \mathbb{E}^\theta (1 - e^{-\sigma}) \in [0, 1)$, then

(i) Formulas (4.18) and (4.30) define a resolvent R_λ of a Feller process.
(ii) The corresponding excursion measure $\hat{\mathbb{P}}$ is equal to $\bar{\mathbb{P}}^\theta$.
(iii) The corresponding Feller process has a zero sojourn at 0 if, and only if, $\mathbb{E}^\theta (1 - e^{-\sigma}) = 1$, in which case $m = 0$ and

$$\lambda R_\lambda f(0) = \frac{\langle \theta, V_\lambda f \rangle}{\langle \theta, V_\lambda 1 \rangle}. \quad (4.31)$$

4.6 Itô's Excursion Theory

It can be checked that the R_λ can be obtained as a uniform limit of resolvents of holding and jumping processes. Hence, according to Theorem 4.5.1, the process with the resolvent R_λ is a weak limit of holding and jumping processes. A nontrivial fact here is that R_λ is indeed a resolvent of a Feller process. We stress that the proof of an approximation by holding and jumping processes given in [14] is constructive and does not appeal to Theorem 4.5.1.

In what follows we discuss the construction of excursion measures other than $\bar{\mathbb{P}}^\theta$. For instance, since excursions of a Brownian motion start from 0 with probability one, its excursion measure falls in this category. Observe that if $\hat{\mathbb{P}}$ is the excursion measure of a process Z, then for any fixed $s > 0$ the conditional distribution of $(u(t+s))_{t \geq 0}$ given $\{\sigma(u) > s\}$ can be constructed with the help of the Ionescu-Tulcea theorem. Indeed, for a fixed $s > 0$, define a measure θ_s on the Borel subsets of $\mathbb{R} \setminus \{0\}$ via

$$\theta_s(\cdot) = \hat{\mathbb{P}}\{u(s) \in \cdot \, , \, \sigma(u) > s\}. \tag{4.32}$$

It follows from the compatibility conditions in combination with Remark 4.6.3 that

$$\hat{\mathbb{P}}\{u(s) \in A_0, u(s+t_1) \in A_1, \ldots, u(s+t_n) \in A_n \, , \, \sigma(u) > s\}$$
$$= \int_{A_0} \theta_s(\mathrm{d}x_0) \int_{A_1} \bar{P}_{t_1}(x_0, \mathrm{d}x_1) \ldots \int_{A_n} \bar{P}_{t_n - t_{n-1}}(x_{n-1}, \mathrm{d}x_n)$$

for all $n \in \mathbb{N}$, all $0 < t_1 < \cdots < t_n$ and any Borel sets $A_0, A_1, \ldots, A_n$ with $0 \notin A_0$. Hence, the measure $\hat{\mathbb{P}}\{(u(t+s))_{t \geq 0} \in \cdot \, , \, \sigma(u) > s\}$ is equal to $\bar{\mathbb{P}}^{\theta_s}$. This suggests that knowing measures θ_s for all $s > 0$ and a semigroup of the killed process should allow us to construct a unique excursion measure $\hat{\mathbb{P}}$ satisfying condition (4.32). Although this is true, it is far from being obvious that excursions have the right limits at 0. To formulate rigorous results, we need a definition.

Definition 4.6.2 A family of finite measures $(\theta_s)_{s>0}$ on the Borel subsets of $\mathbb{R} \setminus \{0\}$ is called *entrance law* for (P_t^0), if the following two conditions hold true. First,

$$\theta_s P_t^0 = \theta_{t+s}, \quad t \geq 0, \quad s > 0,$$

that is,

$$\langle \theta_s, P_t^0 f \rangle = \langle \theta_{t+s}, f \rangle, \quad t \geq 0, \quad s > 0$$

for each bounded measurable $f : \mathbb{R} \setminus \{0\} \to \mathbb{R}$. Second,

$$\langle \theta_s, V_1 1 \rangle \leq 1, \quad s > 0.$$

The second condition is equivalent to

$$\mathbb{E}^{\theta_s}(1 - e^{-\sigma}) \leq 1, \quad s > 0.$$

Lemma 4.6.1 *Let $(\theta_t)_{t>0}$ be an entrance law.*

(I) *The function $t \mapsto \theta_t(\mathbb{R} \setminus \{0\})$ is nonincreasing on $(0, \infty)$.*
(II) *If $\sup_{t>0} \theta_t(\mathbb{R} \setminus \{0\}) < \infty$, then there exists a finite measure θ such that $\theta_t = \theta P_t^0$ for all $t > 0$.*

The first claim follows from the definition. The second statement is proved at the bottom of p. 141 in [14].

Theorem 4.6.2 *The following claims hold true:*

(i) *Suppose $\mathbf{A}_1$–$\mathbf{A}_5$. Then the collection of measures defined in (4.32) is an entrance law.*
(ii) *Suppose $\mathbf{A}_1$ and $\mathbf{A}_2$ and let $(\theta_t)_{t>0}$ be an entrance law. Assume that either $(\theta_t(\mathbb{R} \setminus \{0\}))_{t>0}$ is unbounded or $\sup_{s>0} \mathbb{E}^{\theta_s}(1 - e^{-\sigma}) < 1$. Then there is a unique excursion measure $\hat{\mathbb{P}}$ satisfying conditions $\mathbf{A}_3$–$\mathbf{A}_5$ and (4.32).*

Part (i) can be found on p. 138 in [14]. Part (ii) is Theorem 4.7 on p. 160 of the same reference.

Summarizing the discussion above, any Feller extension of the minimal process can be described uniquely provided that the information is available about one of the objects listed below:

(a) The resolvent $R_\lambda f(x)$ for $f \in C_0(\mathbb{R})$ and $x \in \mathbb{R}$.
(b) The resolvent $R_\lambda f(0)$ for $f \in C_0(\mathbb{R})$.
(c) The excursion measure $\hat{\mathbb{P}}$.
(d) the entrance law $(\theta_t)_{t>0}$.

In general, it is easier to verify the definition of the entrance law than the fact that a function (a prospective resolvent) R_λ satisfies conditions of the Hille-Yosida theorem. On the other hand, a resolvent approach is more convenient for proving limit theorems. Excursion measures and Itô's excursion theory are helpful for understanding structural properties of the processes in focus.

Given next is the result which can be found on p. 140 in [14].

Theorem 4.6.3 *Any entrance law for $(P_t^0)_{t \geq 0}$ can be uniquely decomposed as the sum of two entrance laws*

$$\theta_t = \rho_t + \theta P_t^0, \quad t > 0,$$

4.6 Itô's Excursion Theory

where θ is a sigma-finite measure, and the measure ρ_t satisfies

$$\lim_{t \to 0+} \rho_t(\mathbb{R} \setminus [-x, x]) = 0$$

for any $x > 0$.

The entrance laws $(\rho_t)_{t>0}$ and $(\theta P_t^0)_{t>0}$ from Theorem 4.6.3 are called *continuous entrance law* and *jump entrance law*, respectively. They can also be characterized as follows:

$$\theta_t^c(\cdot) := \rho_t = \hat{\mathbb{P}}\{u(t) \in \cdot, \ u(0) = 0, \ t < \sigma(u)\},$$

$$\theta_t^j(\cdot) := \hat{\mathbb{P}}\{u(t) \in \cdot, \ u(0) \neq 0, \ t < \sigma(u)\};$$

see p. 156 in [14]. According to Theorem 4.6.2, there is the one-to-one correspondence between excursion measures and entrance laws. In view of this, the restrictions of $\hat{\mathbb{P}}$ to the sets $\{u(0) = 0\}$ and $\{u(0) \neq 0\}$ are excursion measures compatible with the minimal semigroup. Moreover, for any excursion measure $\hat{\mathbb{P}}$ its restriction $\hat{\mathbb{P}}|_{u(0) \neq 0}$ is equal to $\bar{\mathbb{P}}^\theta$, where a sigma-finite measure θ satisfies $\mathbb{E}^\theta(1 - e^{-\sigma}) \leq 1$, and the entrance law corresponding to $\hat{\mathbb{P}}|_{u(0) \neq 0}$ is $(\theta P_t^0)_{t>0}$.

Corollary 4.6.1

(I) Let θ be a measure satisfying $\mathbb{E}^\theta(1 - e^{-\sigma}) < \infty$ and $(\rho_t)_{t>0}$ a continuous entrance law which corresponds to the characteristic measure $\hat{\mathbb{P}}^c$. Then

$$\theta_t = \alpha \rho_t + \beta \theta P_t^0, \quad t > 0,$$

is an entrance law for any $\alpha, \beta \geq 0$ such that

$$\alpha \hat{\mathbb{P}}^c(1 - e^{-\sigma}) + \beta \mathbb{E}^\theta(1 - e^{-\sigma}) \leq 1.$$

This entrance law corresponds to the characteristic measure $\hat{\mathbb{P}} = \alpha \hat{\mathbb{P}}^c + \beta \bar{\mathbb{P}}^\theta$.

(II) Assume that there exists a unique continuous entrance law $(\rho_t)_{t>0}$ with the characteristic measure $\hat{\mathbb{P}}^c$ satisfying $\hat{\mathbb{P}}^c(1 - e^{-\sigma}) = 1$. Then any entrance law $(\theta_t)_{t>0}$ can be uniquely represented by

$$\theta_t = \alpha \rho_t + \beta \theta P_t^0, \tag{4.33}$$

where $\alpha, \beta \geq 0$, $\alpha + \beta \leq 1$, and the measure θ satisfies $\mathbb{E}^\theta(1 - e^{-\sigma}) = 1$. The corresponding characteristic measure is

$$\hat{\mathbb{P}} = \alpha \hat{\mathbb{P}}^c + \beta \bar{\mathbb{P}}^\theta. \tag{4.34}$$

The corresponding Feller extension has a zero sojourn at 0 if, and only if, $\alpha + \beta = 1$.

Theorem 4.6.4 *For $\alpha \in (1, 2)$, let U_α denote a symmetric α-stable Lévy process with càdlàg paths. There is a unique Feller extension of the minimal process generated by U_α with a continuous entrance law and a zero sojourn at 0. The extension is the process U_α itself.*

Proof The process U_α itself has a continuous entrance law. Indeed, it was mentioned in Sect. 2.3.1 that U_α changes sign infinitely often on exiting 0. Since the paths of U_α are càdlàg, excursions exit 0 continuously, that is,

$$\mathbb{P}\{U_\alpha(t+) = 0 \text{ for any } t \geq 0 \text{ such that } U_\alpha(t) = 0\} = 1.$$

The process U_α has a zero sojourn at 0 because its one-dimensional distributions are absolutely continuous with respect to Lebesgue measure. This completes the proof of existence.

Now we prove uniqueness. The distribution of a Feller extension Z of the minimal process is determined uniquely by its resolvent. According to Lemma 4.5.1, this resolvent is uniquely determined by the path preceding the first visit to 0 and the value $R_\lambda^Z f(0)$. Hence, we have to show that the value $R_\lambda^Z f(0)$ is determined uniquely whenever Z has a continuous entrance law and a zero sojourn at 0, and f is any function from $C_0(\mathbb{R})$.

According to Theorem 4.2 on p. 157 in [14], it suffices to prove that

$$\lim_{x \to 0} \frac{V_\lambda f(x)}{\lambda V_\lambda 1(x)} = R_\lambda^{U_\alpha} f(0).$$

Recall the definition of X_ε in (2.53). Put $\varepsilon = |x|$, $X_\varepsilon(0) = x$, and $\zeta = \operatorname{sgn} x$. Then formula (2.71) reads

$$\frac{V_\lambda f(x)}{\lambda V_\lambda 1(x)} = R_\lambda^{X_\varepsilon} f(x) = \mathbb{E}^x \int_0^\infty e^{-\lambda t} f(X_\varepsilon(t)) dt. \tag{4.35}$$

Since $\mathbb{E}|\zeta| = 1 < \infty$ whenever $x \neq 0$, the equalities

$$\lim_{x \to 0} \mathbb{E}^x \int_0^\infty e^{-\lambda t} f(X_\varepsilon(t)) dt = \mathbb{E} \int_0^\infty e^{-\lambda t} f(U_\alpha(t)) dt = R_\lambda^{U_\alpha} f(0)$$

4.6 Itô's Excursion Theory

are secured by part (b) of Theorem 2.3.1 with $\kappa \equiv 0$ in combination with the Skorokhod representation theorem and the Lebesgue dominated convergence theorem. □

In contrast to U_α with $\alpha \in (1, 2)$, uniqueness of a continuous entrance law fails for the case $\alpha = 2$, as demonstrated in the next example.

Example 4.6.5 Reflected Brownian motions on $[0, \infty)$ and $(-\infty, 0]$ are Feller extensions with continuous entrance laws of a Brownian motion killed at 0. Let $\hat{\mathbb{P}}^\pm$ be their excursion measures and $(\rho_t^\pm)$ their entrance laws. According to Theorem 1.1 on p. 110 in [14],

$$\rho_t^\pm(dx) = \frac{1}{t^{3/2}\pi^{1/2}} |x| e^{-x^2/(2t)} \mathbb{1}_{(0,\infty)}(\pm x) dx.$$

It is known (see, for instance, Example 3(b) on p. 152 in [14]) that the excursion measure of any Feller extension with a continuous entrance law of a Brownian motion killed at 0 can be represented as

$$\hat{\mathbb{P}} = p\hat{\mathbb{P}}^+ + q\hat{\mathbb{P}}^-,$$

where $p, q \geq 0$, $p + q \leq 1$. The corresponding continuous entrance law admits a decomposition

$$\rho_t = p\rho_t^+ + q\rho_t^-.$$

The extension has a zero sojourn at 0 if, and only if, $p + q = 1$, which is equivalent to the fact that this extension is a skew Brownian motion with permeability parameter $\gamma = p - q$. In particular, if $p = q = 1/2$, then the extension is a standard Brownian motion.

We close this section with the definition of a local time. Let Z be a Feller process satisfying assumptions $\mathbf{A}_1$ and $\mathbf{A}_2$ and L a continuous additive functional of Z (the definition can be found, for instance, in [14]). Assume that L may increase only at times when Z is equal to 0. Then, for all $\lambda > 0$ and all $x \in \mathbb{R}$, the strong Markov property entails

$$\mathbb{E}^x \int_{[0,\infty)} e^{-\lambda t} dL(t) = \mathbb{E}^x \int_{[\sigma,\infty)} e^{-\lambda t} dL(t) = c_\lambda \mathbb{E}^x e^{-\lambda \sigma},$$

where $c_\lambda := \mathbb{E}^0 \int_{[0,\infty)} e^{-\lambda t} dL(t)$, and σ is the first hitting time of 0 by Z.

Definition 4.6.3 A continuous additive functional L which satisfies

$$\mathbb{E}^x e^{-\sigma} = \mathbb{E}^x \int_{[0,\infty)} e^{-t} dL(t), \quad x \in \mathbb{R},$$

is called the *Blumenthal-Getoor local time at* 0.

The Blumenthal-Getoor local time at 0 exists and is unique; see pp. 91–93 in [14].

Theorem 4.6.5 *Let Z be a Feller extension of the minimal process constructed in Theorem 4.6.1 and the process τ be as given in (4.20). Then the process $(\varphi(t))_{t \geq 0}$ defined by*

$$\varphi(t) := \inf\{s \geq 0 \,:\, \tau(s) > t\}, \quad t \geq 0, \tag{4.36}$$

is the Blumenthal-Getoor local time of Z at 0.

The proof can be found in Theorem 2.3 on p. 185 in [14].

Bibliography

1. Andreoletti, P., Debs, P.: Simple random walk on $\mathbb{Z}^2$ perturbed on the axes (renewal case). Electron. J. Probab. **28**(Paper no. 110), 25 (2023)
2. Anulova, S.V.: Processes with a Lévy generating operator in a half-space. Math. USSR-Izv. **13**(1), 9–51 (1979)
3. Anulova, S.V.: On stochastic differential equations with boundary conditions in a half-plane. Math. USSR-Izv. **18**(3), 423–437 (1982)
4. Atar, R., Budhiraja, A.: Diffusion limits in the quarter plane and non-semimartingale reflected Brownian motion. arXiv preprint arXiv: 2403.00320 (2024)
5. Barlow, M., Burdzy, K., Kaspi, H., Mandelbaum, A.: Variably skewed Brownian motion. Electron. Comm. Probab. **5**, 57–66 (2000)
6. Barlow, M., Pitman, J., Yor, M.: On Walsh's Brownian motions. In: Séminaire de Probabilités, XXIII. Lecture Notes in Mathematics, vol. 1372, pp. 275–293. Springer, Berlin (1989)
7. Baxter, J.R., Chacon, R.V.: The equivalence of diffusions on networks to Brownian motion. In: Conference in Modern Analysis and Probability (New Haven, Connecticut, 1982). Contemporary Mathematics, vol. 26, pp. 33–48. American Mathematical Society, Providence, RI (1984)
8. Bayraktar, E., Zhang, J., Zhang, X.: Walsh diffusions as time changed multi-parameter processes. arXiv preprint arXiv:2204.07101 (2022)
9. Belkin, B.: A limit theorem for conditioned recurrent random walk attracted to a stable law. Ann. Math. Statist. **41**, 146–163 (1970)
10. Bertoin, J.: Lévy Processes. Cambridge University, Cambridge (1996)
11. Bhattacharya, R., Waymire, E.C.: Continuous Parameter Markov Processes and Stochastic Differential Equations. Springer, Cham (2023)
12. Billingsley, P.: Convergence of Probability Measures, 2nd edn. Wiley, New York (1999)
13. Bingham, N.H., Goldie, C.M., Teugels, J.L.: Regular Variation. Cambridge University, Cambridge (1987)
14. Blumenthal, R.M.: Excursions of Markov Processes. Birkhäuser, Boston (1992)
15. Blumenthal, R.M., Getoor, R.K.: Markov Processes and Potential Theory. Academic Press, New York (1968)
16. Boas, R.P.: A Primer of Real Functions, 4th edn. Mathematical Association of America, Washington, DC (1996)
17. Bobrowski, A., Komorowski, T.: Diffusion approximation for a simple kinetic model with asymmetric interface. J. Evol. Equ. **22**(Paper no. 42), 26 pp. (2022)
18. Bobrowski, A., Ratajczyk, E.: Approximation of skew Brownian motion by snapping-out Brownian motions. Mathematische Nachrichten **298**(3), 829–848 (2024)

19. Bobrowski, A., Ratajczyk, E.: From snapping out Brownian motions to Walsh's spider processes on star-like graphs. arXiv preprint arXiv:2406.16800 (Accepted to Documenta Mathematica, 2024)
20. Bobrowski, A., Ratajczyk, E.: A kinetic model approximation of Walsh's spider process on the infinite star-like graph. arXiv preprint arXiv:2409.15467 (2024)
21. Bogdan, K., Komorowski, T., Marino, L.: Anomalous diffusion limit for a kinetic equation with a thermostatted interface. Probab. Theory Related Fields **189**(1–2), 721–769 (2024)
22. Bogdanskii, V., Pavlyukevich, I., Pilipenko, A.: Limit behaviour of random walks on $\mathbb{Z}^m$ with two-sided membrane. ESAIM Probab. Stat. **26**, 352–377 (2022)
23. Bohun, V., Marynych, A.: On the local time of a recurrent random walk on $\mathbb{Z}^2$. Theory Probab. Math. Statist. **105**, 69–78 (2021)
24. Borovkov, A.A.: Stochastic processes in queueing theory. Springer, New York (1976)
25. Brémont, J.: On homogeneous and oscillating random walks on the integers. Probab. Surv. **20**, 87–112 (2023)
26. Brooks, J.K., Chacon, R.V.: Diffusions as a limit of stretched Brownian motions. Adv. in Math. **49**(2), 109–122 (1983)
27. Buraczewski, D., Dyszewski, P., Iksanov, A., Marynych, A.: Random walks in a strongly sparse random environment. Stochastic Process. Appl. **130**(7), 3990–4027 (2020)
28. Buraczewski, D., Dyszewski, P., Iksanov, A., Marynych, A., Roitershtein, A.: Random walks in a moderately sparse random environment. Electron. J. Probab. **24**(Paper no.69), 44 pp. (2019)
29. Buraczewski, D., Dyszewski, P., Kołodziejska, A.: Weak quenched limit theorems for a random walk in a sparse random environment. Electron. J. Probab. **29**(Paper no. 7), 30 pp. (2024)
30. Caputo, P., Faggionato, A., Gaudillière, A.: Recurrence and transience for long range reversible random walks on a random point process. Electron. J. Probab. **14**(Paper no. 90), 2580–2616 (2009)
31. Carmona, P., Petit, F., Yor, M.: Some extensions of the arc sine law as partial consequences of the scaling property of Brownian motion. Probab. Theory Related Fields **100**(1), 1–29 (1994)
32. Chaleyat-Maurel, M., El Karoui, N., Marchal, B.: Réflexion discontinue et systèmes stochastiques. Ann. Probab. **8**(6), 1049–1067 (1980)
33. Costantini, C.: The Skorohod oblique reflection problem in domains with corners and application to stochastic differential equations. Probab. Theory Related Fields **91**(1), 43–70 (1992)
34. Darling, D.A.: The influence of the maximum term in the addition of independent random variables. Trans. Amer. Math. Soc. **73**, 95–107 (1952)
35. Davis, B.: Reinforced random walk. Probab. Theory Related Fields **84**(2), 203–229 (1990)
36. Davis, B.: Weak limits of perturbed random walks and the equation $Y_t = B_t + \alpha \sup\{Y_s : s \leq t\} + \beta \inf\{Y_s : s \leq t\}$. Ann. Probab. **24**(4), 2007–2023 (1996)
37. DeBlassie, R.D., Toby, E.H.: Reflecting Brownian motion in a cusp. Trans. Am. Math. Soc. **339**(1), 297–321 (1993)
38. Dolgopyat, D., Kosygina, E.: Scaling limits of recurrent excited random walks on integers. Electron. Commun. Probab. **17**(Paper no. 35), 14 pp. (2012)
39. Doney, R.A.: Moments of ladder heights in random walks. J. Appl. Probab. **17**(1), 248–252 (1980)
40. Doney, R.A., Kyprianou, A.E.: Overshoots and undershoots of Lévy processes. Ann. Appl. Probab. **16**(1), 91–106 (2006)
41. Dong, C., Iksanov, A., Pilipenko, A.: On a discrete approximation of a skew stable Lévy process (2024). https://arxiv.org/abs/2302.07298
42. Dong, C., Iksanov, A., Pilipenko, A.: On multidimensional locally perturbed standard random walks. Lithuanian Math. J. **64**(3), 287–301 (2024)

43. Dupuis, P., Ishii, H.: On Lipschitz continuity of the solution mapping to the Skorokhod problem, with applications. Stochastics Stochastics Rep. **35**(1), 31–62 (1991)
44. Durrett, R., Kesten, H., Lawler, G.: Making money from fair games. In: Random walks, Brownian motion, and interacting particle systems. Progress in Probability, vol. 28, pp. 255–267. Birkhäuser Boston, Boston, MA (1991)
45. Dynkin, E.B.: Theory of Markov processes. Pergamon Press, Oxford (1961)
46. Embrechts, P., Hofert, M.: A note on generalized inverses. Math. Methods Oper. Res. **77**(3), 423–432 (2013)
47. Enriquez, N., Kifer, Y.: Markov chains on graphs and Brownian motion. J. Theoret. Probab. **14**(2), 495–510 (2001)
48. Erdős, P., Taylor, S.J.: Some problems concerning the structure of random walk paths. Acta Math. Acad. Sci. Hungar. **11**, 137–162 (1960)
49. Ethier, S.N., Kurtz, T.G.: Markov processes: Characterization and convergence. Wiley, New York (1986)
50. Feller, W.: Fluctuation theory of recurrent events. Trans. Am. Math. Soc. **67**, 98–119 (1949)
51. Feller, W.: Generalized second order differential operators and their lateral conditions. Illinois J. Math. **1**, 459–504 (1957)
52. Feller, W.: An introduction to probability theory and its applications, vol. II. Wiley, New York (1966)
53. Freidlin, M., Sheu, S.-J.: Diffusion processes on graphs: stochastic differential equations, large deviation principle. Probab. Theory Related Fields **116**(2), 181–220 (2000)
54. Freidlin, M.I., Wentzell, A.D.: Diffusion processes on graphs and the averaging principle. Ann. Probab. **21**(4), 2215–2245 (1993)
55. Freidlin, M.I., Wentzell, A.D.: Diffusion processes on an open book and the averaging principle. Stochastic Process. Appl. **113**(1), 101–126 (2004)
56. Galakhov, E.I., Skubachevskiĭ, A.L.: On Feller semigroups generated by elliptic operators with integro-differential boundary conditions. J. Differential Equations **176**(2), 315–355 (2001)
57. Gikhman, I.I., Skorokhod, A.V.: Introduction to the theory of random processes. Dover Publications, Inc., Mineola, NY (1996)
58. Gnedenko, B.V., Kolmogorov, A.N.: Limit distributions for sums of independent random variables. Addison-Wesley Publishing Co., Cambridge (1954)
59. Gradshteyn, I.S., Ryzhik, I.M.: Table of integrals, series, and products, 7th edn. Elsevier/Academic Press, Amsterdam (2007)
60. Gut, A.: Stopped random walks: Limit theorems and applications, 2nd edn. Springer, New York (2009)
61. Hajri, H.: Stochastic flows related to Walsh Brownian motion. Electron. J. Probab. **16**(Paper no. 58), 1563–1599 (2011)
62. Hajri, H., Touhami, W.: Itô's formula for Walsh's Brownian motion and applications. Statist. Probab. Lett. **87**, 48–53 (2014)
63. Harrison, J., Lemoine, A.: Sticky Brownian motion as the limit of storage processes. J. Appl. Probab. **18**(1), 216–226 (1981)
64. Harrison, J.M., Reiman, M.I.: Reflected Brownian motion on an orthant. Ann. Probab. **9**(2), 302–308 (1981)
65. Harrison, J.M., Shepp, L.A.: On skew Brownian motion. Ann. Probab. **9**(2), 309–313 (1981)
66. Helland, I.S.: Convergence to diffusions with regular boundaries. Stochastic Process. Appl. **12**(1), 27–58 (1982)
67. Ikeda, N., Watanabe, S.: Stochastic differential equations and diffusion processes. North-Holland Publishing Co., Amsterdam (1981)

68. Iksanov, A.: Renewal theory for perturbed random walks and similar processes. Birkhäuser/Springer, Cham (2016)
69. Iksanov, A., Pilipenko, A.: A functional limit theorem for locally perturbed random walks. Probab. Math. Statist. **36**(2), 353–368 (2016)
70. Iksanov, A., Pilipenko, A.: On a skew stable Lévy process. Stochastic Process. Appl. **156**, 44–68 (2023)
71. Iksanov, A., Pilipenko, A., Povar, B.: Functional limit theorems for random walks perturbed by positive alpha-stable jumps. Bernoulli **29**(2), 1638–1662 (2023)
72. Ishikawa, Y.: A remark on the existence of a diffusion process with nonlocal boundary conditions. J. Math. Soc. Japan **42**(1), 171–184 (1990)
73. Itô, K.: Poisson point processes attached to Markov processes. In: Proceedings of the Sixth Berkeley symposium on mathematical statistics and probability (University California, Berkeley, California, 1970/1971). Probability theory, vol. III, pp. 225–239. University California, Berkeley (1972)
74. Itô, K., McKean Jr., H.P.: Brownian motions on a half line. Illinois J. Math. **7**, 181–231 (1963)
75. Itô, K., McKean Jr., H.P.: Diffusion processes and their sample paths. Reprint of the 1974 edition. Springer, Berlin (1996)
76. Kabluchko, Z., Marynych, A.: Renewal shot noise processes in the case of slowly varying tails. Theory Stoch. Process. **21**(2), 14–21 (2016)
77. Kallenberg, O.: Foundations of modern probability. Springer, New York (1997)
78. Karatzas, I., Yan, M.: Semimartingales on rays, Walsh diffusions, and related problems of control and stopping. Stochastic Process. Appl. **129**(6), 1921–1963 (2019)
79. Kasahara, Y.: A limit theorem for sums of i.i.d. random variables with slowly varying tail probability. J. Math. Kyoto Univ. **26**(3), 437–443 (1986)
80. Kaspi, H., Mandelbaum, A.: Lévy bandits: multi-armed bandits driven by Lévy processes. Ann. Appl. Probab. **5**(2), 541–565 (1995)
81. Keilson, J., Wellner, J.A.: Oscillating Brownian motion. J. Appl. Probability **15**(2), 300–310 (1978)
82. Kemperman, J.H.B.: The oscillating random walk. Stochastic Process. Appl. **2**, 1–29 (1974)
83. Kesten, H., Spitzer, F.: Ratio theorems for random walks. I. J. Analyse Math. **11**, 285–322 (1963)
84. Kołodziejska, A.: On favourite sites of a random walk in moderately sparse random environment (2024). Arxiv preprint https://arxiv.org/abs/2407.01206
85. Kolokoltsov, V.N.: Markov processes, semigroups and generators. Walter de Gruyter & Co., Berlin (2011)
86. Kononchuk, P.P., Kopytko, B.I.: On a diffusion process on a half-line with Feller-Wentzel boundary condition that corresponds to reflection and jumps. Theory Stoch. Process. **16**(2), 69–76 (2010)
87. Kopytko, B., Shevchuk, R.: One-dimensional diffusion processes with moving membrane: partial reflection in combination with jump-like exit of process from membrane. Electron. J. Probab. **25**(Paper no. 41), 21 pp. (2020)
88. Kostrykin, V., Potthoff, J., Schrader, R.: Brownian motions on metric graphs. J. Math. Phys. **53**(9), 095206, 36 (2012)
89. Kosygina, E., Mountford, T., Peterson, J.: Convergence of random walks with Markovian cookie stacks to Brownian motion perturbed at extrema. Probab. Theory Related Fields **182**(1–2), 189–275 (2022)
90. Kosygina, E., Zerner, M.P.W.: Excited random walks: results, methods, open problems. Bull. Inst. Math. Acad. Sin. (N.S.) **8**(1), 105–157 (2013)

91. Kulik, A.M.: A limit theorem for diffusions on graphs with variable configuration. arXiv preprint math/0701632 (2007)
92. Kushner, H.J.: Heavy traffic analysis of controlled queueing and communication networks. Springer, New York (2001)
93. Kwon, Y.: The submartingale problem for Brownian motion in a cone with nonconstant oblique reflection. Probab. Theory Related Fields **92**(3), 351–391 (1992)
94. Kwon, Y., Williams, R.J.: Reflected Brownian motion in a cone with radially homogeneous reflection field. Trans. Am. Math. Soc. **327**(2), 739–780 (1991)
95. Lambert, A.: The genealogy of continuous-state branching processes with immigration. Probab. Theory Related Fields **122**(1), 42–70 (2002)
96. Lambert, A., Simatos, F.: The weak convergence of regenerative processes using some excursion path decompositions. Ann. Inst. Henri Poincaré Probab. Stat. **50**(2), 492–511 (2014)
97. Lejay, A.: On the constructions of the skew Brownian motion. Probab. Surv. **3**, 413–466 (2006)
98. Lejay, A.: The snapping out Brownian motion. Ann. Appl. Probab. **26**(3), 1727–1742 (2016)
99. Lindvall, T.: Weak convergence of probability measures and random functions in the function space $D(0, \infty)$. J. Appl. Probability **10**, 109–121 (1973)
100. Lions, P.-L., Sznitman, A.-S.: Stochastic differential equations with reflecting boundary conditions. Comm. Pure Appl. Math. **37**(4), 511–537 (1984)
101. Löbus, J.-U., Portenko, M.I.: On a class of perturbed generators of stable processes. Teor. Ĭmovīr. Mat. Stat. **52**, 102–111 (1995)
102. Mandelbaum, A.: Continuous multi-armed bandits and multiparameter processes. Ann. Probab. **15**(4), 1527–1556 (1987)
103. Mandrekar, V., Pilipenko, A.: On a Brownian motion with a hard membrane. Statist. Probab. Lett. **113**, 62–70 (2016)
104. Matzavinos, A., Roitershtein, A., Seol, Y.: Random walks in a sparse random environment. Electron. J. Probab. **21**(Paper No. 72), 20 pp. (2016)
105. Meerschaert, M.M., Scheffler, H.-P.: Limit theorems for continuous-time random walks with infinite mean waiting times. J. Appl. Probab. **41**(3), 623–638 (2004)
106. Meerschaert, M.M., Sikorskii, A.: Stochastic models for fractional calculus, 2nd edn. De Gruyter, Berlin (2019)
107. Méndez, V., Fedotov, S., Horsthemke, W.: Reaction-transport systems. Springer, Heidelberg (2010)
108. Menshikov, M., Petritis, D.: Markov chains in a wedge with excitable boundaries. In:Analytic methods in applied probability. American Mathematical Society Translations: Series 2, vol. 207, pp. 141–164. American Mathematical Society, Providence, RI (2002)
109. Menshikov, M.V., Petritis, D., Wade, A.R.: Heavy-tailed random walks on complexes of half-lines. J. Theoret. Probab. **31**(3), 1819–1859 (2018)
110. Mijatović, A., Uribe Bravo, G.: Limit theorems for local times and applications to SDEs with jumps. Stochastic Process. Appl. **153**, 39–56 (2022)
111. Mijatović, A., Vysotsky, V.: Stationary entrance markov chains, inducing, and level-crossings of random walks (2018). Arxiv preprint https://arxiv.org/pdf/1808.05010.pdf
112. Mikulevičius, R.: The existence of solutions of a martingale problem. Lithuanian Math. J. **17**(4), 538–550 (1977)
113. Mikulevičius, R.: On uniqueness of solutions of the martingale problem. Lithuanian Math. J. **18**(2), 202–209 (1978)
114. Minlos, R.A., Zhizhina, E.A.: A limit diffusion process for an inhomogeneous random walk on a one-dimensional lattice. Uspekhi Mat. Nauk **52**(2(314)), 87–100 (1997)
115. Nakao, S.: On the pathwise uniqueness of solutions of one-dimensional stochastic differential equations. Osaka Math. J. **9**, 513–518 (1972)

116. Ngo, H.-L., Peigné, M.: Limit theorem for perturbed random walks. Theory Stoch. Process. **24**(2), 61–78 (2019)
117. Ngo, H.-L., Peigné, M.: Limit theorem for reflected random walks. In: Thermodynamic formalism. Lecture Notes in Mathematics, vol. 2290, pp. 205–233. Springer, Cham (2021)
118. Osypchuk, M.M., Portenko, M.I.: One type of singular perturbations of a multidimensional stable process. Theory Stoch. Process. **19**(2), 42–51 (2014)
119. Paulin, D., Szász, D.: Locally perturbed random walks with unbounded jumps. J. Stat. Phys. **141**(6), 1116–1130 (2010)
120. Pavlyukevich, I., Pilipenko, A.: Walsh's Brownian motion and Donsker scaling limits of perturbed random walks. ALEA Lat. Am. J. Probab. Math. Stat. **21**(2), 1669–1707 (2024)
121. Pazy, A.: Semigroups of linear operators and applications to partial differential equations. Springer, New York (1983)
122. Pilipenko, A.: On the Skorokhod mapping for equations with reflection and possible jump-like exit from a boundary. Ukrainian Math. J. **63**(9), 1415–1432 (2012)
123. Pilipenko, A.: An introduction to stochastic differential equations with reflection, vol. 1. Universitätsverlag Potsdam, Potsdam (2014)
124. Pilipenko, A.: A functional limit theorem for excited random walks. Electron. Commun. Probab. **22**(Paper No. 39), 9 pp. (2017)
125. Pilipenko, A., Khomenko, V.: On a limit behavior of a random walk with modifications upon each visit to zero. Theory Stoch. Process. **22**(1), 71–80 (2017)
126. Pilipenko, A., Prykhodko, O.O.: On a limit behaviour of a random walk penalised in the lower half-plane. Theory Stoch. Process. **25**(2), 81–88 (2020)
127. Pilipenko, A., Prykhodko, Y.: On the limit behavior of symmetric random walks with membranes. Theory Probab. Math. Statist. **85**, 93–105 (2012)
128. Pilipenko, A., Prykhodko, Y.: Limit behavior of a simple random walk with non-integrable jump from a barrier. Theory Stoch. Process. **19**(1), 52–61 (2014)
129. Pilipenko, A., Prykhodko, Y.: On the limit behavior of a sequence of Markov processes perturbed in a neighborhood of the singular point. Ukrainian Math. J. **67**(4), 564–583 (2015)
130. Pilipenko, A., Prykhodko, Y.: A limit theorem for singular stochastic differential equations. Mod. Stoch. Theory Appl. **3**(3), 223–235 (2016)
131. Pilipenko, A., Sakhanenko, L.: On a limit behavior of a one-dimensional random walk with non-integrable impurity. Theory Stoch. Process. **20**(2), 97–104 (2015)
132. Pilipenko, A., Sarantsev, A.: Boundary approximation for sticky jump-reflected processes on the half-line. Electron. J. Probab. **29**(Paper no. 32), 21 pp. (2024)
133. Poisat, J., Simenhaus, F.: A limit theorem for the survival probability of a simple random walk among power-law renewal obstacles. Ann. Appl. Probab. **30**(5), 2030–2068 (2020)
134. Poisat, J., Simenhaus, F.: Localization of a one-dimensional simple random walk among power-law renewal obstacles. Ann. Appl. Probab. **34**(4), 4137–4192 (2024)
135. Portenko, M.I.: One class of transformations of a symmetric stable processes. Theory Stoch. Process. **3**(19)(3–4), 373–387 (1997)
136. Portenko, N.I.: Generalized diffusion processes. American Mathematical Society, Providence (1990)
137. Portenko, N.I.: Generalized diffusion processes. In: Proceedings of the Third Japan-USSR symposium on probability theory, pp. 500–523. Springer, Berlin (1976)
138. Raimond, O., Schapira, B.: Excited Brownian motions as limits of excited random walks. Probab. Theory Related Fields **154**(3–4), 875–909 (2012)
139. Ramanan, K.: Reflected diffusions defined via the extended Skorokhod map. Electron. J. Probab. **11**(Paper no. 36), 934–992 (2006)

140. Revuz, D., Yor, M.: Continuous martingales and Brownian motion, 3rd edn. Springer, Berlin (1999)
141. Rogers, L.C.G.: Itô excursion theory via resolvents. Z. Wahrsch. Verw. Gebiete **63**(2), 237–255 (1983)
142. Rogozin, B.A., Foss, S.G.: Recurrency of an oscillating random walk. Theor. Probab. Appl. **23**(1), 155–162 (1978)
143. Saisho, Y.: Stochastic differential equations for multidimensional domain with reflecting boundary. Probab. Theory Related Fields **74**(3), 455–477 (1987)
144. Salisbury, T.: Construction of right processes from excursions. Probab. Theory Related Fields **73**(3), 351–367 (1986)
145. Samorodnitsky, G., Taqqu, M.: Stable non-Gaussian random processes. Chapman & Hall, New York (1994)
146. Sato, K.-i.: Lévy processes and infinitely divisible distributions. Cambridge University, Cambridge (2013)
147. Sato, K.-i., Ueno, T.: Multi-dimensional diffusion and the Markov process on the boundary. J. Math. Kyoto Univ. **4**, 529–605 (1964/65)
148. Skorokhod, A.V.: Limit theorems for stochastic processes. Theor. Probab. Appl. **1**(3), 261–290 (1956)
149. Skorokhod, A.V.: Stochastic equations for diffusion processes in a bounded region. Theor. Probab. Appl. **6**(3), 264–274 (1961)
150. Skorokhod, A.V.: Stochastic equations for diffusion processes in a bounded region. II. Theor. Probab. Appl. **7**(1), 3–23 (1962)
151. Skorokhod, A.V.: Studies in the theory of random processes. Addison-Wesley Publishing, Reading (1965)
152. Skorokhod, A.V.: Anatolii V. Skorokhod—selected works. Springer, Cham (2016)
153. Spitzer, F.: Principles of random walk, 2nd edn. Springer, New York (1976)
154. Stone, C.: Limit theorems for random walks, birth and death processes, and diffusion processes. Illinois J. Math. **7**, 638–660 (1963)
155. Straka, P., Henry, B.I.: Lagging and leading coupled continuous time random walks, renewal times and their joint limits. Stochastic Process. Appl. **121**(2), 324–336 (2011)
156. Stroock, D., Varadhan, S.: Diffusion processes with boundary conditions. Comm. Pure Appl. Math. **24**, 147–225 (1971)
157. Stroock, D., Varadhan, S.R.: Multidimensional diffusion processes. Springer, Berlin (1979)
158. Szász, D., Telcs, A.: Random walk in an inhomogeneous medium with local impurities. J. Statist. Phys. **26**(3), 527–537 (1981)
159. Taira, K.: On the existence of Feller semigroups with boundary conditions. Mem. Am. Math. Soc. **99**(475), 65 pp. (1992)
160. Tanaka, H.: Stochastic differential equations with reflecting boundary condition in convex regions. Hiroshima Math. J. **9**(1), 163–177 (1979)
161. Tóth, B.: Generalized Ray-Knight theory and limit theorems for self-interacting random walks on $\mathbf{Z}^1$. Ann. Probab. **24**(3), 1324–1367 (1996)
162. Tsirelson, B.: Triple points: from non-Brownian filtrations to harmonic measures. Geom. Funct. Anal. **7**(6), 1096–1142 (1997)
163. Varadhan, S., Williams, R.: Brownian motion in a wedge with oblique reflection. Comm. Pure Appl. Math. **38**(4), 405–443 (1985)
164. Ventcel', A.D.: Semigroups of operators that correspond to a generalized differential operator of second order. Dokl. Akad. Nauk SSSR (N.S.) **111**, 269–272 (1956)
165. Ventcel', A.D.: On boundary conditions for multi-dimensional diffusion processes. Theor. Probability Appl. **4**, 164–177 (1959)

166. Vo, T.D.: The oscillating random walk on $\mathbb{Z}$. J. Theoret. Probab. **36**(4), 2426–2447 (2023)
167. Vo, T.D., Peigné, M.: A functional limit theorem for lattice oscillating random walks. ALEA Lat. Am. J. Probab. Math. Stat. **20**(2), 1433–1457 (2023)
168. Walsh, J.B.: A diffusion with a discontinuous local time. Astérisque **52**(53), 37–45 (1978)
169. Watanabe, S.: On stable processes with boundary conditions. J. Math. Soc. Japan **14**, 170–198 (1962)
170. Watanabe, S.: On stochastic differential equations for multi-dimensional diffusion processes with boundary conditions. J. Math. Kyoto Univ. **11**, 169–180 (1971)
171. Watanabe, S.: Construction of diffusion processes with Wentzell's boundary conditions by means of Poisson point processes of Brownian excursions. In: Probability theory (Papers, VIIth Semester, Stefan Banach Internat. Math. Center, Warsaw, 1976). Banach Center Publication, vol. 5, pp. 255–271. PWN, Warsaw (1979)
172. Watanabe, S.: Itô's theory of excursion point processes and its developments. Stochastic Process. Appl. **120**(5), 653–677 (2010)
173. Werner, W.: Some remarks on perturbed reflecting Brownian motion. In: Séminaire de Probabilités, XXIX. Lecture Notes in Mathematics, vol. 1613, pp. 37–43. Springer, Berlin (1995)
174. Whitt, W.: Some useful functions for functional limit theorems. Math. Oper. Res. **5**(1), 67–85 (1980)
175. Whitt, W.: Stochastic-process limits: An introduction to stochastic-process limits and their application to queues. Springer, New York (2002)
176. Williams, R.J.: Reflected Brownian motion in a wedge: semimartingale property. Z. Wahrsch. Verw. Gebiete **69**(2), 161–176 (1985)
177. Williams, R.J.: Semimartingale reflecting Brownian motions in the orthant. In: Stochastic networks. The IMA Volumes in Mathematics and Its Applications, vol. 71, pp. 125–137. Springer, New York (1995)
178. Yano, K.: Convergence of excursion point processes and its applications to functional limit theorems of Markov processes on a half-line. Bernoulli **14**(4), 963–987 (2008)
179. Yano, K.: Functional limit theorems for processes pieced together from excursions. J. Math. Soc. Japan **67**(4), 1859–1890 (2015)

Index

B
Boundedness in probability, 112
Brownian motion
    holding and jumping, 226
    oscillating, 139, 201
    reflected, 3
    skew, 36
        oscillating, 123
        stochastic differential eq., 37
    snapping out, 47
    sticky reflected, 104
    stretched, 8
    Walsh, 43
        Markovian charact., 43
        martingale charact., 44
    with jump-type exit from zero, 25

C
Continuity of functionals
    addition, 214
    composition, 214
    concatenation, 213
    inversion, 218

D
Descending ladder
    epochs, 109
    height, 109, 126
Domain of attraction, 220
    normal, 221

E
Entrance law, 233
    continuous, 235
    jump, 235

G
Generalized inverse, 216
    overshoot, 217
    undershot, 217

L
Lindley model, 12, 79
    with refilling, 80
Local time
    Blumenthal-Getoor, 237
    symmetric semimartingale, 17

M
Markov chain
    entrance, 125
    exit, 125
Markov process
    convergence theorems, 223
    extension, 225
    Feller, 222
    Feller extension, 225
    minimal, 225
    resolvent, 222
    semigroup, 222
    strong, 222
Moran's model of a dam, 105

O

1-arithmetic distribution, 78

P

Permeability parameter, 36
Point of growth, 96

R

Random walk
    among power-law renewal obstacles, 9
    aperiodic, 181
        strongly, 184
    excited, 8
    oscillating, 80
    reflected, 81
    in a sparse random environment, 8
    standard, 1
    stretched, 8
    with freezing, 81
    with membrane, 74
    with repulsive half-plane, 80
Regenerative
    model, 107
    sequence, 107

S

Skew stable process, 50
    equation with local time, 62
Skorokhod reflection
    generalized map, 21
    generalized problem, 19
    map, 12
    problem, 13
Solution to stochastic differential eq.
    with local time (weak), 60
    with reflection at 0, 16
Stable
    distribution, 219
        Lévy-Khintchine formula, 220
        spectrally negative, 219
        spectrally positive, 219
        symmetric, 219
    process, 221
Switch model (two-stage model), 91
    gap, 91
    noise, 91
    regulator, 91

T

Theorem
    continuous mapping, 208
    Donsker in C, 210
    Donsker in D, 213
    Prohorov, 207
    Skorokhod representation, 208
Tightness, 207
Topology
    locally uniform, 209
    product, 211
    Skorokhod J_1 on $D([0, \infty), \mathbb{R}^d)$, 212
    Skorokhod J_1 on $D([0, T], \mathbb{R}^d)$, 211

W

Wiener measure, 210

The manufacturer's authorised representative in the EU is Springer Nature Customer Service Centre GmbH, Europaplatz 3, 69115 Heidelberg, Germany. If you have any concerns regarding our products, please contact ProductSafety@springernature.com

Printed and bound by CPI Group (UK) Ltd, Croydon, CR0 4YY
26/03/2026
02078972-0007